Mammalian Cell Culture

The Use of Serum-Free Hormone-Supplemented Media

Mammalian Cell Culture

The Use of Serum-Free Hormone-Supplemented Media

Edited by

Jennie P. Mather

The Population Council and The Rockefeller University
New York, New York

Plenum Press · New York and London

Library of Congress Cataloging in Publication Data

Main entry under title:

Mammalian cell culture.

Bibliography: p.
Includes index.
1. Cell culture. 2. Culture media (Biology). 3. Mammals—Cytology. I. Mather, Jennie P., 1948- . II. Title: Serum-free hormone supplemented media. [DNLM: 1. Culture media. 2. Cells, Cultured. QS 530 M265]
QH585.M36 1984 599′.087 84-1985
ISBN-13: 978-1-4615-9363-8 e-ISBN-13: 978-1-4615-9361-4
DOI: 10.1007/978-1-4615-9361-4

Softcover reprint of the hardcover 1st edition 1984
A Division of Plenum Publishing Corporation
233 Spring Street, New York, N.Y. 10013

Contributors

David Barnes • Department of Biological Sciences, University of Pittsburgh, Pittsburgh, Pennsylvania 15260

Theodore R. Breitman • Laboratory of Tumor Cell Biology, National Cancer Institute, National Institutes of Health, Bethesda, Maryland 20205

Brian I. Carr • Division of Cytogenetics and Cytology and Department of Medical Oncology, City of Hope National Medical Center, Duarte, California 91010

Paul V. Cherington • Laboratory of Neoplastic Disease Mechanisms, Dana-Farber Cancer Institute, Boston, Massachusetts 02115

J. E. Estes • Department of Pharmacology, Cancer Cell Biology Program, Cancer Research Center, University of North Carolina School of Medicine, Chapel Hill, North Carolina 27514

Izumi Hayashi • Division of Cytogenetics and Cytology and Department of Medical Oncology, City of Hope National Medical Center, Duarte, California 91010

Hiromichi Hemmi • Laboratory of Tumor Cell Biology, National Cancer Institute, National Institutes of Health, Bethesda, Maryland 20205

P. H. Howe • Department of Pharmacology, Cancer Cell Biology Program, Cancer Research Center, University of North Carolina School of Medicine, Chapel Hill, North Carolina 27514

Douglas M. Jefferson • Department of Molecular Pharmacology, Albert Einstein College of Medicine, Bronx, New York 10461

Beverly R. Keene • Laboratory of Tumor Cell Biology, National Cancer Institute, National Institutes of Health, Bethesda, Maryland 20205

E. B. Leof • Department of Pharmacology, Cancer Cell Biology Program, Cancer Research Center, University of North Carolina School of Medicine, Chapel Hill, North Carolina 27514

Jennie P. Mather • The Population Council and The Rockefeller University, New York, New York 10021

W. J. Pledger • Department of Pharmacology, Cancer Cell Biology Program, Cancer Research Center, University of North Carolina School of Medicine, Chapel Hill, North Carolina 27514

Lola M. Reid • Department of Molecular Pharmacology, Albert Einstein College of Medicine, Bronx, New York 10461

Ginette Serrero-Davé • Cancer Center, University of California, San Diego, La Jolla, California 92093. *Present address:* W. Alton Jones Cell Science Center, Lake Placid, New York 12946

Mary Taub • Biochemistry Department, State University of New York at Buffalo, Buffalo, New York 14214

Stephen D. Wolpe • The Population Council, New York, New York 10021

Preface

The advantages of obtaining a completely defined environment for the growth of cells *in vitro* were recognized very early in the history of cell culture (Lewis and Lewis, 1911). Continued interest in the nutritional requirements of cells *in vitro* and in providing an optimal environment for cells led to the development of the complex nutrient mixtures available today in many media (Waymouth, 1972; Ham, 1965).

However, serum remained an essential component of medium for the growth of most cell types in culture. The question of what factor (or factors) in serum was essential for cell growth and survival remained unanswered for several decades. Initially, experiments were designed to purify the "active component" of serum for the growth of cells in culture. These experiments identified fetuin (Fisher *et al.,* 1958) and nonsuppressible insulinlike activity (Temin *et al.,* 1972) as important components of serum. However, the complexity of serum and the very low levels of active components in serum hindered progress in identifying and isolating serum factors.

In retrospect, the analytical approach can be seen to be extremely difficult due to the complex interactions of various hormones and growth factors and the requirements of cells for several very different types of hormones for growth and function. However, these experiments and the development of hormone-dependent cell lines (Clark *et al.,* 1972; Armelin, 1973; Nishikawa *et al.*, 1975) led to the hypothesis that the role of serum in cell culture was to provide a complex of hormones that were growth-stimulatory for given cell types (Sato, 1975). The first experimental evidence in support of this hypothesis was provided when the GH_3 pituitary cell line was grown in serum-free medium supplemented with hormones, growth factors, and transferrin (Hyashi and Sato, 1976).

Subsequently, the development of hormone-supplemented serum-free conditions for the growth of several cell lines originating from different tissues (Bottenstein *et al*, 1979; Mather and Sato, 1979; Barnes and Sato, 1981) allowed several generalizations concerning the growth of cells in serum-free medium. (1) Serum can be replaced by a mixture of hormones, growth factors, transport proteins, vitamins, and attachment factors. (2) The requirements for these factors differ for different cell types. (3) The requirements for cell lines derived from the same cell type are similar or identical for cells from different species. (4) The factors used to grow a cell line derived from a particular cell type can be used for primary cultures of that cell type. These primary cultures will frequently exhibit improved growth and/or function compared with those in serum-supplemented culture. (5) Serum-free culture can be used to avoid fibroblast overgrowth in primary cultures and select for specific cell types. (6) Functional cell lines can be established in serum-free medium from primary cultures of cell types not previously established in culture.

Since the first report of the growth of GH_3 cells in 1976, this approach has been widely used by investigators in a number of areas (Sato *et al.*, 1982). It should be emphasized, however, that the use of hormone-supplemented serum-free culture is more than a new technique. It involves some very basic changes in how we view *in vitro* cultures of eukaryotic cells. These cells can no longer be viewed simply as a microorganism (Puck, 1972), but rather as part of a complex interactive system involving matrix substrate, cell–cell contact, and cell–cell communication via soluble factors. It is also apparent that many aspects of cell behavior *in vitro* that led some investigators to consider this an artifactual system are, in fact, responses to an inappropriate environment (i.e., serum and plastic surfaces). With the use of appropriate hormone-supplemented media and substrates, it can be seen that cell lines, even those that have been in culture for many years, can express cell-specific functions and respond to hormones in a physiologic manner. This is, perhaps, the greatest power of defined cell culture. The synthetic approach, by defining step-by-step the conditions required for cell growth or function, gives deep insight into the complexity of the regulation of cell physiology *in vivo*, which is impossible to obtain by other means.

This book makes no attempt to cover the entire field of hormone-supplemented cell culture. Rather, the contributions have been chosen to illustrate several of the diverse areas of research that have benefited from this new view of cell physiology. By giving in-depth coverage to these areas and the potential of using serum-free culture to define new areas of research, it is hoped that the reader will gain a deeper appreciation of the possibilities of this approach.

Studies on the role of hormones in controlling the initiation of cell proliferation and cell cycle (Chapters 1 and 2) give more insights into the physiologic regulation of cell growth than was possible using other methods of cell synchronization such as amino acid starvation. The use of serum-free culture has also allowed studies of differentiation of cells such as adipocytes (Chapter 3) and myelomonocytes (Chapter 4) in defined conditions. Cell-produced products can be more easily purified from serum-free culture. This is especially useful in monoclonal antibody production (Chapter 5). The regulation of physiologic function of cells can be studied (Chapter 6). The complex interactions of cells *in vivo* in response to extraneous agents such as chemical carcinogens (Chapter 7), hormones (Chapter 6), or factors produced locally in an organ (Chapter 8) can be studied using *in vitro* defined culture systems. The role of defined attachment factors (Chapter 9) or isolated biomatrix (Chapter 10) can be studied *in vitro* in defined systems to increase our understanding of the role of the basal lamina and cell–substrate interactions *in vivo*.

The complex nature of hormone action and interactions *in vitro* is immediately apparent from these studies. Thus, a hormone may act directly on a cell *in vitro* to stimulate progression through the cell cycle or specific cell function. A hormone might also act indirectly by stimulating the secretion of a cell-produced attachment factor or inhibit the production of enzymes such as collagenase, which might degrade a cell-produced attachment factor. The increased levels of such factors might then lead to increased cell growth or function. A hormone might also act by increasing the secretion of a transport protein such as transferrin, which would then transport a required nutrient into the cell and increase cell growth. A hormone might increase the production of a growth factor, which would then act to stimulate growth in that cell or inhibit the production of a growth inhibitor. Given these and other possible modes of action and interaction discussed in these chapters, it is apparent that the regulation of cell growth and function by a few hormones, growth factors, transport factors, and attachment factors can be a very complex matter indeed. However, these problems can be approached using defined culture. The next few decades should prove to be a period of exciting advances in our understanding of cell physiology *in vitro* and *in vivo*.

It is appropriate to express the appreciation of the author for many conversations with Dr. Gordon Sato. Without his vision, generosity, and support, much of the work in this field, and this book, would not have been possible. I would also like to thank Ms. Catherine Galloway for her editorial assistance.

Jennie P. Mather

References

Armelin, H. A., 1973, Pituitary extracts and steroid hormones in the control of 3T3 cell growth, *Proc. Natl. Acad. Sci. U.S.A.* **70**:2702–2706.

Barnes, D., and Sato, G., 1981, Serum-free cell culture: A unifying approach, *Cell* **22**:69–655.

Bottenstein, J. E., Sato, G. H., and Mather, J. P., 1979, Growth of neuroepithelial derived cell lines in serum-free hormone-supplemented media, in: *Hormones and Tissue Culture* (G. Sato and R. Ross, eds.), Cold Spring Harbor Press, Cold Spring Harbor, New York, pp. 531–544.

Clark, J. L., Jones, K. L., Gospodarowicz, D., and Sato, G. H., 1972, Growth response to hormones by a new rat ovary cell line, *Nature New Bio.* **236:**180–181.

Fisher, H. W., Puck, T. T., and Sato, G., 1958, Molecular growth requirements of single mammalian cells: The action of fetuin in promoting cell attachment to glass, *Proc. Natl. Acad. Sci. U.S.A.* **44:**4–10.

Ham, R. G., 1965, Clonal growth of mammalian cells in a chemically defined synthetic medium, *Proc. Natl. Acad. Sci. U.S.A.* **53:**288-293.

Hayashi, I., and Sato, G., 1976, Replacement of serum by hormones permits the growth of cell in a defined medium, *Nature* (*London*) **159:**132–134.

Lewis, M. R., and Lewis, W. H., 1911, The cultivation of tissues from chick embryos in solutions of NaCl, $CaCl_2$, KCl, and $NaHCO_3$, *Anat. Rec.* **5:**277–293.

Mather, J. P., and Sato, G., 1979, The growth of mouse melanoma cells in serum-free hormone supplemented medium, *Exp. Cell Res.* **120:**191–200.

Nishikawa, K., Armelin, H. A., and Sato, G., 1975, Control of ovarian cell growth in culture by serum and pituitary factors, *Proc. Natl. Acad. Sci. U.S.A.* **72:**483–487.

Puck, T. T., 1972, *The Mammalian Cells as a Microorganism,* Holden-Day, Inc., San Francisco.

Sato, G., 1975, The role of serum in cell culture, in: *Biochemical Action of Hormones,* Volume III (G. Litwak, ed.), Academic Press, Inc., New York, pp. 391–396.

Sato, G. H., Pardee, A. B., and Sirbasku, D. A., 1982, *Growth of Cell in Hormonally Defined Media,* Cold Spring Harbor Laboratory, Cold Spring Harbor, New York.

Temin, H., Pierson, R. W., Jr., and Dulak, N. C., 1972, The role of serum in the control of multiplication of avian and mammalian cells in culture, in: *Growth, Nutrition and Metabolism of Cells in Culture,* Volume I (G. H. Rothblat and V. J. Cristafalo, eds.), Academic Press, New York, pp. 49–81.

Waymouth, C., 1972, Construction of Tissue Culture Media, in: *Growth, Nutrition and Metabolism of Cells in Culture,* Volume I (G. H. Rothblatt and V. J. Cristofalo, eds.), Academic Press, New York, pp. 11–49.

Contents

CHAPTER 1

Serum Factor Requirements for the Initiation of Cellular Proliferation

W. J. PLEDGER, J. E. ESTES, R. H. HOWE, and E. B. LEOF

CHAPTER 2

Regulation of Fibroblast Growth by Multiple Growth Factors in Serum-Free Medium

PAUL V. CHERINGTON

CHAPTER 3

Growth and Differentiation of Preadipocyte Cell Lines in Serum-Free Medium

GINETTE SERRERO-DAVÉ

CHAPTER 4

Growth and Differentiation of Human Myelomonocyte Leukemia Cell Lines in Serum-Free Medium

THEODORE R. BREITMAN, BEVERLY R. KEENE, and HIROMICHI HEMMI

CHAPTER 5

In Vitro Immunization and Growth of Hybridomas in Serum-Free Medium

STEPHEN D. WOLPE

Chapter 6

Kidney Cell Cultures in Hormonally Defined Serum-Free Medium

MARY TAUB

Chapter 7

Rat Hepatocytes in Culture: A Model for Studies of Growth Control during Experimental Chemical Hepatocarcinogenesis

IZUMI HAYASHI and BRIAN I. CARR

CHAPTER 8

Intratesticular Regulation: Evidence for Autocrine and Paracrine Control of Testicular Function

JENNIE P. MATHER

CHAPTER 9

Attachment Factors in Cell Culture

DAVID BARNES

CHAPTER 10

Cell Culture Studies Using Extracts of Extracellular Matrix to Study Growth and Differentiation in Mammalian Cells

LOLA M. REID and DOUGLAS M. JEFFERSON

CHAPTER 1

Serum Factor Requirements for the Initiation of Cellular Proliferation

W. J. PLEDGER, J. E. ESTES, P. H. HOWE, and E. B. LEOF

Introduction

The regulatory events in the G_0/G_1 portion of the cell cycle are believed to control, at least in part, the rate of cellular proliferation. When quiescent density-arrested fibroblasts are stimulated to initiate proliferation, for example, by the addition of fresh serum, they begin DNA synthesis after a defined G_0/G_1 phase lag. Such stimulation of density-arrested cells provides a population of cells that can be useful in the investigation to elucidate the biochemical events that regulate the traverse of G_0/G_1. Serum is a complex heterogeneous mixture of proteins, hormones, growth factors, and other components. This complexity has made it difficult to characterize the necessary events leading to proliferation because other cellular reactions, not required for the control of proliferation, are also stimulated by the addition of serum.

We have developed a fibroblast culture protocol using a defined medium in order to investigate the required biochemical events that lead from quiescent growth arrest to the cellular commitment to proliferate. Using the minimum required serum components for traverse of G_0/G_1, the portion of the G_0/G_1 phase requiring various factors

W. J. PLEDGER, J. E. ESTES, P. H. HOWE, and E. B. LEOF • Department of Pharmacology, Cancer Cell Biology Program, Cancer Research Center, University of North Carolina School of Medicine, Chapel Hill, North Carolina 27514.

can be determined as well as the biochemical events involved in the regulation of the cell cycle. Transformation of cells leads to alteration of normal serum requirements and growth control (Scher *et al.*, 1979). The comparison of normally required serum components and the biochemical reactions they regulate to that of transformed cells may lead to a clearer understanding of the transformation process.

Multiple Serum Components Required for Cellular Proliferation

The growth of cultured cells in hormonally defined medium has demonstrated that multiple growth factors control cellular proliferation (Barnes and Sato, 1980). The studies performed by Jimenez de Asua and associates clearly illustrated that the regulatory control of G_0/G_1 is affected by multiple growth factors (Jimenez de Asua *et al.*, 1977; Otto *et al.*, 1981). Platelet-derived growth factor (PDGF), a polypeptide stored in platelets and released into serum during clotting, has been shown to be a potent mitogen for cultured fibroblast cells (Balk, 1971; Ross *et al.*, 1974; Kohler and Lipton, 1974). Platelet-derived growth factor or other mitogens such as fibroblast growth factor (FGF), calcium phosphate (Stiles *et al.*, 1979a) and a macrophage-derived growth factor (Wharton *et al.*, 1982a) have been shown to stimulate the processes involved in the initiation of cell cycle traverse and cellular proliferation. Even though platelet-poor plasma does not stimulate density-arrested BALB/c-3T3 cells to undergo proliferative growth, it functions synergistically with PDGF to give maximum growth stimulation of the quiescent cell population (Pledger *et al.*, 1977; Vogel *et al.*, 1978).

In order to study the synergistic interactions between PDGF and plasma or plasma-derived factors we developed a system where density-arrested BALB/c-3T3 cells were transiently treated with PDGF, then transferred to medium with plasma but without PDGF (Pledger *et al.*, 1977). Under these conditions the density-arrested cells did not initiate DNA synthesis unless they received both the transient exposure to PDGF and also medium containing plasma: separately, the PDGF exposure or the plasma-supplemented medium did not stimulate DNA synthesis. Following a transient exposure to PDGF, density-arrested BALB/c-3T3 cells responded to the addition of plasma-derived growth factors by initiating DNA synthesis after a minimum lag of 12 hours.

Progression through G_1 in response to plasma has been shown to proceed by an ordered sequence of events, since progression in G_1 to specific growth arrest points can be demonstrated by incubating PDGF-treated cells in plasma for varying periods of time under conditions that did not permit DNA synthesis to occur (Pledger *et al.*, 1978). With such techniques, two specific arrest points have been described: the "V" point, which is operationally defined as being six hours prior to DNA synthesis, and the "W" point, which is located shortly before the beginning of S phase. Growth arrest at the "V" point has been observed in BALB/c-3T3 cells grown in medium supplemented with platelet-poor plasma derived from hypopituitary animals (Stiles *et al.*, 1979a), in medium deficient in amino acids (Stiles *et al.*, 1979b) or by increasing intracellular cAMP by the addition of cholera toxin and isobutylmethylxanthine to the growth-stimulating medium (Leof *et al.*, 1982a).

Previously reported data have suggested that several plasma-derived factors may be required for progression of density-arrested BALB/c-3T3 cells through the G_0/G_1 phase. Stiles *et al.* (1979a) demonstrated that somatomedin C and one or more of the components in hypophysectomized rat plasma were necessary for the progression of BALB/c-3T3 cells. Clemmons *et al.* (1981) and Clemmons and VanWyk (1981) reported that human fibroblast cells responded to growth stimulation (i.e., PDGF) in somatomedin-C-deficient medium because they produced sufficient somatomedin C for growth. Whether this is a general mechanism by which some cell types respond to proliferative signals is at present unknown. In general, defined growth medium has required multiple growth factors.

Epidermal Growth Factor (EGF) and Somatomedin C Can Replace the G_0/G_1 Progression Activity of Platelet Poor Plasma

Leof *et al.* (1982b) demonstrated that the addition of epidermal growth factor (EGF) and somatomedin-C-supplemented plasma-free medium to density-inhibited BALB/c-3T3 cells, which had been given a transient exposure to PDGF, resulted in a somatomedin C and EGF concentration-dependent increase in the percentage of cells that initiated DNA synthesis (Fig. 1A). The simultaneous presence of nanogram concentrations (10 to 20 ng/ml) of both growth factors in the culture medium was capable of replacing the progression activity of 5% PPP, as determined by the number of PDGF-treated cells that synthesized DNA, whereas 20 ng/ml of either factor alone resulted in only 5% to 25% of the cells undergoing DNA synthesis (Fig. 1A,B). The variation in the percentage of PDGF-treated cells that enter the S phase in medium supplemented with only EGF or somatomedin C increased

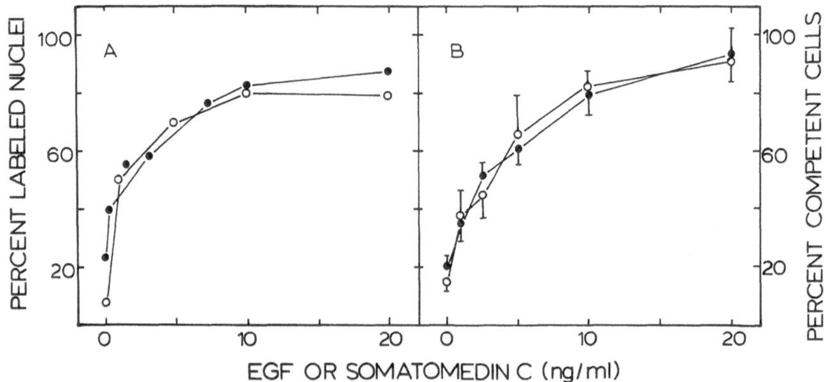

Figure 1. (A) BALB/c-3T3 cells were grown to confluence in 96-well Nunc microtiter plates in 0.2 ml Dulbecco-Vogt modified Eagle medium (DME) containing 10% calf serum. Two to three days after the cells reached confluence (4 to 5 days after initial plating), the spent medium was removed and 0.2 ml DME containing 5 μg/ml partially purified PDGF was added for 5 h. The cultures were then washed and DME containing 5 μCi/ml [^{3}H]thymidine plus 20 ng/ml EGF and varying concentrations of somatomedin C (●——●) or media containing 20 ng/ml somatomedin C and varying concentrations of EGF (○——○) were added. The cells were fixed 24 h later and processed for autoradiography. Platelet poor plasma (PPP) (5%) addition to competent cells allowed 90% of the cells to enter S phase. (B) Confluent cultures of BALB/c-3T3 cells were exposed to 0.7 μg/well PDGF further purified by Biogel P150 chromatography or 50 ng/well FGF for 5.5 h. The cultures were then washed and exposed to DME containing various concentrations of EGF or somatomedin C as previously described for (A). Individual points represent the mean of four experiments (two using Biogel PDGF, one using CM-Sephadex PDGF, and one using FGF). The data are plotted as the percent of competent cells that entered S phase (error bars represent the SEM) and in all cases 100% was defined as the percentage of PDGF- or FGF-treated cells that initiated DNA synthesis in DME supplemented with 5% PPP (ranged from 26% to 90% of the culture). It is demonstrated that identical results are observed, regardless of both the manner and the total number of cells which had previously been made competent.

with the number of previous cell passages and was related to the deterioration of proper growth control leading to the appearance of spontaneous transformants. Insulin, when used at concentrations of 10^{-6} to 10^{-5}M, could be substituted for somatomedin C (data not shown). At these hyperphysiological concentrations, insulin has been shown to bind to somatomedin C receptors in a wide variety of tissues and is thought to function as a somatomedinlike peptide (VanWyk *et al.*, 1975).

Since the experiments illustrated in Fig. 1A were performed with PDGF that had been purified on CM-Sephadex (Antoniades *et al.*,

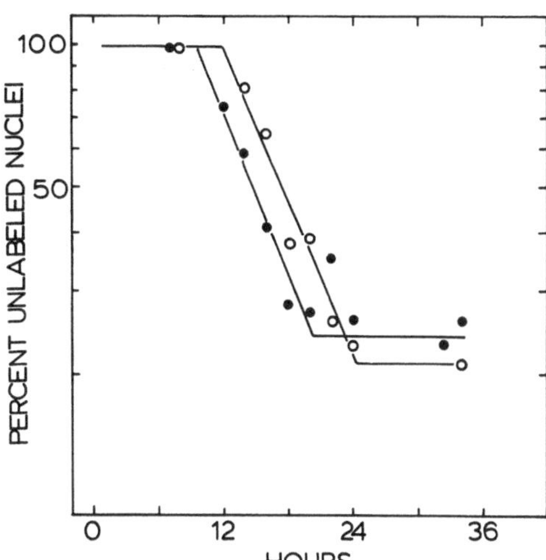

Figure 2. Quiescent density-arrested BALB/c-3T3 cells were exposed to 5 μg/well partially purified PDGF for 5 h. The cells were washed and a 1 : 1 mixture of DME and Ham's F12 media containing 5 μCi/ml [^{3}H]thymidine and either 5% PPP (○——○) or 10 ng/ml EGF plus 20 ng/ml somatomedin C (●——●) was added. At the indicated times, wells were fixed by the addition of 1 M ascorbic acid and the percentage of labeled cells was determined by autoradiography. Lines were drawn by linear regression analysis and had r values of 0.979 (●——●) and 0.974 (○——○).

1979), the same procedures were repeated using PDGF that had been further purified by chromatography on Biogel P150 in 1 M acetic acid (Antoniades *et al.*, 1979). Purified FGF, another polypeptide that initiates the growth of quiescent BALB/c-3T3 cells, was also used to repeat the experimental protocol of Fig. 1A. Figure 1B shows the results of these experiments. Epidermal growth factor (EGF) and somatomedin C were capable of replacing the progression activity of medium supplemented with 5% PPP independent of the particular growth factor used to initiate cellular proliferation. The ability of EGF and somatomedin C to replace the progression activity of plasma was independent of the percent of density-arrested cells that were stimulated to initiate cell cycle traverse. It should be pointed out that the BALB/c-3T3 cells not treated with PDGF were not stimulated to undergo DNA synthesis when transferred to fresh medium containing either plasma or EGF and somatomedin C.

EGF and Somatomedin C Reduce the Minimum G_1 Transit Time

The addition of plasma-supplemented medium to PDGF-treated BALB/c-3T3 cells brought about the initiation of DNA synthesis after a minimum lag of 12 h (Fig. 2). In a similar experiment, Leof *et al.* (1982b) showed when cells were transiently treated with PDGF, washed,

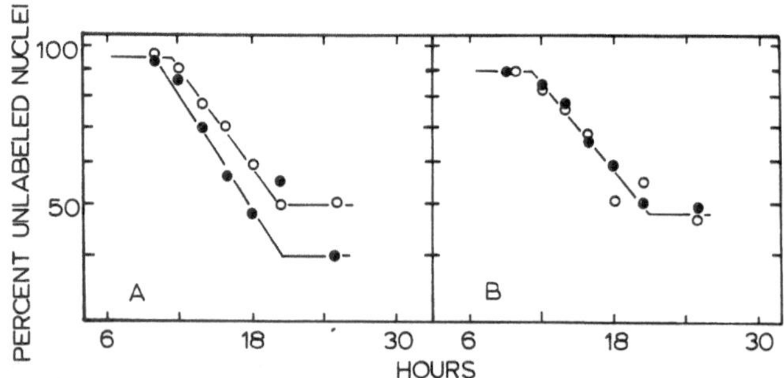

Figure 3. Quiescent density-arrested BALB/c-3T3 cells were made competent by a 5-h exposure to 5 μg/well PDGF. The cultures were then washed and transferred to DME supplemented with 5 μCi/ml [^{3}H]thymidine containing 5% PPP or 5% PPP plus various growth factors. At the indicated times, the cells were fixed and the percentage of labeled nuclei was determined. (A) Cellular response to PPP supplemented with 10^{-5} M insulin (●——●) or 15 ng/ml somatomedin C (○——○); (B) cellular response to unsupplemented PPP (●——●) or PPP supplemented with 15 ng/ml EGF (○——○). The lines were drawn by linear regression analysis. Correlation values for (A) 0.993 (●——●) and 0.994 (○——○); (B) r values 0.993 (●——●) and 0.936 (○——○).

and then transferred to medium supplemented with only EGF and somatomedin C, a decrease in the minimum lag time necessary for G_1 progression and initiation of DNA synthesis was observed (Fig. 2). As can be seen in Fig. 2, only 10 h were required after the addition of EGF and somatomedin-C-supplemented medium until DNA synthesis began. However, the maximum number of cells that synthesized DNA and the rate at which the cells entered S phase were the same in conditions of either PPP or EGF and somatomedin-C-supplemented medium. Insulin at concentrations greater than 10^{-6} M could also replace somatomedin C in these experiments. A decrease in the minimum lag time preceding S phase had been noted by Phillips and Cristofalo (1981) when serum was replaced with purified growth factors for the stimulation of quiescent WI-38 cells.

Availability of Free Somatomedin C Controls Minimum G_1 Transit Time

In order to explain this reduced minimum lag time (compared with PPP) following the addition of EGF and somatomedin-C-supplemented plasma-free medium to PDGF-treated cells, the experiment illustrated in Fig. 3 was performed. Density-inhibited BALB/c-3T3 cells were exposed to PDGF for 5 h. The PDGF-treated cells were then transferred to medium that had been supplemented with PPP or PPP plus either somatomedin C, insulin, or EGF. As can be seen in Fig. 3, the addition of insulin to PPP-supplemented medium allowed cells to enter S phase after a lag of 10 h as was observed using EGF and somatomedin-C-supplemented plasma-free medium (Fig. 2). In contrast to these findings, the addition of somatomedin C or EGF to PPP-supplemented medium allowed competent cells to enter DNA synthesis after a minimum lag of 12 h, a time similar to that found using medium supplemented with PPP alone. The difference in availability of free peptide as compared with peptide bound to plasma proteins may explain the differences in minimum G_1 lag times.

The high concentration of insulin used in Fig. 3 has been shown to be capable of mimicking the action of somatomedin C by binding to the somatomedin C receptor. Insulin differs from somatomedin C in that it does not bind to the sometomedin C plasma-binding proteins or other binding proteins found in plasma. The addition of small quantities of free somatomedin C to plasma could have resulted in the formation of a somatomedin-C-binding protein complex. Therefore, since plasma does not affect the 10-h lag time found in medium containing insulin, this observation is consistent with the hypothesis that the availability of free somatomedin C as compared with the total serum somatomedin C (including bound somatomedin C) was responsible for the different minimum lag times before entrance into S phase and directly regulates cell cycle traverse. Moreover, this also rules out the possibility that a plasma component is directly inhibiting or delaying the G_1 phase. Whether plasma also retards the availability of EGF cannot at the present time be directly deduced from these experiments; however, the factor that insulin by itself is ineffective in causing entry into DNA synthesis but causes entry at 10 h in the presence of plasma suggests that the additional plasma factor (presumably EGF) is freely available.

Somatomedin C Regulates Progression of Late G_1 Phase and Commitment to DNA Synthesis

Stiles *et al.* (1979b) have shown that plasma-supplemented medium deficient in essential amino acids brought about the inhibition of progression through G_1 at a point that preceded DNA synthesis by 6 h. This arrest point apparently bisected the plasma-dependent 12-h lag time before DNA synthesis. When the nutritionally arrested cells were

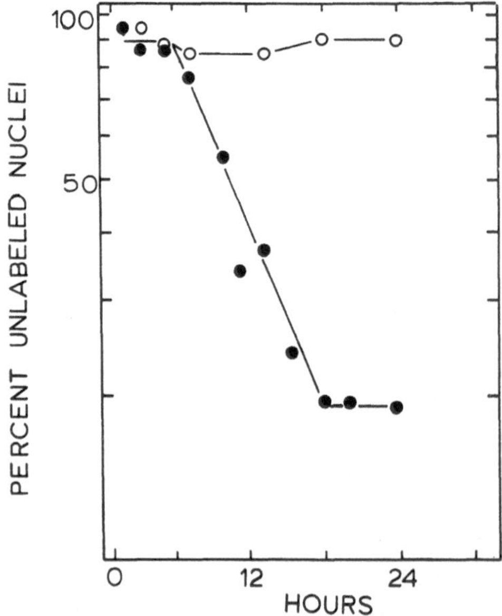

Figure 4. Quiescent density-arrested BALB/c-3T3 cells were placed in DME lacking group B amino acids and supplemented with 5% dialyzed PPP for 34 h. The cultures were arrested at the V point as described by Stiles *et al.* (1979b) after which they were washed with DME containing the full complement of amino acids and placed in DME supplemented with 5 μCi/ml [^{3}H]thymidine (O——O) or DME containing 5 μCi/ml [^{3}H]thymidine and 5% PPP (●——●). At the indicated times, wells were fixed by the addition of 1 M ascorbic acid and the percentage of labeled nuclei was determined. The r value was 0.972.

returned to complete medium, plasma was required for progression through the final 6 h of G_1 and entry into S phase.

In order to determine which factor or factors were required for progression through late G_1, cells were treated with PDGF and placed in dialyzed plasma-supplemented medium deficient in amino acids essentially as described by Stiles *et al.* (1979b) (Fig. 4). After the PDGF-treated cultures had been arrested in the plasma-supplemented amino-acid-deficient medium, the cells were washed and transferred to fresh medium containing all essential amino acids and supplemented with the appropriate growth factors. Figure 5 shows that Dulbecco-Vogt modified Eagle medium (DME) supplemented with 20 ng/ml somatomedin C was equivalent to 5% PPP in permitting progression from the nutrient arrest point to DNA synthesis. The progression activity of the somatomedin C could also be obtained with 10^{-5} M insulin (data not shown). Addition of unsupplemented DME or DME containing 10 to 20 ng/ml EGF or FGF was incapable of sustaining progression through G_1 and entry into S phase (Fig. 4 and data not shown). These data extend the findings of Wharton *et al.* (1981), who showed that somatomedin C alone would sustain progression from a point 4 h prior to DNA synthesis.

Regardless of the method of arrest (discussed earlier) at the 6-h pre-DNA synthesis "V" point, further traverse, 6 h, of G_1 (V point to

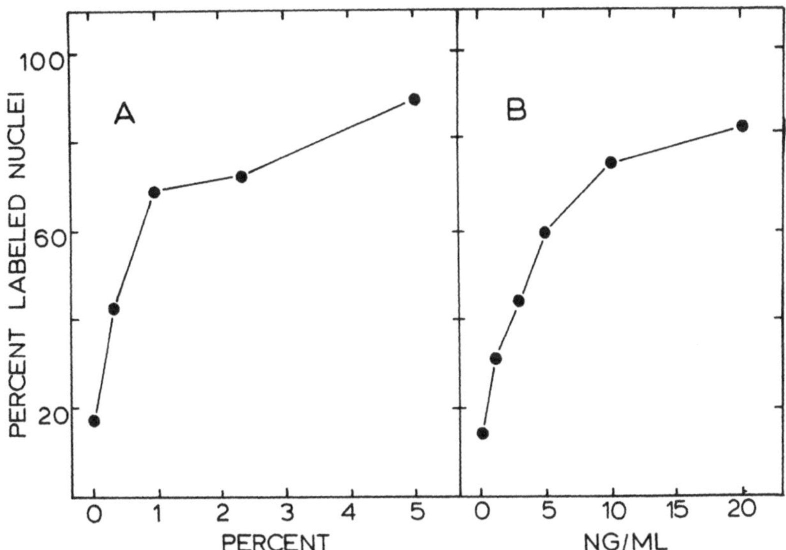

Figure 5. Confluent cultures were arrested at the "V" point as described in Fig. 4. The cells were washed and DME containing 5 μCi/ml [^{3}H]thymidine and various concentrations of either (A) PPP or (B) somatomedin C was added. Twenty-four hours later the plates were fixed and processed for autoradiography.

S phase) and commitment to DNA synthesis could be obtained in serum-free DME medium containing only nanogram quantities of somatomedin C.

EGF is Required during Traverse of Early G_1

EGF in combination with somatomedin C (or hyperphysiological concentrations of insulin) could completely replace the G_0/G_1 progression activity of 5% PPP for density-arrested BALB/c-3T3 cells that have been previously treated with PDGF (Leof *et al.*, 1982b). Since Stiles *et al.* (1979a) had shown that plasma from hypophysectomized rats containing reduced quantities of somatomedin C (and presumably normal EGF levels) allowed PDGF-treated cells to undergo progression only to the "V" point, it seemed likely that EGF played a critical role in the early G_1 progression of PDGF-treated cells either by acting alone or by acting in conjunction with low concentrations of somatomedin C. To discriminate between these possibilities, the experiments described below were undertaken in density-arrested BALB/c-3T3 cells that had

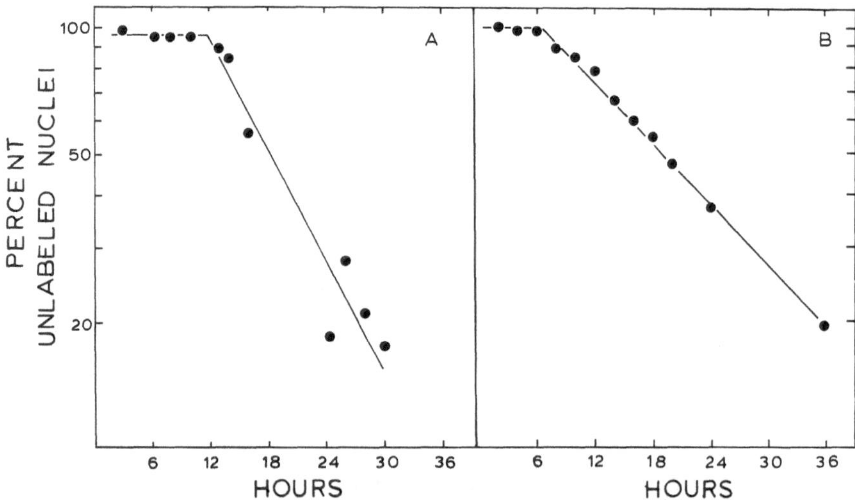

Figure 6. (A) Density-arrested BALB/c-3T3 cells were treated for 5 h at 37 °C with 5 μg/well partially purified PDGF. The cultures were then washed with DME and transferred to DME supplemented with 20 ng/ml EGF. After an 11-h incubation at 37 °C the cultures were then refed with DME supplemented with 5% PPP and 5 μCi/ml [^{3}H]thymidine (6.7 Ci/mmole). At the indicated times DNA synthesis was stopped by the addition of 60 μl/well 1 M ascorbic acid. The plates were washed with phosphate-buffered saline and fixed by two ten-minute applications of 100% methanol. The curve was drawn according to linear regression analysis and had a *r* value of −0.960. (B) Quiescent BALB/c-3T3 cells were treated for 3 h with DME supplemented with 2 μg/well PDGF that was further purified by Biogel P-150 chromatography in 1 M acetic acid (11). The plates were washed with DME and placed in DME containing 5 ng/ml EGF and 0.75 ng/ml somatomedin C. After a 12-h incubation at 37 °C, 5% plasma-supplemented DME plus 5 μCi/ml [^{3}H]thymidine was added. At the indicated times DNA synthesis was stopped by the addition of 1 M ascorbic acid and processed for autoradiography as described for Fig. 1A. Linear regression analysis indicated an *r* value of −0.998.

been previously exposed to PDGF. In some experiments insulin (1.0×10^{-8} M) was used as a substitute for somatomedin C.

In the first experiment, density-inhibited BALB/c-3T3 cells were exposed to PDGF for 5 h, after which the cells were rinsed with DME medium and transferred to fresh unsupplemented DME or DME supplemented with 1 to 50 ng/ml EGF. The cultures were incubated at 37 °C for 10 to 12 h in these media and then transferred to DME containing 5% PPP (0 time). Cells treated in this manner entered S phase after a minimum lag of 12 h after the addition of PPP (Fig. 6A). This lag time was similar to that found after immediate plasma addition to PDGF-treated cells. EGF alone was therefore not sufficient to provide the necessary proliferative signal(s) to allow PDGF-treated cells to progress temporally in the cell cycle. Incubation of the PDGF-treated cells for 36 to 48 h in medium containing EGF also did not allow the cells to progress into the G_1 phase (data not shown).

Figure 6B shows that the addition of a subnanogram concentration of somatomedin C to EGF-supplemented medium in an otherwise identical experiment (Fig. 6A) permitted traverse of PDGF-treated BALB/c-3T3 cells through the early portions of G_1 but not into S phase. After density-arrested cells were exposed to PDGF for 5 h, they were incubated in DME supplemented with EGF (5 to 20 ng/ml) and subnanogram concentration of somatomedin C (0.25 to 0.75 ng/ml) for 12 h. The cells were then washed and placed in DME containing 5% PPP and [^{3}H]thymidine. At the times indicated, cells were fixed and the percent of labeled nuclei determined. Following the addition of plasma-supplemented medium, these cells entered S phase after a minimum lag of only 6 h. This interval is similar to that previously reported in other types of "V" point arrest.

The "V" point arrest produced by incubating PDGF-treated cells in medium with EGF plus subnanogram concentrations of somatomedin C differed from other types of "V" point arrest; however, since recovery of G_0/G_1 progression could not be fully restored by the addition of higher dosages of somatomedin C (10 to 20 ng/ml) or by these dosages of somatomedin C plus EGF (10 ng/ml). The addition of these elevated concentrations of somatomedin C to the cultures treated with PDGF then incubated 12 h in EGF plus subnanogram concentrations of somatomedin C allowed only half as many cells to progress to DNA synthesis as when 5% PPP was added. This result is clearly different from other types of "V" point arrest where somatomedin C alone is fully as effective as 5% PPP (Stiles *et al.*, 1979b; Leof *et al.*, 1982a). The poor recovery from this growth arrest may be due to the prolonged incubation of the cell cultures arrested in mid G_0/G_1 in medium containing only limiting growth factors and no plasma. Apparently, arrest at the "V" point in limited growth factors left the arrested cells in a state that required an additional serum component(s).

Summary and Conclusions

Thus far we have shown that for PDGF-treated density-arrested BALB/c-3T3 cells:

1. EGF (10 ng/ml) plus somatomedin C (15 ng/ml) or insulin (10^{-5} M) can replace the progression activity of 5% PPP.

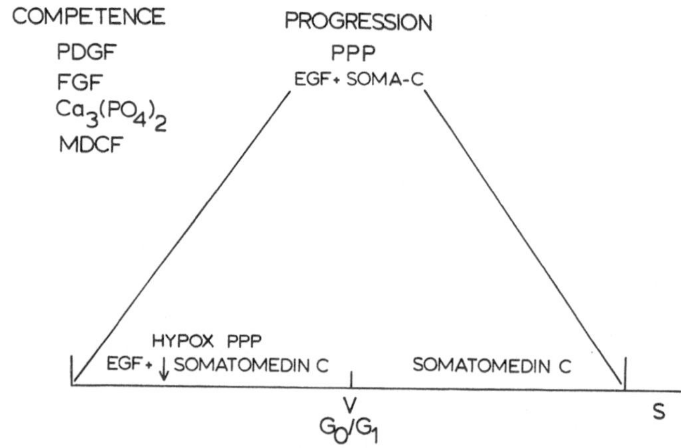

Figure 7. A schematic model that depicts the growth factor requirements of BALB/c-3T3 cells for discrete regions of G_0/G_1 is illustrated. The arrow down indicates that suboptimal serum levels of somatomedin C is required.

2. Following arrest at the "V" point by incubation in limited amino acid concentrations or by inclusion of cholera toxin and isobutylmethylxanthine in the incubation medium or incubation in hypopituitary plasma, somatomedin C (10 to 20 ng/ml) or insulin (10^{-5} M) alône will allow cells to continue the last 6 h of G_1 and initiate DNA synthesis.
3. EGF (10 ng/ml) plus subnanogram amounts of somatomedin C (0.25 to 0.75 ng/ml) or low levels of insulin (1.0×10^{-8} M) allow competent cells to traverse the early portion of G_1 and arrest at a point(s) 6 h prior to S phase; however, once cells are treated in this manner, EGF (10 ng/ml) plus somatomedin C (15 ng/ml) or insulin (10^{-5} M) supplemented medium is unable to reinitiate the total population to complete cell cycle traverse.
4. EGF is required only during the first six h of the 12-h G_0/G_1 lag time.

A model consistent with these data and that shows the portions of the G_0/G_1 phase regulated by somatomedin C and EGF is illustrated in Fig. 7. The model suggests unique functions for EGF and somatomedin C. Only somatomedin C is required for late G_1 traverse. EGF, on the other hand, is not required for traverse of the last 6 h of G_1 and will supplement subnanogram concentrations of somatomedin C in early G_1. Even though subnanogram amounts of somatomedin C (when present with EGF) are sufficient for early G_1 traverse, higher concentrations are necessary for traverse of the last 6 h of G_1. The greater requirement for somatomedin C during late G_1 may reflect an increased requirement for a function that this peptide regulates.

It seems unlikely that residual EGF in the cultures can explain the requirement for only somatomedin C during late G_0/G_1, as has been demonstrated following certain types of "V" point arrest. When PDGF-treated cells were exposed to either EGF for 12 h or EGF and somatomedin C for less than 6 h, the cells did not undergo DNA synthesis when transferred to medium containing only somatomedin C (data not shown). If significant quantities of EGF had remained in the cultures after these periods, the addition of 10 to 20 ng/ml of somatomedin C should have stimulated cell cycle, traverse since this concentration of somatomedin C and EGF can completely replace the requirement for plasma. However, this did not occur.

By using the minimum growth factor requirements for the stimulation of cellular proliferation, we can now investigate required biochemical regulatory events and synergistic interactions of growth factor. For example, PDGF has been shown to induce specific proteins (Pledger *et al.*, 1981) and mRNAs (Cochran and Stiles, 1982). Recently, Wharton *et al.* (1982) and Leof *et al.* (1982c) demonstrated that cAMP potentiated the PDGF-induced down regulation of EGF receptors. This almost total down regulation of EGF receptors brought about the abrogation of EGF reguirement during G_0/G_1 traverse. New studies designed to elucidate normal growth control and the alteration of these control events by transformation can now be carried out in the defined medium without aberrant effects from other serum components.

Acknowledgments. This work was supported by NCI Grants CA24193 and CA30479. W.J.P. was supported by JFRA-32 from the A.C.S.

References

Antoniades, H. N., Scher, C. D., and Stiles, C. D., 1979, Purification of human platelet-derived growth factor, *Proc. Natl. Acad. Sci. U.S.A.* **76:**1809.

Balk, S. D., 1971, Calcium as a regulator of the proliferation of normal, but not transformed chicken fibroblasts in a plasma-containing media, *Proc. Natl. Acad. Sci. U.S.A.* **68:**271.

Barnes, D., and Sato, G., 1980, Methods for growth of cultured cells in serum-free medium, *Anal. Biochem.* **102:**255.

Clemmons, D. R., and VanWyk, J. J., 1981, Somatomedin C and platelet-derived growth factor stimulate human fibroblast replication, *J. Cell Physiol.* **106:**361.

Clemmons, D. R., Underwood, L. E., and VanWyk, J. J., 1981, Hormonal control of immunoreactive somatomedin production by human fibroblasts, *J. Clin. Invest.* **67:**10.

Cochran, B. H., and Stiles, C. D., 1982, Molecular cloning of gene sequences regulated by platelet-derived growth factor, *J. Cell Biol.* **95:**198.

Jimenez de Asua, L., O'Farrell, M. K., Clingan, D., and Rudland, P. S., 1977, Temporal sequence of hormonal interactions during the prereplicative phase of quiescent cultured 3T3 fibroblasts, *Proc. Natl. Acad. Sci. U.S.A.* **74:**3845.

Kohler, N., and Lipton, A., 1974, Platelets as a source of a fibroblast growth-promoting activity, *Exp. Cell Res.* **87:**297.

Leof, E. B., Wharton, W., O'Keefe, E., and Pledger, W. J., 1982a, Elevated intracellular concentrations of cyclic AMP inhibited serum-stimulated density-arrested BALB/c-3T3 cells in mid-G_1, *J. Cell. Biochem.* **19:**93.

Leof, E. B., Wharton, W., VanWyk, J. J., and Pledger, W. J., 1982b, Epidermal growth factor (EGF) and somatomedin C regulate G_1 progression in competent BALB/c-3T3 cells, *Exp. Cell. Res.* **141:**107.

Leof, E. B., Olashaw, N. E., Pledger, W. J., and O'Keefe, E. J., 1982c, Cyclic AMP potentiates down regulation of epidermal growth factor receptors by platelet-derived growth factor, *Biochem. Biophys. Res. Commun.* **109:**83.

Otto, A. M., Natoli, C., Richmond, K. M., Iacobelli, S., and Jimenez de Asua, L., 1981, Glucocorticoids inhibit the stimulatory effect of epidermal growth factor on the initiation of DNA synthesis, *J. Cell Physiol.* **107:**155.

Phillips, P. D., and Cristofalo, V. D., 1981, Growth regulation of WI-38 cells in a serum-free medium, *Exp. Cell Res.* **134:**297.

Pledger, W. J., Stiles, C. D., Antoniades, H. N., and Scher, C. D., 1977, Induction of DNA synthesis in BALB/c-3T3 cells by serum components: Reevaluation of the commitment process, *Proc. Natl. Acad. Sci. U.S.A.* **74:**448.

Pledger, W. J., Stiles, C. D., Antoniades, H. N., and Scher, C. D., 1978, An ordered sequence of events is required before BALB/c-3T3 cells become committed to DNA synthesis, *Proc. Natl. Acad. Sci. U.S.A.* **75:**2839.

Pledger, W. J., Hart, C. A., Locatell, K. L., and Scher, C. D., 1981, Platelet-derived growth factor-modulated proteins: Constitutive synthesis by a transformed cell line, *Proc. Natl. Acad. Sci. U.S.A.* **78:**4358.

Ross, R., Glomser, J., Kariya, B., and Harker, L., 1974, A platelet-dependent serum factor that stimulates the proliferation of arterial smooth muscle cells *in vitro, Proc. Natl. Acad. Sci. U.S.A.* **71:**1207.

Scher, C. D., Pledger, W. J., Martin, P., Antoniades, H., and Stiles, C. D., 1979, Transforming viruses directly reduce the cellular growth requirement for a platelet derived growth factor, *J. Cell. Physiol.* **97:**371.

Stiles, C. D., Capone, G. T., Scher, C. D., Antoniades, H. N., VanWyk, J. J., and Pledger, W. J., 1979a, Dual control of cell growth by somatomedins and platelet-derived growth factor, *Proc. Natl. Acad. Sci. U.S.A.* **76:**1279.

Stiles, C. D., Isberg, R. R., Pledger, W. J., Antoniades, H. N., and Scher, C. D., 1979b, Control of BALB/c-3T3 cell cycle by nutrients and serum factor: Analysis using platelet-derived growth factor and platelet-poor plasma, *J. Cell Physiol.* **99:**395.

VanWyk, J. J., Underwood, L. E., Baseman, J. B., Hintz, R. L., Clemmons, D. R., and Marshall, R. N., 1975, Explorations of the insulin-like and growth-promoting properties of somatomedin by membrane receptor assays, *Adv. Metab. Disord.* **8:**127.

Vogel, A. E., Raines, B., Rivest, M. J., and Ross, R., 1978, Coordinate control of 3T3 cell proliferation by platelet-derived growth factor and plasma components, *Proc. Natl. Acad. Sci. U.S.A.* **75:**2810.

Wharton, W., VanWyk, J. J., and Pledger, W. J., 1981, Inhibition of BALB/c-3T3 cells in late G_1: Commitment to DNA synthesis controlled by somatomedin C, *J. Cell Physiol.* **107:**31.

Wharton, W., Gillespie, G. Y., Russell, S. W., and Pledger, W. J., 1982a, Mitogenic activity elaborated by macrophage-like cell lines act as competence factors for BALB/c-3T3 cells, *J. Cell. Physiol.* **110:**93.

Wharton, W., Leof, E. B., Pledger, W. J., and O'Keefe, E. J., 1982b, Modulation of the epidermal growth factor receptor by platelet-derived growth factor and choleragen: Effects on mitogenesis, *Proc. Natl. Acad. Sci. USA* **79:**5567.

CHAPTER 2

Regulation of Fibroblast Growth by Multiple Growth Factors in Serum-Free Media

PAUL V. CHERINGTON

Introduction

In vitro studies of cellular growth regulation are based on observations that monolayer cultures of cell lines derived from normal tissue reversibly arrest growth at high density (Todaro and Green, 1963; Todaro *et al.*, 1965). What originally made this model of growth control particularly interesting were demonstrations that agents that induced tumors *in vivo* also induced overgrowth of quiescent cell monolayers into piles of cells called foci. Early demonstrations of *in vitro* transformation included studies of carcinogenic hydrocarbons (Chen and Heidelberger, 1969), Rous sarcoma virus (Temin and Rubin, 1958), and the DNA[1] tumor viruses SV40 (Todaro and Green, 1964), and polyoma (Stoker and MacPherson, 1961). Thus, an apparent correlation existed

[1] The abbreviations used include: DNA, deoxyribonucleic acid; RNA, ribonucleic acid; EGF, epidermal growth factor; FGF, fibroblast growth factor; PMA, phorbol myristate acetate; PDGF, platelet-derived growth factor; TGF, transforming growth factor; $PGF_{2\alpha}$, prostaglandin $F_{2\alpha}$; BSA, bovine serum albumin; DME, Dulbecco's modified Eagles medium; MEM, minimal essential medium; MSA, multiplication-stimulating activity; IGF, insulinlike growth factor; FDGF, fibroblast-derived growth factor; BHK, baby hamster kidney, CHEF, Chinese hamster embryo fibroblast; PGE_1, prostaglandin E_1; cAMP, adenosine 3′,5′-cyclic monophosphate; HDL, high-density lipoprotein; IgM, immunoglobulin M; IgG, immunoglobulin G; and CnBr-EGF, cyanogen-bromide-cleaved EGF.

PAUL V. CHERINGTON • Laboratory of Neoplastic Disease Mechanisms, Dana-Farber Cancer Institute, Boston, Massachusetts 02115.

between aberrant growth in monolayer cultures and aberrant growth of tissue *in vivo*. The exact relationship between the properties of transformed cell lines *in vitro* and the invasive and metastatic behavior of tumors *in vivo* is not resolved. However, the cellular changes that permit growth of tumor cells *in vivo* and *in vitro* under conditions normally not suited for cell proliferation are likely to be related.

Quiescent cell populations *in vivo* and *in vitro*, with rare exceptions, are arrested in the cell cycle between mitosis (M phase) and DNA synthesis (S phase), that is, in the G_1 phase. Cells that have arrested growth during the G_1 phase are in a state sometimes referred to as G_0 to indicate a physiological difference between cycling cells in G_1 and quiescent cells. Once cells enter S phase they are committed to mitosis and will not arrest at any point in the cell cycle prior to cell division provided that the medium contains sufficient nutritional supplements. Thus, the biochemical events in G_1 that determine ability of different normal and transformed cells to regulate transit of G_1 phase to S phase are of particular interest in a discussion of growth regulation. The reader is referred to several reviews on the regulation of the animal cell cycle (Prescott, 1976; Baserga, 1976; Leffert and Koch, 1977; Pardee *et al.*, 1978). An excellent introduction to the literature on cell transformation in tissue culture is provided by Pollack (1981).

The following points are discussed in Sections 2 and 3 of this chapter: (1) the characteristics of the density-arrested state and how they relate to serum and/or nutrient deprivation, and (2) relaxed growth regulation *in vitro* as related to relaxed requirements for serum or growth factors. The serum-free medium supplements needed to study the altered medium requirements for growth after transformation are discussed in Sections 4, 5, and 6. Finally, studies that have used serum-free media to study the altered growth requirements of transformed cells are described in Section 7. What will become clear in these sections is that events in G_1 regulated by growth factors and hormones are altered in transformed cells so that the requirements for one or more of these factors are reduced or eliminated. Thus, using serum-free growth conditions, a range of relaxed serum requirements for growth of transformed cells can be further defined as the elimination of one or more growth factor requirements. Such resolution is not possible in serum-containing media.

In Vitro Models of Growth Regulation—Why Do Cells Arrest Their Growth?

A proliferating cell, with rare exceptions, is committed to mitosis by the time it initiates DNA synthesis. If the conditions for continued growth are not optimal, the postmitotic cell arrests its progress through the cell cycle prior to DNA synthesis. The earliest events leading to DNA synthesis and another cycle of growth occur well before mitosis, for example, the maturation of mitotic centers (Brooks *et al.*, 1980) or the induction of "competence" (Scher *et al.*, 1979a). Although most of the events leading to DNA synthesis occur in G_1 in most cell lines, a G_1 period is not necessarily required for DNA synthesis. A Chinese hamster cell line has been studied that expressed no G_1 period; DNA synthesis occurred immediately following mitosis (Liskay, 1977; Liskay and Prescott, 1978). Such cells must complete all preparation for DNA synthesis in the preceding cell cycle thus obviating the "need" for a G_1 period prior to S phase. Most cells studied in culture, however, take longer to prepare for DNA synthesis and express a G_1 phase.

The most commonly used models of arrested cell populations include cell lines grown to high density and sparse cultures deprived of adequate serum supply, for example, 0.25% serum instead of 5% to 10%. Cells deprived of serum arrest growth because the growth factors and nutrients needed for G_1 transit that are provided by serum are no longer present. A small amount of serum is usually provided because serum provides factors necessary for survival as well as stimulation of DNA synthesis (see Section 4). Depriving cells of an essential amino acid, for example, isoleucine or glutamine, can result in a G_1-arrested population, but often such nutrient-deprived populations exhibit kinetics of entry into S phase that are different from serum-deprived or high-density cells. Thus, nutrient-deprived cells are thought to be arrested in a different state.

Deoxyribonucleic acid synthesis and cell proliferation that occur after serum or growth factor stimulation can be assayed in a number of ways, but the most common methods are outlined below. Stimulation of DNA synthesis is most often determined by assaying the incorporation of tritiated thymidine into acid precipitable material. After stimulation of cells with serum or growth factors, the rate of incorporation of pulses of tritiated thymidine, or cumulative incorporation over longer labeling periods, may be determined using liquid scintillation counting (Rossow, *et al.*, 1979). The proportion of the population entering S phase, that is, the labeling index, may be determined by fixing tritiated thymidine labelled cells in the culture dish and counting the number of nuclei using autoradiography (Rossow *et al.*, 1979; Stein and Yanishevsky, 1979). Nuclear DNA may be stained with a fluorescent dye and measured on a per nucleus basis cytofluorometrically (Gray and Coffino, 1979). This technique determines the proportion of the cell population in each phase of the cell division cycle. Such data are very useful for determining the proportion of cells arrested at a given point in the cell cycle under different conditions and for monitoring progress through the cell cycle following mitogenic stimulation. Cell proliferation is most commonly measured by counting cell number using a Coulter counter (Cherington and Pardee, 1980).

Whether cells arrested in G_0 are really outside the cell cycle in a biochemical state that is distinct from G_1 (Baserga, 1976) or whether these cells are cycling infinitely slowly (Smith and Martin, 1973) is not resolved. Nonetheless, the rate of entry into S phase is greatly reduced, if not abolished, under some conditions, and such cells are considered to be arrested in this discussion. Some differences between arrested and growing cells include the following: (1) G_0 cells take longer to reach S phase after stimulation than cycling cells take to reach S phase after mitosis (Martin and Stein, 1976; Pardee *et al.*, 1978), (2) human fibroblasts enter a "deeper" G_0 state with time based on decreasing transcriptional and translational activity (Abelson *et al.*, 1974; Rudland *et al.*, 1975; Rossini *et al.*, 1976; Robinson *et al.*, 1976), and (3) certain drug-sensitive processes occur in quiescent cells preparing for DNA synthesis that do not occur in growing cells. For example, DNA synthesis was inhibited by the amino acid analogue histidinol in quiescent Swiss mouse 3T3 cells after serum stimulation, but DNA synthesis was not inhibited in cells progressing from mitosis to S phase (Yen *et al.*, 1978).

Several mechanisms have been proposed to explain why cells grown to high density in culture arrest their growth. The mechanisms proposed include the following: (1) depletion of serum-derived growth factors or depletion of serum- or medium-derived nutrients such as amino acids or glucose that are necessary for cells to enter S phase (Holley *et al.*, 1976; Moses *et al.*, 1978; Vogel *et al.*, 1980), (2) cell–cell contact resulting in inhibition of entry into S phase, i.e., "contact inhibition" (Whittenberger and Glaser, 1977), and (3) rounding of cells at high density (Folkman and Moscona, 1978).

Serum-derived growth factors and the nutrient supply are important variables limiting cell growth at high density. Stoker and Rubin (1967) and Holley and Kiernan (1968) pointed out that the final density reached by cultures of baby hamster kidney cells or mouse 3T3 cells was determined by the amount of serum and nutrients provided. In contrast to what the term "contact inhibition" of growth suggested, the final density was not a function of cell–cell contact, but was a function of nutrient and mitogen supply. In support of this conclusion, different cell lines deplete different medium constituents first. A normal mouse-embryo-derived cell line arrested at high-cell density due to depletion of serum-derived growth factors; supplying fresh serum alone or purified growth factors such as epidermal growth factor (EGF), fibroblast growth factor (FGF), or the tumor promoter phorbol myristate acetate (PMA) stimulated DNA synthesis in those cells (Moses *et al.*, 1978). A chemically transformed derivative of this cell line also arrested growth at high-cell density, but fresh medium, lacking serum or growth factors, stimulated these cells demonstrating that these cells depleted medium-derived nutrients before serum-derived growth factors.

Cell–cell contact and cell surface components may contribute to the density-arrested state, but serum-derived growth factors are ultimately growth limiting for normal cells. Whittenberger and Glaser (1977) have shown that plasma membrane fractions of Swiss 3T3 cells reversibly arrested 3T3 cell growth. This inhibition was partially reversed by increased serum (Whittenberger *et al.*, 1978). Thus, any inhibition of growth mediated by cell–cell contact is overcome by serum.

Folkman and Moscona (1978) reported that they manipulated the growth rate of endothelial cells by altering the cell shape. Cells were placed on substrata that contained films of poly(2-hydroxyethyl) methacrylate [poly(HEMA)] of different thicknesses. The degree to which cells flattened on such substrata was inversely related to the film thickness. The cell growth rate was directly related to the extent to which the cells were allowed to flatten. They also observed that cells in a dense monolayer were rounded much like sparse cells on poly(HEMA). They proposed that the cell shape determined, directly or indirectly, the growth rate. This proposal helped to explain the observations made years earlier by Dulbecco (1970) that cells at the edge of a wound or scratch in a confluent monolayer had a higher labeling index than cells in the middle of the monolayer. Cells at the wound edge spread into the wound, thus flattening and initiating DNA synthesis.

Serum-derived growth factors can overcome growth restriction by cell–cell contact or cell shape. Cells that spread from a wound in a monolayer maintain cell–cell contacts, but entered S phase (Dulbecco, 1970; Folkman and Moscona, 1978), perhaps because flatter cells on an appropriate substratum were more responsive to serum factors (Clarke *et al.*, 1970). Increased serum concentration stimulated growth in density-arrested cells (Todaro *et al.*, 1965; Temin, 1971). Thus, the ability of the cell to respond to the available growth factors can regulate the growth state of that cell.

Neoplastic Transformation and Relaxed Growth Regulation

The term "transformation" regarding animal cell cultures has become synonymous with properties such as altered cellular morphology and cytoskeleton (Pollack *et al.*, 1975), decreased growth requirement for serum, decreased anchorage dependence, or increased density at which growth stops, that is, saturation density (Todaro *et al.*, 1965; Dulbecco, 1970; Clarke *et al.*, 1970). All or some of these growth regulatory phenomena may be affected by a given oncogenic agent such as DNA- or RNA-containing tumor viruses, or carcinogens, or they may be selected for by manipulation of the culture conditions. The correlation between transformed phenotype *in vitro* and tumorigenic potential *in vivo* is never absolute; many cell lines displaying abnormal growth regulation *in vitro* are not tumorigenic (Stiles and Kawahara, 1978). The purpose of this discussion will be to point out the different transformed phenotypes expressed *in vitro* as they relate to growth regulatory processes.

Many transformed cell lines do not arrest growth in G_1 on serum deprivation (Pardee and James, 1975; Bartholomew *et al.*, 1976; Cherington *et al.*, 1979; Dubrow *et al.*, 1979). The transforming agent in such cases tends to be DNA tumor viruses such as SV40 or polyoma. Although no requirements for DNA synthesis stimulating growth factors are expressed in cell lines lacking G_1 growth regulation, growth may be retarded throughout the cell cycle on serum deprivation due to deficiency of nutritional factors supplied by serum. These factors include iron and its transport protein, transferrin (Pardee *et al.*, 1980), biotin (Young *et al.*, 1979), and unsaturated fatty acids (Messmer and Young, 1977).

Transformed cell lines often retain the ability to regulate growth in G_1 on serum deprivation (Holley *et al.*, 1976; Moses *et al.*, 1978; Cherington *et al.*, 1979; Dubrow *et al.*, 1979). This indicates that these cells require one or more serum-derived growth factors, for example, insulin or the related somatomedins (Cherington *et al.*, 1979). However, the serum concentration required for growth may be diminished. The decreased serum requirement is probably a result of the loss of the requirement for an individual growth factor such as EGF (Cherington *et al.*, 1979) or PDGF (Scher *et al.*, 1978). Cell lines transformed by RNA or DNA tumor viruses or chemicals or "spontaneously" transformed cell lines selected *in vitro* have been described that retained G_1 growth regulation while losing a specific growth factor requirement (Scher *et al.*, 1978; Cherington *et al.*, 1979; Rockwell *et al.*, 1980; Scher *et al.*, 1982).

The loss of a requirement for an individual growth factor could occur via more than one mechanism. The process triggered by that growth factor could become constitutive or an endogenous supply of the growth factor, or its functional equivalent, could be produced by the cell thus obviating the need for exogenous supplies. In the former case, persistent activation of an enzyme, such as a protein kinase, or the production of a stable form of a labile protein are both possible. A candidate for such a regulatory protein is an antigen of approximately 54,000 daltons that in some systems is present at low levels in growth-controlled cells and at higher levels in virally transformed cells. The levels of this protein are regulated by its rate of degradation rather than its rate of synthesis (Oren *et al.*, 1981). Kinetic analysis of the entry of cells into S phase after partial or complete inhibition of protein synthesis has indicated that a growth regulatory protein (or proteins) is more stable in transformed cells (Rossow *et al.*, 1979; Medrano and Pardee, 1980; Campisi *et al.*, 1982).

Some transformed cells produce and release proteins into the medium that act as growth factors and confer a transformed phenotype on the producing cell or neighboring cells. DeLarco and Todaro (1978, 1980) discovered a material secreted by cells transformed by murine or feline sarcoma viruses that was originally called sarcoma growth factor. Sarcoma growth factor represents one example of a more general group of secreted materials known as transforming growth factors (TGF). The TGF reported by DeLarco and Todaro was mitogenic and it bound to EGF receptors (DeLarco and Todaro, 1980). However, it was more active than EGF since it stimulated growth in soft agar and conferred a transformed morphology on cells exposed to it (DeLarco and Todaro, 1978).

Transforming growth factor activity may be composed of more than one component. Studying TGF activity contained in medium conditioned by chemically transformed AKR mouse cells, Moses *et al.* (1981) used ion exchange chromatography to dissociate the EGF receptor binding activity from the soft agar growth-stimulating activity. In another study, acid/ethanol extracts of normal and neoplastic cells or tissue contained TGF activity. The EGF receptor binding and two distinct soft agar growth-stimulating activities of these TGFs were separated using high-pressure liquid chromatography (Roberts *et al.*, 1982). Epidermal growth factor potentiated the soft agar growth-stimulating activity of certain extracted TGFs; the ability of EGF to potentiate was related to the endogenous EGF content of the extracts (Roberts *et al.*, 1982).

Transforming growth factors have growth factor activity in serum-free media. The growth of a rat cell line (rat-1) in a defined medium containing transferrin, insulin, and TGF was nearly as rapid as growth in 10% serum (Kaplan *et al.*, 1982). The TGF in this work was partially purified from medium conditioned by Kirsten murine sarcoma virus-transformed rat kidney cells. This material also disrupted actin cables and stimulated anchorage-independent growth (Ozanne *et al.*, 1980), thus conferring a transformed phenotype on normal cells.

Secretion of TGFs is not evident in all transformed cells (Cherington and Pardee, 1982). Either some transformed cells do not secrete these endogenous growth factors or they secrete only sufficient quantities to stimulate the cell producing them. Concentrations reached in the medium may not be sufficient to stimulate neighboring cells or cells in heterologous cultures.

Replacing Serum with Defined Medium Components to Study Growth Regulation *in Vitro*

The use of serum-free or serum-depleted conditions to study factors that stimulate DNA synthesis in growth-arrested cells predates our ability to grow and passage cell lines in the absence of serum. This is because agents that stimulate DNA synthesis, for example, growth factors and hormones, are not sufficient to support the transit of a cell from S phase through mitosis and through subsequent cell division cycles. Therefore, the different contributions by serum to the medium must be considered when attempting to create serum-free cultures. Bottenstein *et al.* (1979) and Barnes and Sato (1980) have presented detailed discussions of factors to consider when preparing serum-free media. Some contributions of serum particularly relevant to growth-control studies are briefly discussed below.

Some of the more important contributions of serum to the culture conditions are the following: (1) basement membrane components such as fibronectin that promote attachment and spreading of cells on the growth surface (Yamada and Olden, 1978), (2) low-molecular-weight nutrients such as lipids, amino acids, vitamins, and trace minerals that are not present in the basal nutrient medium or are present in suboptimal concentrations (Ham and McKeehan, 1979), (3) transport proteins such as transferrin that facilitate the uptake of nutrients and minerals required by the cells (Rudland *et al.*, 1977; Ham and McKeehan, 1979), and (4) growth factors or hormones that stimulate DNA synthesis. Proteins such as albumin or fetuin in serum conceivably also play important roles in tissue culture. They can bind toxic substances, stabilize labile, essential substances in the medium, and they possess some pH-buffering capacity (Ham and McKeehan, 1979).

Adhesion-Promoting Components

Anchorage-dependent cells must attach and spread on the substratum before growth can occur. Although tissue culture plastic is specially treated to permit attachment of animal cells, serum facilitates attachment for many cell types. Creating a positive charge on the culture surface by coating with polylysine, for example, increases the efficiency with which cells adhered to the plastic (McKeehan and Ham, 1976). Such a treatment, however, did not necessarily promote spreading and optimal growth of animal cells. Cell-derived fibronectin or plasma fibronectin (cold insoluble globulin) promoted spreading and growth in the absence of serum on substrata of plastic, or on matrices of gelatin, fibrin, or collagen (Grinnell and Minter, 1978; Grinnell and Feld, 1979; Chiquet *et al.*, 1979; Grinnell *et al.*, 1980). Some cell types attached and spread efficiently in the absence of serum or exogenous fibronectin because they were capable of secreting enough endogenous fibronectin (Grinnell and Feld, 1979; Cherington and Pardee, 1980). Pretreating the growth surface with serum or allowing cells to attach in the presence of serum and subsequently shifting to serum-free media were also effective (Barnes and Sato, 1980), although contaminating factors provided by these procedures must be considered.

Extracellular matrix material can replace growth factors such as FGF in promoting cell proliferation (Gospodarowicz *et al.*, 1982). This could occur by promoting attachment and thus altering cell shape and sensitivity to growth factors present. Alternatively, the extracellular matrix could harbor tightly adherent growth factors such as FGF or PDGF (Smith *et al.*, 1982).

Nutrients and Trace Minerals

Serum contains a range of low-molecular-weight nutrients that are required for growth or for maximal stimulation of DNA synthesis by mitogens. Most of these nutrients, for example, glucose, pyruvate, phosphate, amino acids, some vitamins, and bulk ions such as sodium, potassium, calcium, and magnesium, are contained in standard media. Some nutrients required for optimal DNA synthesis and cell proliferation are not in all standard nutrient media. Dulbecco's modified Eagles medium, for example, has no biotin, vitamin B_{12}, fatty acid, or purine source (Ham and McKeehan, 1979). Each of these compounds is required for optimal growth and stimulation of DNA synthesis by purified growth factors.

Biotin and fatty acids normally provided by serum were required by Swiss mouse 3T3 cells for maximal stimulation of DNA synthesis by growth factors such as prostaglandin $F_{2\alpha}$ ($PGF_{2\alpha}$) and insulin (Hatten *et al.*, 1977; O'Farrell *et al.*, 1979). In addition, delipidized serum was biotin deficient and did not support the growth of 3T3 cells or SV40-transformed 3T3 cells; reconstitution of delipidized serum with fatty acids and biotin allowed growth of these cells (Hatten *et al.*, 1977). Also, supplementing media containing low (0.2%) serum with fatty acids and biotin allowed the growth of SV40-transformed 3T3 cells (Messmer and Young, 1977). Rockwell *et al.* (1980) grew SV40-transformed 3T3 cells without serum by providing *cis*-unsaturated fatty acid, BSA, insulin, and transferrin and an adequate nutrient medium containing biotin.

Vitamin B_{12} is also important for the stimulation of DNA synthesis by serum or growth factors. Mierzejewski and Rozengurt (1977) had to supplement serum-containing media with vitamin B_{12} for optimal stimulation of DNA synthesis in 3T6 cells. Stimulation of Swiss 3T6 cells by EGF and insulin (Mierzejewski and Rozengurt, 1976) and stimulation of both 3T3 and 3T6 cells by $PGF_{2\alpha}$ and insulin (O'Farrell *et al.*, 1979) was vitamin B_{12} dependent. Hypoxanthine and other purines increased the rate of DNA synthesis stimulated by growth factors or serum in baby hamster kidney cells (Clarke and Smith, 1973; Brooks, 1975), and 3T3 cells (Schor and Rozengurt, 1973; O'Farrell *et al.*, 1979).

Ham and McKeehan (1979) have shown that for cloning or continuous passage of animal cells in the absence of serum, or in the presence of limiting concentrations of serum proteins, a variety of trace minerals were required that were not normally included in basal nutrient media. These minerals included iron, zinc, and selenium. In addition, copper, manganese, molybdenum, and vanadium may be required by animal cells. Normally, such trace minerals are provided in sufficient quantities in serum itself, or are present as trace contaminants in most serum-free media. However, when stringent conditions of serum-free cloning are imposed or when cell lines are passaged in serum-free media, the requirements for these elements may be revealed.

Dr. Richard Ham and his collaborators have developed basal nutrient media that contain many of the above-mentioned components, which are omitted from standard media such as DME or MEM. A readily available medium designed by Ham (1965) is called F12, and has been very useful by itself or when mixed with DME in designing serum-free growth media for a variety of cell types (Hayashi and Sato, 1976; Bottenstein *et al.*, 1979; Barnes and Sato, 1980). More recently developed media (MCDB 104, 105, 107, 108, and 402) for serum-free growth incorporate much of the recent knowledge concerning the nutrient and trace element requirements of animal cells (McKeehan *et al.*, 1977, 1978; McKeehan and McKeehan, 1980a; Shipley and Ham, 1981; Walthall and Ham, 1981). McKeehan and McKeehan (1980b) have recently suggested that serum-derived growth factors merely make nutrients and ions more available to the cell; given the appropriate medium composition, a cell would not require growth factors. Further evidence is required to generalize this concept to all cells.

Transport Proteins

Several proteins in blood serve as carriers for compounds essential for cell growth and survival. Hemoglobin transports oxygen, albumin binds a wide range of essential compounds such as steroid hormones or vitamins, and transferrin binds iron and facilitates its transport into cells. Transferrin and iron are essential for growth of cells in culture as well as *in vivo*. Most serum-free media reported to date utilize transferrin (see Bottenstein *et al.*, 1979; Barnes and Sato, 1980, for examples).

Transferrin facilitates iron transport throughout the cell cycle, but does not regulate growth. Transformed cells retained their requirement for transferrin although requirements for factors that regulated entry into S phase were often relaxed (Cherington *et al.*, 1979; McClure *et al.*, 1979; Young *et al.*, 1979). Depriving cells of factors that regulated growth and stimulated DNA synthesis resulted in the arrest of the cell population in G_1 (Pardee *et al.*, 1980). In contrast, transferrin and iron deprivation resulted in cell death or growth arrest throughout the cell cycle. Omission of transferrin did not significantly decrease the stimulation of DNA synthesis in quiescent fibroblasts by growth factors (Rudland *et al.*, 1977; Cherington and Pardee, 1980).

Factors That Stimulate DNA Synthesis

Most mammalian factors that stimulate growth of cell cultures also stimulate DNA synthesis. These factors include hormones that were well characterized when their mitogenic actions were first observed, for example, insulin (Hollenberg and Cuatrecasas, 1975; Vaheri *et al.*, 1973) and $PGF_{2\alpha}$ (Jimenez de Asua *et al.*, 1975, 1977a). Mitogenic activities have also been detected in and occasionally purified from serum, plasma, pituitary glands, salivary glands, and platelets or from media conditioned by cell cultures. Certain factors act synergistically to stimulate DNA synthesis, indicating distinct but complementary modes of action; other combinations of factors are merely additive at suboptimal doses, suggesting that these factors are stimulating the same intracellular events that lead to DNA synthesis.

Combinations of Growth Factors Are Required in Defined Media to Support Cell Proliferation

Growth factors stimulate DNA synthesis in resting cells and keep growing cells from entering the resting (G_0) state. They are insufficient to stimulate sustained proliferation of a cell culture without the appropriate nutritional factors. In addition, no single growth factor is sufficient to stimulate maximal DNA synthesis in a normal, G_0-arrested population. Combinations of factors are needed. Armelin (1973) made one of the earliest reports in this regard. He observed that a combination of supplements, hydrocortisone and a contaminant in a pituitary extract that probably was FGF, stimulated more DNA synthesis and proliferation of Swiss mouse 3T3 cells than any of several factors alone. Many reports have described combinations of growth factors that stimulated DNA synthesis and, in some studies, cell proliferation in media containing low concentrations of serum or serum depleted by cell growth (see reviews by Gospodarowicz and Moran, 1976; Scher *et al.*, 1979b; Rudland and Jimenez de Asua, 1979). A variety of hormones and growth factors of use in defined media were identified in these studies. In addition, the complexity of events occurring in the prereplicative phase (between G_0 and S phase) was indicated by the fact that more than one stimulus was required.

The work of Rutherford and Ross (1976) allowed the classification of serum-derived growth factors into two or more categories. They found that whole human or monkey serum obtained from clotted blood could be functionally divided into two classes of growth-promoting components, a platelet-derived activity and the plasma-derived activities. Neither platelet-poor plasma nor the platelet-derived activity stimulated growth of monkey fibroblasts or smooth muscle cells alone. They had to be present together. Pledger *et al.* (1977) extended the studies of Rutherford and Ross (1976). They showed that BALB/c-3T3 cells, grown to high-density quiescence, entered S phase with a 12-h lag after the addition of platelet-poor plasma only if the cells had first been exposed to platelet-derived growth factor (PDGF) for 4 h. These cells were not able to synthesize DNA even with plasma present unless they were "competent," that is, exposed to PDGF. The event(s) stimulated by PDGF persisted after the removal of the hormone unlike the responses induced by other factors such as insulin. This may be because PDGF adheres tenaciously to tissue culture plastic and extracellular matrix material such as collagen (Smith *et al.*, 1982).

A number of laboratories have adopted defined conditions with which to study stimulation of DNA synthesis by combinations of growth factors. One class of factors that are universally required includes insulin and the related somatomedins. Insulin, with other factors, has been used in serum-free media to initiate DNA synthesis in quiescent cultures of mouse 3T6 cells (Mierzejewski and Rozengurt, 1976), 3T3 cells (Dicker and Rozengurt, 1978), BALB/c-3T3 cells and C3H/10T/2 cells (Pledger *et al.*, 1982), and Chinese hamster CHEF/18 fibroblasts (Cherington and Pardee, 1980). Insulin has also been incorporated into defined culture media developed for the growth of a variety of cell types (see Barnes and Sato, 1980).

The somatomedins, which include somatomedin C, somatomedin A, MSA, and the insulinlike growth factors (IGFs), can replace insulin for growth stimulation of some cell lines at one hundredth the concentration required for insulin. For example, Stiles *et al.* (1979) demonstrated that nanomolar concentrations of somatomedins were as active as 200 nM insulin in stimulating BALB/c-3T3 cell DNA synthesis in the presence of PDGF and plasma from hypophysectomized rats (somatomedin free). The somatomedins may be more stable in growth media than insulin. In some cases this could explain the high concentrations of insulin required for growth stimulation. Another explanation concerns which class of receptor is mediating somatomedin, or IGF, and insulin growth stimulation. The type 1 IGF receptor appears to mediate the mitogenic response of many cell lines to somatomedins or insulin, although high concentrations of insulin are required because insulin binds to this receptor with low affinity (Czech, 1982). Blocking the high-affinity insulin receptor with specific antibody did not block high-affinity IGF binding or the mitogenic response of human fibroblasts to IGF or insulin (King *et al.*, 1980). Thus, the insulin receptor did not mediate the mitogenic response in this system. In a rat hepatoma cell line, however, where insulin stimulated growth via its own high-affinity receptor, subnanomolar concentrations were active (Koontz and Iwashashi, 1981; Czech, 1982).

A wide range of factors acts synergistically with insulin to stimulate DNA synthesis. These include FGF (Gospodarowicz and Moran, 1974; Stiles *et al.*, 1979), EGF (Mierzejewski and Rozengurt, 1976; Cherington and Pardee, 1980), $PGF_{2\alpha}$ (Jimenez de Asua *et al.*, 1977b; O'Farrell *et al.*, 1979), glucocorticoids (Thrash and Cunningham, 1973; Gospodarowicz and Moran, 1974), fibroblast-derived growth factor (FDGF) (Bourne and Rozengurt, 1976), vasopressin (Rozengurt *et al.*, 1979), the tumor promoter PMA (Dicker and Rozengurt, 1978; Frantz *et al.*, 1979), and thrombin (Zetter *et al.*, 1977; Cherington and Pardee, 1980).

Fibroblast growth factor (Gospodarowicz, 1975) replaced PDGF in stimulating BALB/c-3T3 cell growth in the presence of platelet-poor plasma (Stiles *et al.*, 1979) or EGF and somatomedin C (Pledger *et al.*, 1982). Calcium phosphate precipitates and scratching the cell monolayer ("wounding") also obviated the need for PDGF if platelet-poor plasma was present (Stiles *et al.*, 1979).

Epidermal growth factor isolated from human urine or mouse submaxillary glands stimulates DNA synthesis and growth in a variety of cell types (reviewed by Carpenter and Cohen, 1979), especially if insulin, other growth factors, or serum are present. In serum-free conditions, EGF stimulated DNA synthesis synergistically with somatomedin C and PDGF in BALB/c-3T3 cells (Leof *et al.*, 1980; Pledger *et al.*, 1982), and it was synergistic with insulin in Swiss 3T3 cells (Dicker and Rozengurt, 1978), 3T6 cells (Mierzejewski and Rozengurt, 1976), Chinese hamster embryo diploid fibroblasts (Cherington and Pardee, 1980), BHK cells (Cherington and Pardee, unpublished data), and C3H/10T1/2 cells (Pledger *et al.*, 1982).

Platelet-derived growth factor decreased the effective EGF concentration required for stimulation of DNA synthesis in mouse C3H/10T1/2 cells (Pledger *et al.*, 1982). Epidermal growth factor (10 ng/ml) and insulin (5×10^{-6} M) were sufficient to stimulate DNA synthesis in 40% of the cells in quiescent cultures. If these cells were pretreated for 3 h with 50 ng/ml PDGF prior to shifting to medium containing EGF and insulin, 40% stimulation was achieved at 1 ng/ml EGF. Thus, PDGF increased the sensitivity of these cells to EGF tenfold so that they were responsive to concentrations of EGF commonly measured in plasma and serum (Carpenter and Cohen, 1979).

Proteases such as trypsin or pronase have been shown to stimulate DNA synthesis and cell division in quiescent cells (Burger, 1970; Sefton and Rubin, 1970; Noonan, 1976; Hovi and Vaheri, 1976). Thrombin also stimulated DNA synthesis and cell growth in culture (Chen and Buchanan, 1975; Blumberg and Robbins, 1975; Zetter *et al.*, 1977; Gospodarowicz *et al.*, 1978; Carney *et al.*, 1978). Thrombin had a greater substrate specificity than other proteolytic mitogens (Magnusson *et al.*, 1975), and had limited targets that are on the surface of cells responsive to it (Blumberg and Robbins, 1975; Carney and Cunningham, 1978a,b;

Glenn and Cunningham, 1979). Thrombin thus could be useful in determining the identity and role of the cell surface target that is responsible for growth stimulation by proteases and possibly by other growth factors.

Thrombin potentiates the growth response of a variety of cell types to growth factors. In serum-containing media, the mitogenic effect of FGF, and to a lesser extent EGF, on human vascular endothelial cells was increased by thrombin (Gospodarowicz *et al.*, 1978). Zetter *et al.* (1977) observed potentiation of EGF, $PGF_{2\alpha}$, insulin, and FGF by thrombin, with low serum present, in a variety of mouse and rabbit cell lines that responded weakly to these factors alone. Pohjanpelto (1979) observed a synergy between insulin, platelet extract, and thrombin in human fibroblasts in the absence of serum. Growth of a Chinese hamster diploid fibroblast line (CHEF/18) was maximal at 3 nM thrombin when insulin was present. Thrombin acted synergistically with insulin to stimulate DNA synthesis and proliferation of these cells when studied in serum-free media, and appeared to replace EGF activity. Transformed cell lines related to the thrombin-responsive hamster cell line had lost requirements for both EGF and thrombin (Cherington and Pardee, 1980). Thus, EGF and thrombin mechanisms for stimulating growth of these cells may be related. A subclone of CHEF/18 was isolated which was tenfold more sensitive to thrombin (Low *et al.*, 1982).

Vasopressin and the tumor promoter PMA have potentiated DNA synthesis stimulated by a variety of growth factors in Swiss mouse 3T3 cells in serum-free media. These factors included EGF, insulin, and an FDGF produced by SV40-transformed BHK cells (Dicker and Rozengurt, 1978; Rozengurt, 1980; Rozengurt *et al.*, 1979). Vasopressin and PMA were not synergistic with one another (Dicker and Rozengurt, 1979; 1980), suggesting a similar mode of action. Other similarities between vasopressin and PMA action have also been reported. Both stimulated ouabain-sensitive $^{86}Rb^{+}$ uptake that indicated stimulation of the Na^{+}–K^{+}-ATPase (Rozengurt and Mendoza, 1980). Epidermal growth factor and insulin showed much less enhancement of $^{86}Rb^{+}$ influx (Rozengurt and Heppel, 1975). Both vasopressin and PMA inhibited EGF binding indicating similar perturbations of the plasma membrane. Shoyab *et al.* (1979), Brown *et al.* (1979a), and Lee and Weinstein (1978) observed decreased binding of EGF to mouse and human cell lines when PMA was present. Whether this decreased binding was a result of decreased EGF receptor affinity or decreased receptor number per cell was not resolved by these studies. Dicker and Rozengurt (1980) reported vasopressin inhibition of EGF binding. Epidermal growth factor does not compete for PMA receptors recently identified by Driedger and Blumberg (1980). Oxytocin, which was 1000 times less mitogenic than vasopressin, was also reportedly less effective at inhibiting EGF binding (Dicker and Rozengurt, 1980). Frantz *et al.* (1979) found that PMA was also synergistic with FGF, EGF, platelet extract, and insulin in stimulating DNA synthesis in BALB/c-3T3 cells in serum-free medium.

Prostaglandin $F_{2\alpha}$ was synergistic with insulin in stimulating density-arrested Swiss mouse 3T3 cells (Jimenez de Asua *et al.*, 1977b). When the growth factors were added directly to the density-arrested cultures without removing the depleted, serum-containing medium, it appeared that insulin (50 ng/ml) increased the rate at which the population entered S phase, and $PGF_{2\alpha}$ (300 ng/ml) regulated the length of the lag period prior to S phase. This suggests that $PGF_{2\alpha}$ initiated a process leading to DNA synthesis and that insulin determined the rate at which this process occurred. Hydrocortisone (40 ng/ml) or dexamethasone (4 ng/ml) added within 5 h of $PGF_{2\alpha}$ decreased the rate of entry into S phase, in contrast to other studies where glucocorticoids potentiated growth (this section). Prostaglandin E_1 (PGE_1) (50 μg/ml) or the synthetic phosphodiesterase inhibitor SQ-20006 (0.4 mM), both of which increase intracellular cAMP, both decreased the rate of entry into S phase. Whether cAMP was involved in this inhibition is conjectural.

Glucocorticoids potentiated the actions of growth factors in some studies even though they appeared to inhibit $PGF_{2\alpha}$ and insulin stimulation of DNA synthesis in Swiss 3T3 cells (Jimenez de Asua *et al.*, 1977a). Dexamethasone (1 μg/ml) potentiated the growth of BALB/c-3T3 cells in a low-serum medium containing FGF and insulin (Armelin, 1973; Gospodarowicz and Moran, 1974). At 100 ng/ml, dexamethasone potentiated the stimulation of human fibroblast growth by EGF (Baker *et al.*, 1978). This may have been due to increased binding of EGF by these cells; dexamethasone influenced EGF (increased), insulin (increased), and thrombin (decreased) binding to cell surface receptors. Other examples of glucocorticoid potentiation of FGF growth stimulation have been reported (Holley and Kiernan, 1974; Kamely and Rudland, 1976).

The ability of cytoskeleton-disrupting agents to increase the proliferative responses to growth factors is controversial. Teng *et al.* (1977), Friedkin *et al.* (1979), and Otto *et al.* (1979) reported that micromolar concentrations of microtubule-disrupting agents such as colchicine, demecolcine, or vinblastine potentiated DNA synthesis stimulated by insulin, serum, EGF, FDGF, FGF, vasopressin, or $PGF_{2\alpha}$ in confluent, growth-arrested mouse and chick embryo fibroblasts. These results contradicted reports by Baker (1976) and McClain *et al.* (1977) who observed that growth stimulation of subconfluent, serum-deprived mouse neuroblastoma or chick embryo fibroblasts by serum was inhibited by antimitotic agents. A critical difference between the above contradictory studies was the density of the arrested populations; the populations stimulated by perturbing their cytoskeleton were at high density and the populations inhibited by such treatment were sparse. McClain and Edelman (1980) compared the effect of colchicine on DNA synthesis stimulation by serum, insulin, FGF, or EGF in sparse and dense cultures. In general, the sparse cells were inhibited and the dense cells were stimulated by colchicine when growth factors were present. Potentiation of serum stimulation of dense cultures by colchicine was evident but less dramatic.

Mechanism of Growth Factor Action—Events Stimulated by Growth Factors Prior to DNA Synthesis

A range of biochemical events may be observed following stimulation of a quiescent cell population with serum or purified growth factors. Macromolecular synthesis increases over a period of hours during the prereplicative phase, that is, between G_0 and S phase, and this increased synthesis is maintained into S phase (Prescott, 1976; Pardee *et al.*, 1978). Polysome formation correlates with the rate of protein synthesis; whether the level of mRNA or some other event limits polysome formation has been disputed (Levis *et al.*, 1977; Rudland *et al.*, 1975). Recently, Thomas *et al.* (1982) have shown that increased polysome formation alone is not sufficient to stimulate DNA synthesis. Insulin added alone to the spent serum-containing medium of confluent Swiss mouse 3T3 cells did not stimulate DNA synthesis. It did stimulate as much ribosomal protein S6 phosphorylation and polysome formation as mitogenic doses of EGF or $PGF_{2\alpha}$. Combinations of any of these factors (EGF, $PGF_{2\alpha}$, or insulin) resulted in synergistic increases in S6 phosphorylation, polysome formation, and DNA synthesis over those levels observed with each factor alone. Thus, S6 phosphorylation and polysome formation may have contributed to, but were not sufficient for, a full mitogenic response.

Protein phosphorylation is an event that occurs rapidly after exposure of cells to growth factors that could contribute to the stimulation of DNA synthesis. The presence or absence of phosphoproteins or certain protein kinase types depending on the growth state of cultured cells has been reported (reviewed by Pardee *et al.*, 1978).

Recently, specific phosphoproteins associated with growth factor action have been reported. Platelet-derived-growth-factor-stimulated phosphorylation of proteins in fibroblast and glial cell membranes was reported (Ek *et al.*, 1982). Fibroblast growth factor treatment rapidly stimulated the phosphorylation of a 33,000-dalton protein in Swiss 3T3 cells (Nilsen-Hamilton and Hamilton, 1979). Insulin and EGF stimulated the phosphorylation of a 31,000-dalton protein (perhaps the ribosomal protein S6) in 3T3-Ll preadipocytes (Smith *et al.*, 1980).

Tyrosine kinase activity is implicated in growth regulation. Phosphorylation of cellular proteins at tyrosines is increased in growth-factor-treated and virally transformed cells (Cooper and Hunter, 1981). Similar phosphoproteins containing phosphotyrosine appear in Swiss 3T3 cells after EGF, PDGF, or serum stimulation indicating that these mitogens may act, in part, via common pathways of tyrosine kinase activity (Cooper *et al.*, 1982). Tyrosine kinase activity is associated with the EGF (Ushiro and Cohen, 1980; Buhrow *et al.*, 1982) and insulin (Kasuga *et al.*, 1982) receptors, and possibly with the PDGF receptor (Ek and Heldin, 1982). Thus, an early event in growth stimulation appears to be activation of one or more tyrosine kinases. The substrate(s) for such kinase activities that are involved in growth regulation are not known.

The uptake of monovalent and divalent cations, phosphate, and metabolic precursors such as hexose sugars, nucleosides, and amino acids are all influenced after mitogenic stimulation. Rates of glycolysis and levels of cyclic nucleotides are also affected. The nature of these effects and their relative importance in regulating the onset of DNA synthesis have been recently reviewed (Rozengurt, 1979; 1980). McKeehan (McKeehan and McKeehan, 1980b; McKeehan, 1982) has suggested that a major action of serum-derived growth factors is to reduce the extracellular levels of ions and nutrients, in particular Ca^{2+}, K^{+}, and pyruvic acid, which are required for cellular proliferation. For example, EGF decreased the concentration of extracellular calcium required for proliferation of human prostate epithelial cells and human fibroblasts (Lechner and Kaighn, 1979; McKeehan and McKeehan, 1979). The mechanism of this effect is not known, however.

Extracellular divalent cations are required for the stimulation of DNA synthesis in resting cells (Dulbecco and Elkington, 1975; Swierenga *et al.*, 1976; Boynton *et al.*, 1977). More calcium was associated with the cell surface in quiescent than in growing 3T3 cells, although intracellular levels apparently were not very different (Tupper and Zorgniotti, 1977). Epidermal growth factor stimulated Ca^{2+} uptake and Ca^{2+}-dependent phosphatidylinositol turnover in human A-431 cells (Sawyer and Cohen,

1981). Whether calcium fluxes or phospholipid metabolism are involved in DNA synthesis stimulation is conjectural, although calcium has been proposed as a second intracellular messenger in a variety of hormonal actions (Rasmussen, 1970). Magnesium deprivation also results in growth arrest (Rubin, 1975a), but as with calcium, the effect of serum or growth factors on magnesium fluxes has not been determined. Subtoxic doses of zinc, cadmium, and mercury have been reported to stimulate DNA synthesis in resting chick cells (Rubin, 1975b).

Individual growth factors do not necessarily stimulate all of the early events observed after serum stimulation. It appears that multiple factors are required for the stimulation of early events prior to DNA synthesis as well as for the stimulation of DNA synthesis. For example, insulin was less able to stimulate $^{86}Rb^+$ fluxes than serum (Rozengurt and Heppel, 1975) and was only a weak mitogen for 3T3 cells (Scher *et al.*, 1974), BHK cells (Clarke and Stoker, 1971), and chick fibroblasts (Vaheri *et al.*, 1973; Baseman *et al.*, 1974) when added alone. When insulin was combined with a factor that stimulated $^{86}Rb^+$ flux, for example, vasopressin, greatly increased DNA synthesis was observed (Dicker and Rozengurt, 1979; Rozengurt and Mendoza, 1980). Multiple factors were required to stimulate the dual centriole deciliations that occur prior to DNA synthesis after stimulating quiescent cells (Tucker *et al.*, 1979a,b). The absolute importance of these early events, however, was not established.

Receptor cross-linking and aggregation may be involved in growth stimulation. Multimeric IgM monoclonal antibodies to the EGF receptor as well as intact EGF stimulated DNA synthesis and receptor clustering in addition to rapid cellular events including protein phosphorylation (Schreiber *et al.*, 1981a, 1983). Chemically modified EGF (cyanogen-bromide-treated, CnBr-EGF), monovalent F_{ab} fragments of antireceptor IgM, and a monoclonal antireceptor IgG that did not induce receptor clustering did not stimulate DNA synthesis (Schechter *et al.*, 1979; Schreiber *et al.*, 1983), although protein phosphorylation and other rapid responses were stimulated in the absence of receptor aggregation (Schreiber *et al.*, 1981b, 1983; Yarden *et al.*, 1982). Thus, the early events observed in the prereplicative phase were either not involved in DNA synthesis stimulation or alone were insufficient. Cross-linking of CnBr-EGF or F_{ab} fragments of antireceptor IgM using antibodies against EGF or immunoglobulin, respectively, restored the mitogenicity and receptor-clustering activity to these ligands (Schechter *et al.*, 1979; Schreiber *et al.*, 1983). The ability of bivalent anti-insulin receptor antibodies to induce receptor aggregation and biological responses has also been reported (Kahn *et al.*, 1978).

Studying receptor–ligand activity in cells grown in serum-free conditions may give results that are different from serum-grown cells. Wolfe *et al.* (1980) reported that [^{125}I]-EGF binding did not decrease on HeLa cells grown in serum-free medium. This is in contrast to the 80% decrease of binding seen in serum-grown cultures after treatment for 60 minutes or more with EGF (Carpenter and Cohen, 1979). Reevaluation of the steady-state levels of cell surface receptors before and after ligand binding may be necessary as conditions for serum-free culture of particular cell lines are established.

Comparison of the Growth Factor Requirements of Normal and Transformed Cells Grown in Serum-Free Media

Cells transformed *in vitro* have a diminished growth requirement for serum (Clarke *et al.*, 1970; Dulbecco, 1970), and acute treatment of cell monolayers with the SV40-transforming virus can induce growth in the absence of serum (Smith *et al.*, 1971). Therefore, one expects that transformed cell lines will no longer require or will express reduced requirements for growth factors. Studies of individual growth factor requirements are possible if serum levels are reduced or eliminated from the culture medium. For example, Scher *et al.* (1978) demonstrated that a wide range of virally transformed cell lines related to BALB/c-3T3 no longer required PDGF; they grew well in media supplemented only with platelet-poor plasma, whereas BALB/c-3T3 cells grew only if platelet-poor plasma was combined with PDGF. Brown and Holley (1979) were able to show that benzo[a]pyrene-transformed BALB/c-3T3 cells retained an EGF growth requirement, but the effective concentration for stimulating DNA synthesis in confluent cultures was decreased approximately fivefold. A quantitative reduction in the requirement for a particular growth factor may reflect a reduction in the rate of its uptake and destruction rather than change in the biochemical events regulated by that factor (Brown *et al.*, 1979b).

Since plasma- and serum-free media are now available for a number of cell lines used in growth control studies, a more complete analysis of the altered responsiveness of transformed cell lines to growth factors is possible. McClure *et al.* (1979) showed in short-term serum-free cultures that SV40-transformed BALB/c-3T3 cells had lost the requirements expressed by the parent BALB/c-3T3 cell lines for FGF, EGF, and a "gimmel" factor extract of rat submaxillary glands. Subsequent studies of these cell lines, and similarly isolated Swiss 3T3 lines, in long-term, multiple-passage serum-free cultures confirmed these results and also indicated that the transformed lines had an increased requirement for *cis*-unsaturated fatty acids, such as linoleic acid. The transformed lines expressed a requirement for insulin, but the optimal concentration of insulin was reduced (Serrero *et al.*, 1979; Rockwell *et al.*, 1980). Taub *et al.* (1981) also reported a quantitative, but incomplete, reduction of the insulin growth requirement of canine kidney epithelial cells (MCDK) after Moloney sarcoma virus transformation. No other requirements were lost in these transformed cells, although an increased fibronectin requirement was observed.

Scher *et al.* (1982) described a spontaneously arising subclone of BALB/c-3T3 cells (ST3T3) that was tumorigenic, grew to an elevated saturated density, and expressed an altered morphology. Growth-arrested BALB/c-3T3 cells in 0.5% platelet-poor plasma required additions of PDGF, EGF, and insulin for maximal stimulation of DNA synthesis. Under the same conditions, insulin alone maximally stimulated ST3T3 cells; PDGF and EGF had no additional effect on the transformed cells. In addition, proteins synthesized preferentially in response to PDGF in BALB/c-3T3 cells were constitutively synthesized in ST3T3 cells. The plasma used by Scher *et al.* (1982) in these studies could have been eliminated by using supplements described by McClure *et al.* (1979), Serrero *et al.* (1979), and Rockwell *et al.* (1980), perhaps revealing other growth requirements altered in the SV3T3 cell line.

Transformed cell lines that retain a requirement for DNA-synthesis-stimulating factor such as insulin also retain the ability to arrest growth in G_1. Cherington *et al.* (1979) used a serum-free medium containing FGF, EGF, insulin, and transferrin to grow Syrian hamster BHK cells and found that a polyoma virus-transformed derivative cell line (pyBHK) had dispensed with all but the transferrin requirement. When growing populations were shifted into media lacking the DNA-synthesis-stimulating factors required for G_1 transit, that is, FGF, EGF, and insulin, and analyzed with cytofluorometry, the BHK cell line arrested growth in G_1 and the pyBHK line did not. Omission of transferrin resulted in pyBHK cell growth ceasing nonspecifically throughout the cell cycle.

Chinese hamster embryo fibroblasts (CHEF/18), originally described by Sager and Kovac (1978) grew well in a medium containing EGF, insulin, transferrin, and a crude preparation of fetuin (Sager *et al.*, 1979). Subsequent studies determined that 10 ng/ml highly purified thrombin could replace fetuin (Cherington and Pardee, 1980). A spontaneous transformant, CHEF/16, and a chemically transformed derivative of CHEF/18, T30-4, both had lost requirements for EGF and thrombin while retaining a requirement for insulin. These cell lines arrested growth in G_1 on insulin deprivation, unlike the pyBHK line noted above (Cherington *et al.*, 1979). The independent expression of EGF and insulin growth requirements was further demonstrated recently by Powers *et al.* They have shown that certain revertants of SV40-transformed 3T3 cells, which retained anchorage independence, reexpressed an insulin requirement without regaining sensitivity to EGF (Powers, Pollack, Graham, Boersig, and Fisher, unpublished data). Apparently the insulin-regulated events in G_1 are the last to be bypassed or eliminated in these transformed cell lines.

Epidermal growth factor and thrombin may have been acting via overlapping pathways leading to DNA synthesis in CHEF/18 cells since (1) the requirements for these factors were lost coordinately in two separate transformed cell lines, and (2) 100 ng/ml thrombin eliminated the EGF requirement for maximal growth. Epidermal growth factor had a half-maximal effect at approximately 3 to 4 ng/ml with insulin plus 0 or 10 ng/ml thrombin present, but at 100 ng/ml thrombin little or no additional growth stimulation was seen with EGF (Cherington and Pardee, 1980). Insulin, on the other hand, potentiated the growth response to EGF and thrombin regardless of the thrombin concentration. The basis for the overlapping actions of EGF and thrombin are not known. The two growth factors do not bind to or effect the same cell surface targets. Thrombin does not compete for EGF binding sites or destroy them after prolonged incubation with cells (Cherington and Pardee, 1982). Furthermore, EGF does not cause the changes in iodinatable surface proteins observed after thrombin treatment (Cherington, unpublished observations).

The basis for the lost growth requirements for EGF and thrombin in the transformed CHEF cell lines (Cherington and Pardee, 1980) is not clear. The transformed CHEF lines, CHEF/16 and T30-4, had 50% and 90% fewer EGF receptors, respectively, than the normal CHEF/18 cell line. The affinity of the receptors on the three cell lines were the same (Cherington and Pardee, 1982). Transforming growth factors secreted by some transformed cell lines can compete for EGF binding sites in a variety of nontransformed lines (DeLarco and Todaro, 1980). In fact, serum-free medium conditioned by the TGF-secreting cell line 3B11-IC (DeLarco and Todaro, 1980) contained material that competed for EGF receptors on CHEF/18 cells and eliminated the CHEF/18 growth requirement for EGF and thrombin in a defined medium (Cherington and Pardee, 1982). However, medium conditioned by T30-4 did not diminish EGF binding to the parent CHEF/18 cell line or replace the EGF and thrombin growth requirement. In addition,

CHEF/18 cells cocultured with T30-4 did not bind less EGF than equivalent numbers of cells in separate cultures. Thus, T30-4 did not appear to secrete a TGF that interfered with EGF receptors on neighboring cells or that stimulated growth of neighboring cells. It is possible that T30-4 cells produced an endogenous factor that acted on the same cells that produced it, but that was present in the medium at concentrations too low to influence other cells in the culture. Alternatively, an intracellular event triggered by both EGF and thrombin may have become constitutive in growth-factor-independent cells in the absence of a secreted TGF. If a constitutive intracellular activity caused, indirectly, the reduction of EGF receptor number, then decreased receptor number would be secondary to the mechanism bypassing the EGF and thrombin requirements.

Shipley and Moses (1982) recently described a serum-free medium that supported the growth of AKR-2B mouse cells. This medium consisted of MCDB 402 without linoleic acid (Shipley and Ham, 1981) supplemented with EGF, insulin and ethanolamine. A chemically transformed derivative cell line that previously had been shown to produce TGFs with EGF receptor binding activity (Moses *et al.*, 1981) no longer required EGF or ethanolamine for growth; a combination of insulin and linoleic acid was sufficient. Thus, the secretion of a TGF by this cell line may have effectively replaced the EGF requirement for growth in a defined medium. Kaplan *et al.* (1982) have shown that TGFs released by virally transformed rat kidney cells will, when combined with insulin and transferrin in a serum-free medium, support rapid growth of the untransformed rat-1 cell line.

Sager *et al.* (1982) attempted to relate growth factor requirements with transformation-related growth characteristics *in vitro* and tumorigenicity in athymic nude mice. Analysis of subclones derived from hybrids formed between the growth-factor-dependent, nontumorigenic line CHEF/18 and the EGF/thrombin-independent, tumorigenic line CHEF/16 indicated that anchorage independence, tumorigenicity, and growth factor independence were coordinately expressed. Mutagenesis of CHEF/18 with alkylating agents, however, generated subclones that did not always support the correlation between growth factor requirements and tumorigenicity; several tumorigenic clones retained EGF or thrombin requirements. In addition, some mutant clones, either tumorigenic or nontumorigenic, did not grow in the serum-free medium suitable for the parent cell line. This is not surprising, however, since mutagenesis could have changed the growth requirements of the cells rendering the medium used unsuitable. Others have reported that additional nutritional requirements were expressed following transformation (Rockwell *et al.*, 1980). It is possible that cells in the CHEF cell system lost growth factor requirements after, rather than coordinately with, acquiring tumor-forming potential. The multistep nature of transformation in this system has been described recently (Smith and Sager, 1982).

McClure *et al.* (1982) used a colony-formation assay to compare the growth requirements of SV40-transformed rat embryo cells with the requirements of the parent cell line. This assay had the potential to stringently test a number of growth variables, for example, initiation of growth or attachment as reflected by the number of colonies and the rate of growth as reflected by the size of the colonies after a fixed time period. The medium included hydrocortisone, insulin, transferrin, high-density lioprotein (HDL), EGF, vasopressin, and fibronectin. In terms of percent colony formation, all lines required transferrin and all but two transformed lines required HDL. The transformed lines appeared to have a somewhat relaxed requirement for fibronectin. Omission of either EGF, vasopressin, insulin, or hydrocortisone individually did not greatly decrease the plating efficiency of any line, including the untransformed parent line, although the colony size was in some cases reduced. However, omission of both EGF and vasopressin together resulted in low plating efficiency of the parent line, whereas the transformed lines displayed normal plating efficiencies. Perhaps EGF and vasopressin had overlapping modes of growth stimulation in these cells, as indicated for tumor promoter and vasopressin stimulation of DNA synthesis in 3T3 cells (Dicker and Rozengurt, 1980) and for EGF and thrombin growth stimulation of CHEF/18 cells (Cherington and Pardee, 1980).

Conclusion

Animal cell proliferation is regulated by events leading to DNA synthesis; whether or not a cell proceeds with DNA synthesis or arrests growth in G_1 depends on the extracellular conditions. Studying the conditions required for optimal cell growth *in vitro*, using serum-free media, will lead to a better understanding of how the combined effects of substratum, nutritional factors, growth-promoting factors, and hormones interact to limit growth. This interaction promises to be complex; conditions altering cell shape probably alter sensitivity to serum-derived growth factors (Folkman and Moscona, 1978), matrix proteins alter growth factor requirements (Gospodarowicz *et al.*, 1982), one growth factor can increase sensitivity to another growth factor (Pledger *et al.*,

1982), or decrease the requirement for a given nutrient or ion (McKeehan and McKeehan, 1980b). Defining the culture conditions will also permit a meaningful analysis of the biochemical effects of specific medium components that limit growth.

Programmed differentiation is coupled to growth arrest. Defining conditions that regulate growth may also help determine what regulates the transition from a reversible, growth-arrested state to an irreversible, terminally differentiated state. For example, serum-free media will facilitate the study of extracellular conditions that permit adipocyte conversion following growth arrest of preadipocytes (Scott *et al.*, 1982). Similarly, defined conditions for the growth of keratinoctyes will allow further study for the factors regulating growth and the programmed differentiation that follows growth arrest in these cells (Green, 1979).

Recent advances in the molecular analysis of cell transformation allow the transfection of untransformed cell lines or primary cell cultures with well-characterized viral genes (Carmichael *et al.*, 1982; Rassoulzadegan *et al.*, 1982; Land *et al.*, 1983; Ruley, 1983) and cellular DNA (Shih *et al.*, 1979; Cooper *et al.*, 1980; Land *et al.*, 1983; Ruley, 1983). The use of serum-free growth conditions for cells transfected with well-characterized genetic material will help determine the relationship between expression of particular transforming genes and altered requirements for growth.

ACKNOWLEDGMENTS. The author is grateful to Ms. Roselle Sherman for her help in preparing the manuscript and to Dr. Claire Fearon for her critical review of the manuscript. Some of the work described in this chapter was conducted while the author was working in Dr. Arthur Pardee's laboratory at the Dana-Farber Cancer Institute in Boston, Massachusetts under the support of a National Research Service Award (#5 T32 GM07306-05) made to the Department of Pharmacology, Harvard Medical School, Boston, Massachusetts. Currently, the author is supported by a Damon Runyon-Walter Winchell Cancer Fund Fellowship, DRG-503.

References

Abelson, H. T., Johnson, L. R., Penman, S., and Green, H., 1974, Changes in RNA in relation to growth of the fibroblast: II. The lifetime of mRNA, rRNA and tRNA in resting and growing cells, *Cell* **1**:161–165.

Armelin, H. A., 1973, Pituitary extracts and steroid hormones in the control of 3T3 cell growth, *Proc. Natl. Acad. Sci. U.S.A.* **70**:2702–2706.

Baker, J. B., Barsh, G. S., Carney, D. H., and Cunningham, D. D., 1978, Dexamethasone modulates binding and action of epidermal growth factor in serum-free cell culture, *Proc. Natl. Acad. Sci. U.S.A.* **75:**1882–1886.

Baker, M. E., 1976, Colchicine inhibits mitogenesis in C1300 neuroblastoma cells that have been arrested in G_0, *Nature* **262:**785–786.

Barnes, D., and Sato, G., 1980, Methods for growth of cultured cells in serum-free medium, *Anal. Biochem.* **102:**255–270.

Bartholomew, J. C., Yokota, H., and Ross, P., 1976, Effect of serum on the growth of Balb 3T3 A31 mouse fibroblasts and a SV40-transformed derivative, *J. Cell. Physiol.* **88:**277–286.

Baseman, J. B., Paolini, D., and Amos, H., 1974, Stimulation by insulin of RNA synthesis in chick fibroblasts, *J. Cell Biol.* **60:**54–64.

Baserga, R., 1976, *Multiplication and Division in Mammalian Cells,* Marcel Dekker, New York.

Blumberg, P. M., and Robbins, P. W., 1975, Effects of proteases on activation of resting chick embryo fibroblasts and on cell surface proteins, *Cell* **6:**137–147.

Bottenstein, J., Hayashi, I., Hutchings, S., Masui, H., Mather, J., McClure, D. B., Ohasa, S., Rizzino, A., Sato, G., Serrero, G., Wolfe, R., and Wu, R., 1979, The growth of cells in serum-free hormone-supplemented media, *Methods Enzymol.* **58:**94–109.

Bourne, H., and Rozengurt, E., 1976, An 18,000 molecular weight polypeptide induces early events and stimulates DNA synthesis in cultured cells, *Proc. Natl. Acad. Sci. U.S.A.* **73:**4555–4559.

Boynton, A. L., Whitfield, J. R., Isaacs, R. J., and Tremblay, R., 1977, The control of human WI-38 cell proliferation by extracellular calcium and its elimination by SV40 virus-induced proliferative transformation, *J. Cell. Physiol.* **92:**241–248.

Brooks, R. F., 1975, The kinetics of serum-induced initiation of DNA synthesis in BHK-21/C13 cells, and the influence of exogenous adenosine, *J. Cell. Physiol.* **86:**369–378.

Brooks, R. F., Bennett, D. C., and Smith, J. A., 1980, Mammalian cell cycles need two random transitions, *Cell* **19:**493–504.

Brown, K. D., and Holley, R. W., 1979, Epidermal growth factor and the control of proliferation of Balb 3T3 benzo[a]pyrene-transformed Balb 3T3 cells, *J. Cell. Physiol.* **100:**139–146.

Brown, K. D., Dicker, P., and Rozengurt, E., 1979a, Inhibition of epidermal factor binding to surface receptors by tumor promoters, *Biochem. Biophys. Res. Commun.* **86:**1037–1043.

Brown, K. D., Yeh, Y.-C., and Holley, R. W., 1979b, Binding, internalization, and degradation of epidermal growth factor by Balb 3T3 and BP 3T3 cells: Relationship to cell density and the stimulation of cell proliferation, *J. Cell. Physiol.* **100:**227–238.

Buhrow, S. A., Cohen, S., and Stavros, J. V., 1982, Affinity labeling of the protein kinase associated with the epidermal growth factor receptor in membrane vesicles from A431 cells, *J. Biol. Chem.* **257:**4019–4022.

Burger, M. M., 1970, Proteolytic enzymes initiating cell division and escape from contact inhibition of growth, *Nature (London)* **277:**170–171.

Campisi, J., Medrano, E. E., Morreo, G., and Pardee, A. B., 1982, Restriction point control of cell growth by a labile protein: Evidence for increased stability in transformed cells, *Proc. Natl. Acad. Sci. U.S.A.* **79:**436–440.

Carmichael, G. G., Schaffhausen, B. S., Dorsky, D. I., Oliver, D. B., and Benjamin, T. L., 1982, Carboxy terminus of polyoma middle-sized tumor antigen is required for attachment to membranes, associated protein kinase activities, and cell transformation, *Proc. Natl. Acad. Sci. U.S.A.* **79:**3579–3583.

Carney, D. H., and Cunningham, D. D., 1978a, Cell surface action of thrombin is sufficient to initiate division of chick cells, *Cell* **14:**811–823.

Carney, D. H., and Cunningham, D. D., 1978b, Role of specific cell surface receptors in thrombin-stimulated cell division, *Cell* 15:1341–1349.

Carney, D. H., Glenn, K. C., and Cunningham, D. D., 1978, Conditions which affect initiation of animal cell division and thrombin, *J. Cell. Physiol.* **95:**13–22.

Carpenter, G., and Cohen, S., 1979, Epidermal growth factor, *Annu. Rev. Biochem.* **48:**193–216.

Chen, L. B., and Buchanan, J. M., 1975, Mitogenic activity of blood components. I. Thrombin and prothrombin, *Proc. Natl. Acad. Sci. U.S.A.* **72:**131–135.

Chen, T. T., and Heidelburger, C., 1969, Quantitative studies on the malignant transformation of mouse prostate cells by carcinogenic hydrocarbons *in vitro, Int. J. Cancer* **4:**166–178.

Cherington, P. V., and Pardee, A. B., 1980, Synergistic effects of epidermal growth factor and thrombin on growth stimulation of diploid Chinese hamster fibroblasts, *J. Cell. Physiol.* **105:**25–32.

Cherington, P. V., and Pardee, A. B., 1982, On the basis for loss of the EGF growth requirement by transformed cells, in: *Growth of Cells in Hormonally Defined Media* (G. H. Sato, A. B. Pardee, and D. A. Sirbasku, eds.), Cold Spring Harbor Laboratory, Cold Spring Harbor, New York, pp. 221–230.

Cherington, P. V., Smith, B. L., and Pardee, A. B., 1979, Loss of epidermal growth factor requirement and malignant transformation, *Proc. Natl. Acad. Sci. U.S.A.* **76:**3937–3941.

Chiquet, M., Puri, E. C., and Turner, D. C., 1979, Fibronectin mediates attachment of chicken myoblasts to a gelatin-coated substratum, *J. Biol. Chem.* **254:**5475–5482.

Clarke, G. D., and Smith, C., 1973, The response of normal and polyoma virus-transformed BHK/21 cells to exogenous purines, *J. Cell. Physiol.* **81:**125–132.

Clarke, G. D., and Stoker, M. G. P., 1971, Conditions affecting the response of cultured cells to serum, in: *Growth Control in Cell Cultures* (G. E. W. Wolstenholme and J. Knight, eds.), Churchill Livingstone, London, pp. 17–32.

Clarke, G. D., Stoker, M. G. P., Ludlow, A., and Thornton, M., 1970, Requirement of serum for DNA systhesis in BHK/21 cells: Effects of density, suspension and virus transformation, *Nature (London)* **227:**798–801.

Cooper, G. M., Okenquist, S., and Silverman, L., 1980, Transforming activity of DNA of chemically transformed and normal cells, *Nature (London)* **284:**418–421.

Cooper, J. A., and Hunter, T., 1981, Similarities and differences between the effects of epidermal growth factor and Rous sarcoma virus, *J. Cell. Biol.* **91:**878–883.

Cooper, J. A., Bowen-Pope, D. F., Raines, E., Ross, R., and Hunter, T., 1982, Similar effects of platelet-derived growth factor and epidermal growth factor on the phosphorylation of tyrosine in cellular proteins, *Cell* **31:**263–273.

Czech, M. P., 1982, Structural and functional homologies in the receptors for insulin and the insulin-like growth factors, *Cell* **31:**8–10.

DeLarco, J. E., and Todaro, G. J., 1978, Growth factors from murine sarcoma virus-transformed cells, *Proc. Natl. Acad. Sci. U.S.A.* **75:**4001–4005.

DeLarco, J. E., and Todaro, G. J., 1980, Sarcoma growth factor (SGF): Specific binding to epidermal growth factor (EGF) membrane receptors, *J. Cell. Physiol.* **102:**267–277.

Dicker, P., and Rozengurt, E., 1978, Stimulation of DNA synthesis by tumor promoter and pure mitogenic factors, *Nature (London)* **276:**723–726.

Dicker, P., and Rozengurt, E., 1979, Synergistic stimulation of early events and DNA synthesis by phorbol esters, polypeptide growth factors, and retinoids in cultured fibroblasts, *J. Supramol. Struct. Cell. Biochem.* **11:**79–93.

Dicker, P., and Rozengurt, E., 1980, Phorbol esters and vasopressin stimulate DNA synthesis by a common mechanism, *Nature* (*London*) **287:**607–612.

Driedger, P. E., and Blumberg, P. M., 1980, Specific binding of phorbol ester tumor promoters, *Proc. Natl. Acad. Sci. U.S.A.* **77:**567–571.

Dubrow, R., Riddle, V. G. H., and Pardee, A. B., 1979, Different responses to drugs and serum of cells transformed by various means, *Cancer Res.* **39:**2718–2726.

Dulbecco, R., 1970, Topoinhibition and serum requirement of transformed and untransformed cells, *Nature* (*London*) **277:**802–806.

Dulbecco, R., and Elkington, J., 1975, Induction of growth in resting fibroblastic cell cultures by Ca^{++}, *Proc. Natl. Acad. Sci. U.S.A.* **72:**1584–1588.

Ek, B., and Heldin, C. -H., 1982, Characterization of a tyrosine-specific kinase activity in human fibroblast membranes stimulated by platelet-derived growth factor, *J. Biol. Chem.* **257:**10486–10492.

Ek, B., Westermark, B., Wasteson, A., and Heldin, C. -H., 1982, Stimulation of tyrosine-specific phosphorylation by platelet derived growth factor, *Nature* (*London*) **295:**419–420.

Folkman, J., and Moscona, A., 1978, Role of cell shape in growth control, *Nature* (*London*) **273:**345–349.

Frantz, C. N., Stiles, C. D., and Scher, C. D., 1979, The tumor promoter 12-0-tetradecanoyl-phorbol-13-acetate enhances the proliferative response of Balb/c-3T3 cells to hormonal growth factors, *J. Cell. Physiol.* **100:**413–424.

Friedkin, M., Legg, A., and Rozengurt, E., 1979, Antitubulin agents enhance the stimulation of DNA synthesis by polypeptide growth factors in 3T3 mouse fibroblasts, *Proc. Natl. Acad. Sci. U.S.A.* **76:**3909–3912.

Glenn, K. C., and Cunningham, D. D., 1979, Thrombin-stimulated cell division involves proteolysis of its cell surface receptor, *Nature* (*London*) **278:**711–714.

Gospodarowicz, D., 1975, Purification of a fibroblast growth factor from bovine pituitary, *J. Biol. Chem.* **250:**2515–2520.

Gospodarowicz, D., and Moran, J. S., 1974, Stimulation of sparse and confluent 3T3 cell populations by a fibroblast growth factor, dexamethasone, and insulin, *Proc. Natl. Acad. Sci. U.S.A.* **71:**4584–4588.

Gospodarowicz, D., and Moran, J. S., 1976, Growth factors in mammalian cell culture, *Annu. Rev. Biochem.* **45:**531–558.

Gospodarowicz, D., Brown, K. D., Birdwell, C. R., and Zetter, B. R., 1978, Control of proliferation of human vascular endothelial cells. Characterization of the responses of human umbilical vein endothelial cells to fibroblast growth factor, epidermal growth factor, and thrombin, *J. Cell. Biol.* **77:**774–788.

Gospodarowicz, D., Cohen, D., and Fujii, D. K., 1982, Regulation of cell growth by the basal lamina and plasma factors: Relevance to embryonic control of cell proliferation and differentiation, in: *Growth of Cells in Hormonally Defined Media* (G. H. Sato, A. B. Pardee, and D. A. Sirbasku, eds.), Cold Spring Harbor Laboratory, Cold Spring Harbor, New York, pp. 95–124.

Gray, J. W., and Coffino, P., 1979, Cell cycle analysis by flow cytometry, *Methods. Enzymol.* **58:**233–248.

Green, H., 1979, The keratinocyte as differentiated cell type, *Harvey Lect.* **74:**101–139.

Grinnell, F., and Feld, M. K., 1979, Initial adhesion of human fibroblasts in serum-free medium: Possible role of secreted fibronectin, *Cell* **17:**117–129.

Grinnell, F., and Minter, D., 1978, Attachment and spreading of baby hamster kidney cells to collagen substrata: Effects of cold-insoluble globulin, *Proc. Natl. Acad. Sci. U.S.A.* **75:**4408–4412.

Grinnell, F., Feld, M., and Minter, D., 1980, Fibroblast adhesion to fibrinogen and fibrin substrata: Requirement for cold-insoluble globulin (plasma fibronectin), *Cell* **19:**517–525.

Ham, R. G., 1965, Clonal growth of mammalian cells in a chemically defined, synthetic medium, *Proc. Natl. Acad. Sci. U.S.A.* **53:**288–293.

Ham, R. G., and McKeehan, W. L., 1979, Media and growth requirements, *Methods Enzymol.* **58:**44–93.

Hatten, M. E., Horowitz, A. F., and Burger, M. M., 1977, The influence of membrane lipids on the proliferation of transformed and untransformed cell-lines, *Exp. Cell. Res.* **107:**31–34.

Hayashi, I., and Sato, G. H., 1976, Replacement of serum by hormones permits growth of cells in a defined medium, *Nature (London)* **259:**132–134.

Hollenberg, M. D., and Cuatrecasas. P., 1975, Insulin and epidermal growth factor. Human fibroblast receptors related to deoxyribonucleic acid synthesis and amino acid uptake, *J. Biol. Chem.* **250:**3845–3853.

Holley, R. W., and Kiernan, J. A., 1968, "Contact inhibition" of cell division in 3T3 cells, *Proc. Natl. Acad. Sci. U.S.A.* **60:**300–304.

Holley, R. W., and Kiernan, J. A., 1974, Control of the initiation of DNA synthesis in 3T3 cells: Serum factors, *Proc. Natl. Acad. Sci. U.S.A.* **71:**2908–2911.

Holley, R. W., Baldwin, J. H., Kiernan, J. A., and Messmer, T. O., 1976, Control of growth of benzo(a)pyrene-transformed 3T3 cells, *Proc. Natl. Acad. Sci. U.S.A.* **73:**3229–3232.

Hovi, T., and Vaheri, A., 1976, Reversible release of chick embryo fibroblast cultures from density dependent inhibition of growth, *J. Cell. Physiol.* **87:**245–252.

Jimenez de Asua, L., Clingan, D., and Rudland, P. S., 1975, Initiation of cell proliferation in cultured mouse fibroblasts by prostaglandin $F_{2\alpha}$, *Proc. Natl. Acad. Sci. U.S.A.* **72:**2724–2728.

Jimenez de Asua, L., Carr, B., Clingan, D., and Rudland, P., 1977a, Specific glucocorticoid inhibition of growth promoting effects of prostaglandin $F_{2\alpha}$ on 3T3 cells, *Nature (London)* **265:**450–452.

Jimenez de Asua, L., O'Farrell, M. K., Clingan, D., and Rudland, P. S., 1977b, Temporal sequence of hormone interactions during the prereplicative phase of quiescent cultured 3T3 fibroblasts, *Proc. Natl. Acad. Sci. U.S.A.* **74:**3845–3849.

Kahn, C. R., Baird, K. L., Jarrett, D. B., and Flier, J. S., 1978, Direct demonstration that receptor crosslinking or aggregation is important in insulin action, *Proc. Natl. Acad. Sci. U.S.A.* **75:**4209–4213.

Kamely, D., and Rudland, P., 1976, Induction of DNA synthesis and cell division in human diploid skin fibroblasts by fibroblast growth factor, *Exp. Cell Res.* **97:**120–126.

Kaplan, P. L., Anderson, M., and Ozanne, B., 1982, Transforming growth factor(s) production enables cells to grow in the absence of serum: An autocrine system, *Proc. Natl. Acad. Sci. U.S.A.* **79:**485–489.

Kasuga, M., Zick, Y., Blith, D. L., Karlsson, F. A., Haring, H. U., and Kahn, C. R., 1982, Insulin stimulation of phosphorylation of the β subunit of the insulin receptor. Formation of both phosphoserine and phosphotyrosine, *J. Biol. Chem.* **257:**9891–9894.

King, G. L., Kahn, C. R., Rechler, M. M., and Nissley, S. P., 1980, Direct demonstration of separate receptors for growth and metabolic activities of insulin and multiplication-stimulating activity (an insulin-like growth factor) using antibodies to the insulin receptor, *J. Clin. Invest.* **66:**130–140.

Koontz, J. W., and Iwahashi, M., 1981, Insulin as a potent, specific growth factor in a rat hepatoma cell line, *Science* **211:**947–949.

Land, H., Parada, L. F., and Weinberg, R. A., 1983, Tumorigenic conversion of primary embryo fibroblasts requires at least two cooperating oncogenes, *Nature* **304:**596–602.

Lechner, J. F., and Kaighn, M. E., 1979, Reduction of the calcium requirement of normal human epithelial cells by EGF, *Exp. Cell Res.* **121:**432–435.

Lee, L. -S., and Weinstein, I. B., 1978, Tumor-promoting phorbol esters inhibit binding of epidermal growth factor to cellular receptors, *Science* **202:**313–315.

Leffert, H. L., and Koch, K. S., 1977, Control of animal cell proliferation, in: *Growth, Nutrition, and Metabolism of Cells in Culture,* Volume III (G. H. Rothblatt and V. J. Cristofalo, eds.), Academic Press, New York, pp. 225–294.

Leof, E. G., Wharton, W., VanWyk, J. J., and Pledger, W. J., 1980, Epidermal growth factor and somatomedin C control G_1 progression of competent BALB/c 3T3 cells: Somatomedin C regulates commitment to DNA synthesis, *J. Cell Biol.* **87:**5a.

Levis, R., McReynolds, L., and Penman, S., 1977, Coordinate regulation of protein synthesis and messenger RNA content during growth arrest of suspension Chinese hamster ovary, *J. Cell. Physiol.* **90:**485–502.

Liskay, R. M., 1977, Absence of a measurable G_2 phase in two Chinese hamster cell lines, *Proc. Natl. Acad. Sci. U.S.A.* **74:**1622–1625.

Liskay, R. M., and Prescott, D., 1978, Genetic analysis of G1 period: Isolation of mutants (or variants) with a G1 period from a Chinese hamster cell line lacking G1, *Proc. Natl. Acad. Sci. U.S.A.* **74:**1622–1625.

Low, D. A., Scott, R. W., Baker, J. B., and Cunningham, D. D., 1982, Cells regulate their mitogenic response to thrombin through release of protease nexin, *Nature (London)* **298:**476–478.

Magnusson, S., Petersen, T. E., Sottrup-Jensen, L., and Claeys, H., 1975, Complete primary structure of prothrombin isolation, structure, and reactivity of ten carboxylated glutamic acid residues and regulation of prothrombin activation by thrombin, in: *Proteases and Biological Control* (E. Reich, D. B. Rifkin, and E. Shaw, eds.), Cold Spring Harbor Laboratory, Cold Spring Harbor, New York, pp. 123–149.

Martin, R. G., and Stein, S., 1976, Resting state in normal and simian virus 40 transformed Chinese hamster lung cells, *Proc. Natl. Acad. Sci. U.S.A.* **73:**1655–1659.

McClain, D. A., and Edelman, G. M., 1980, Density-dependent stimulation and inhibition of cell growth by agents that disrupt microtubules, *Proc. Natl. Acad. Sci. U.S.A.* **77:**2748–2752.

McClain, D. A., D'Eustachio, P., and Edelman, G. M., 1977, Role of surface modulating assemblies in growth control of normal and transformed fibroblasts, *Proc. Natl. Acad. Sci. U.S.A.* **74:**66–670.

McClure, D. B., Rockwell, G., and Sato, G., 1979, The growth requirements of normal and SV40 transformed Balb/c 3T3 cells in serum-free medium, *J. Cell. Biol.* **83:**96a.

McClure, D. B., Hightower, M. J., and Topp, W. C., 1982, Effect of SV40 transformation on the growth factor requirements of the rat embryo line REF52 in serum-free medium, in: *Growth of Cells in Hormonally Defined Media* (G. H. Sato, A. B. Pardee, and D. A. Sirbasku, eds.), Cold Spring Harbor Laboratory, Cold Spring Harbor, New York, pp. 345–364.

McKeehan, W. L., 1982, Growth-factor-nutrient interrelationships in control of normal and transformed cell proliferation, in: *Growth of Cells in Hormonally Defined Media* (G. H. Sato, A. B. Pardee, and D. A. Sirbasku, eds.), Cold Spring Harbor Laboratory, Cold Spring Harbor, New York, pp. 65–74.

McKeehan, W. L., and Ham, R. G., 1976, Stimulation of clonal growth of normal fibroblasts with substrata coated with basic polymers, *J. Cell Biol.* **71:**727–734.

McKeehan, W. L., and McKeehan, K. A., 1979, Epidermal growth factor modulates extracellular Ca^{2+} requirement for multiplication of normal human skin fibroblasts, *Exp. Cell Res.* **123:**397–400.

McKeehan, W. L., and McKeehan, K. A., 1980a, Calcium, magnesium and serum factors in multiplication of normal and transformed human lung fibroblasts, *In Vitro* **16:**475–485.

McKeehan, W. L., and McKeehan, K. A., 1980b, Serum factors modify the cellular requirement for Ca^{2+}, K^{+}, Mg^{2+}, phosphate ions, and 2-oxocarboxylic acids for multiplication of normal fibroblasts, *Proc. Natl. Acad. Sci. U.S.A.* **77:**3417–3421.

McKeehan, W. L., McKeehan, K. A., Hammond, S. L., and Ham, R. G., 1977, Improved medium for clonal growth of human diploid fibroblasts at low concentrations of serum protein, *In Vitro* **13:**399–416.

McKeehan, W. L., Genereux, D. P., and Ham, R. G., 1978, Assay and partial purification of factors from serum that control multiplication of human diploid fibroblasts, *Biochem. Biophys. Res. Commun.* **80:**1013–1021.

Medrano, E. E., and Pardee, A. B., 1980, Prevalent deficiency in tumor cells of cycloheximide-induced cycle arrest, *Proc. Natl. Acad. Sci. U.S.A.* **77:**4123–4126.

Messmer, T. O., and Young, D. V., 1977, The effects of biotin and fatty acids on SV3T3 cell growth in the presence of normal calf serum, *J. Cell. Physiol.* **90:**265–270.

Mierzejewski, K., and Rozengurt, E., 1976, Stimulation of DNA synthesis and cell division in a chemically defined medium: Effect of epidermal growth factor, insulin, and vitamin B^{12} on resting cultures of 3T6 cells, *Biochem. Biophys. Res. Commun.* **73:**271–278.

Mierzejewski, K., and Rozengurt, E., 1977, Vitamin B^{12} enhances the stimulation of DNA synthesis by serum in resting cultures of 3T6 cells, *Exp. Cell Res.* **106:**394–397.

Moses, H. L., Proper, J. A., Volkenant, M. E., Wells, D. J., and Getz, M. J., 1978, Mechanism of growth arrest of chemically transformed cells in culture, *Cancer Res.* **38:**2807–2812.

Moses, H. L., Branum, E. L., Proper, J. A., and Robinson, R. A., 1981, Transforming growth factor production by chemically transformed cells, *Cancer Res.* **41:**2842–2848.

Nilsen-Hamilton, M., and Hamilton, R. T., 1979, Fibroblast growth factor causes an early increase in phosphorylation of a membrane protein in quiescent 3T3 cells, *Nature (London)* **279:**444–446.

Noonan, K. D., 1976, Role of serum in protease-induced stimulation of 3T3 cell division past the monolayer stage, *Nature (London)* **251:**573–576.

O'Farrell, M. K., Clingan, D., Rudland, P. S., and Jimenez de Asua, L., 1979, Stimulation of the initiation of DNA synthesis and cell division in several cultured mouse cell types. Effect of growth-promoting hormones and nutrients, *Exp. Cell Res.* **118:**311–321.

Oren, M., Maltzman, W., and Levine, A. J., 1981, Post-translational regulation of the 54K cellular tumor antigen in normal and transformed cells, *Mol. Cell. Biol.* **1:**101–110.

Otto, A. M., Zumbe, A., Bison, L., Kubler, A. M., and Jimenez de Asua, L., 1979, Cytoskeleton-disrupting drugs enhance effect of growth factors and hormones on initiation of DNA synthesis, *Proc. Natl. Acad. Sci. U.S.A.* **76:**6435–6438.

Ozanne, B., Fulton, R. J., and Kaplan, P. L., 1980, Kirsten murine sarcoma virus transformed cell lines and a spontaneously transformed rat cell-line produce transforming factors, *J. Cell. Physiol.* **105:**163–180.

Pardee, A. B., and James, L. J., 1975, Selective killing of transformed baby hamster kidney (BHK) cells, *Proc. Natl. Acad. Sci. U.S.A.* **72:**4994–4998.

Pardee, A. B., Dubrow, R., Hamlin, J. L., and Kletzien, R. F., 1978, Animal cell cycle, *Annu. Rev. Biochem.* **47:**715–750.

Pardee, A. B., Cherington, P. V., and Medrano, E. E., 1980, On deciding which factors regulate cell growth, in: *Control of Cellular Division and Development*, Part A (D. Cunningham, E. Goldwasser, J. Watson, and C. F. Fox, eds), Alan R. Liss, Inc., New York, pp. 495–502.

Pledger, W. J., Stiles, C. D., Antoniades, H. N., and Scher, C. D., 1977, Induction of DNA synthesis in BALB/c 3T3 cells by serum components: Reevaluation of the commitment process, *Proc. Natl. Acad. Sci. U.S.A.* **74:**4481–4485.

Pledger, W. J., Leof, E. B., Chou, B. B., Olashaw, N., O'Keefe, E. F., Van Wyk, J. J., and Wharton, W. R., 1982, Initiation of cell-cycle traverse by serum-derived growth factors, in: *Growth of Cells in Hormonally Defined Media* (G. H. Sato, A. B. Pardee, and D. A. Sirbasku, eds.) Cold Spring Harbor Laboratory, Cold Spring Harbor, New York, pp. 259–273.

Pohjanpelto, P., 1979, Synergistic effect of thrombin and platelet extract on growth of human fibroblasts, *Thromb. Res.* **14:**353–362.

Pollack, R. (ed.), 1981, *Readings in Mammalian Cell Culture*, 2nd ed., Cold Spring Harbor Laboratory, Cold Spring Harbor, New York.

Pollack, R., Osborn, M., and Weber, K., 1975, Patterns of organization of actin and myosin in normal and transformed cultured cells, *Proc. Natl. Acad. Sci. U.S.A.* **72:**994–998.

Prescott, D. M., 1976, The cell cycle and the control of cellular reproduction, *Adv. Genet.* **18:**99–177.

Rasmussen, H., 1970, Cell communication, calcium ion, and cyclic adenosine, *Science* **170:**404–411.

Rassoulzadegan, M., Cowie, A., Carr, A., Glaichenhaus, N., Kamen, R., and Cuzin, F., 1982, The roles of individual polyoma virus early proteins in oncogenic transformation, *Nature (London)* **300:**713–718.

Roberts, A. B., Anzano, M. A., Frolik, C. A., and Sporn, M. B., 1982, Transforming growth factors: Characteriztion of two classes of factors from neoplastic and non-neoplastic tissues, in: *Growth of Cells in Hormonally Defined Media* (G. H. Sato, A. B. Pardee, and D. A. Sirbasku, eds.), Cold Spring Harbor Laboratory, Cold Spring Harbor, New York, pp. 319–332.

Robinson, J. H., Smith, J. A., and Dee, L. A., 1976, Synthesis and degradation of proteins in cultured, androgen-responsive tumor cells, *Exp. Cell Res.* **102:**117–126.

Rockwell, G. A., Sato, G. H., and McClure, D. B., 1980, The growth requirements of SV40 virus transformed Balb/c-3T3 cells in serum-free monolayer culture, *J. Cell. Physiol.* **103:**323–331.

Rossini, M., Lin, J. C., and Baserga, R., 1976, Effects of prolonged quiescence on nuclei and chromatin of WI-38 fibroblasts, *J. Cell. Physiol.* **88:**1–11.

Rossow, P. W., Riddle, V. G. H., and Pardee, A. B., 1979, Synthesis of labile, serum-dependent protein in early G_1 controls of animal cell growth, *Proc. Natl. Acad. Sci. U.S.A.* **76:**4446–4450.

Rozengurt, E., 1979, Early events in growth stimulation, in: *Surfaces of Normal and Malignant Cells* (R. Hynes, ed.), J. Wiley and Sons, Sussex, England, pp. 323–353.

Rozengurt, E., 1980, Stimulation of DNA synthesis in quiescent cultured cells: Exogenous agents, internal signals and early events, *Curr. Topics Cell. Regul.* **17:**59–88.

Rozengurt, E., and Heppel, 1975, Serum rapidly stimulates ouabain-sensitive $^{86}Rb^+$ influx in quiescent 3T3 cells, *Proc. Natl. Acad. Sci. U.S.A.* **72:**4492–4495.

Rozengurt, E., and Mendoza, S., 1980, Monovalent ion fluxes and the control of cell proliferation in cultured fibroblasts, *Ann. N.Y. Acad. Sci.* **339:**175–190.

Rozengurt, E., Legg, A., and Pettican, P., 1979, Vasopressin stimulation of mouse 3T3 cell growth, *Proc. Natl. Acad. Sci. U.S.A.* **76:**1284–1287.

Rubin, H., 1975a, Central role for magnesium in coordinate control of metabolism and growth in animal cells, *Proc. Natl. Acad. Sci. U.S.A.* **72:**3551–3555.

Rubin, H., 1975b, Nonspecific nature of the stimulus to DNA synthesis in cultures of chick embryo cells, *Proc. Natl. Acad. Sci. U.S.A.* **72:**1676–1680.

Rudland, P. S., and Jimenez de Asua, L., 1979, Actions of growth factors in the cell cycle, *Biochim. Biophys. Acta* **560:**91–133.

Rudland, P. S., Weil, S., and Hunter, A. R., 1975, Changes in RNA metabolism and accumulation of presumptive messenger RNA during transition from the growing to the quiescent state of cultured mouse fibroblasts, *J. Mol. Biol.* **96:**745–766.

Rudland, P. S., Durbin, H., Clingan, D., and Jimenez de Asua, L., 1977, Iron salts and transferrin are specifically required for cell division of cultured 3T6 cells, *Biochem. Biophys. Res. Commun.* **75:**556–562.

Ruley, H. E., 1983, Adenovirus early region 1A enables viral and cellular transforming genes to transform primary cells in culture, *Nature* **304:**602–606.

Rutherford, R. B., and Ross, R., 1976, Platelet factors stimulate fibroblasts and smooth muscle cells quiescent in plasma serum to proliferate, *J. Cell Biol.* **69:**196–203.

Sager, R., and Kovac, P. E., 1978, Genetic analysis of tumorigenesis: I. Expression of tumor-forming ability in hamster hybrid cell lines, *Somatic Cell Genet.* **4:**375–392.

Sager, R., Cherington, P. V., Smith, B. L., and Pardee, A. B., 1979, Growth factor requirements of Chinese hamster cell hybrids, clonal progeny, and tumorigenic derivatives, *Fed. Proc.* 38:635.

Sager, R., Bennett, F., and Smith, B. L., 1982, Altered growth-factor requirements of transformed mutants and tumor derived cell populations of CHEF cell origin, in: *Growth of Cells in Hormonally Defined Media* (G. H. Sato, A. B. Pardee, and D. A. Sirbasku, eds.), Cold Spring Harbor Laboratory, Cold Spring Harbor, New York, pp. 231–241.

Sawyer, S. T., and Cohen, S., 1981, Enhancement of calcium uptake and phoshatidylinositol turnover by epidermal growth factor in A-431 cells, *Biochemistry* **20:**6280–6286.

Schechter, Y., Hernaez, L., Schlessinger, J., and Cuatrecasas, P., 1979, Local aggregation of hormone-receptor complexes is required for activation by epidermal growth factor, *Nature (London)* **278:**835–838.

Scher, C. D., Stathakos, D., and Antoniades, H. N., 1974, Dissociation of cell division stimulating capacity for Balb/c-3T3 from the insulin-like activity in human serum, *Nature (London)* **247:**279–281.

Scher, C. D., Pledger, W. J., Martin, D., Antoniades, H., and Stiles, C. D., 1978, Transforming viruses directly reduce the cellular growth requirement for a platelet-derived growth factor, *J. Cell. Physiol.* **97:**371–380.

Scher, C. D., Stone, M. E., and Stiles, C. D., 1979a, Platelet-derived growth factor prevents G_0 growth arrest, *Nature (London)* **281:**39–392.

Scher, C. D., Shepard, R. C., Antoniades, H. N., and Stiles, C. D., 1979b, Platelet-derived growth factor and the regulation of the mammalian fibroblast cell cycle, *Biochem. Biophys. Acta* **560:**217–241.

Scher, C. D., Hendrickson, S. L., Whipple, A. P., Gottesman, M. M., and Pledger, W. J., 1982, Constitutive synthesis by a tumorigenic cell line of proteins modulated by platelet-derived growth factor, in: *Growth of Cells in Hormonally Defined Media* (G. H. Sato, A. B. Pardee, and D. A. Sirbasku, eds.), Cold Spring Harbor Laboratory, Cold Spring Harbor, New York, pp. 289–303.

Schor, S., and Rozengurt, E., 1973, Enhancement by purine nucleosides and nuclotides of serum-induced DNA synthesis in quiescent 3T3 cells, *J. Cell. Physiol.* **81:**339–346.

Schreiber, A. B., Yarden, Y., Eshhar, Z., and Schlessinger, J., 1981a, Monoclonal antibodies against receptor for epidermal growth factor induce early and delayed effects of epidermal growth factor, *Proc. Natl. Acad. Sci. U.S.A.* **78:**7535–7539.

Schreiber, A. B., Yarden, Y., and Schlessinger, J., 1981b, A non-mitogenic analogue of epidermal growth factor enhances the phosphorylation of endogenous membrane proteins, *Biochem. Biophys. Res. Commun.* **101:**517–523.

Schreiber, A. B., Libermann, T. A., Lax, I., Yarden, Y., and Schlessinger, J., 1983, Biological role of epidermal growth factor-receptor clustering. Investigation with monoclonal anti-receptor antibodies, *J. Biol. Chem.* **258:**846–853.

Scott, R. E., Hoerl, B. J., Wille, J. J., Jr., Florine, D. L., Krawisz, B. R., and Yun, K., 1982, Coupling of proadipocyte growth arrest and differentiation. II. A cell cycle model for the physiological control of cell proliferation. *J. Cell Biol.* **94:**400–405.

Sefton, B. M., and Rubin, H., 1970, Release from density dependent growth inhibition by proteolytic enzymes, *Nature* (*London*) **277:**843–845.

Serrero, G. R., McClure, D. B., and Sato, G. H., 1979, Growth of 3T3 fibroblasts in serum-free, hormone-supplemented media, in: *Hormones and Cell Culture* (R. Ross, and G. H. Sato, eds.), Cold Spring Harbor Laboratory, Cold Spring Harbor, New York, pp. 523–530.

Shih, C., Shilo, B. -Z., Goldfarb, M. P., Dannenberg, A., and Weinberg, R. A., 1979, Passage of phenotypes of chemically transformed cells via transfection of DNA and chromatin, *Proc. Natl. Acad. Sci. U.S.A.* **76:**5714–5718.

Shipley, G. D., and Ham, R. G., 1981, Improved medium and culture conditions for clonal growth with minimal serum protein and for enhanced serum-free survival of Swiss 3T3 cells, *In Vitro* **17:**656–670.

Shipley, G. D., and Moses, H. L., 1982, Growth of nontransformed and transformed mouse AKR-2B cells in serum-free medium. Loss of epidermal growth factor response with transformation, *J. Cell Biol.* **95:**194a.

Shoyab, M., DeLarco, J. E., and Todaro, G. J., 1979, Biologically active phorbol esters specifically alter affinity of epidermal growth factor membrane receptors, *Nature* (*London*) **279:**387–391.

Smith, B. L., and Sager, R., 1982, Multistep origin of tumor-forming ability in Chinese hamster embryo fibroblast cells, *Cancer Res.* **42:**389–396.

Smith, C. J., Rubin, C. S., and Rosen, O. A., 1980, Insulin-treated 3T3-L1 adipocytes and cell-free extracts derived from them incorporate ^{32}P into ribosomal protein S6, *Proc. Natl. Acad. Sci. U.S.A.* **77:**2641–2645.

Smith, H., Scher, C., and Todaro, G., 1971, Induction of cell division in medium lacking serum growth factor by SV40, *Virology* **44:**359–370.

Smith, J. A., and Martin, L., 1973, Do cells cycle?, *Proc. Natl. Acad. Sci. U.S.A.* **70:**1263–1267.

Smith, J. C., Singh, J. P., Lillquist, J. S., Goon, D. S., and Stiles, C. D., 1982, Growth factor adherent to cell substrate are mitogenically active *in situ*, *Nature* (*London*) **296:**154–156.

Stein, G. H., and Yanishevsky, R., 1979, Autoradiography, *Methods Enzymol.* **58:**279–292.

Stiles, C. D., and Kawahara, A., 1978, The growth behavior of virus-transformed cells in nude mice, in: *The Nude Mouse in Experimental and Clinical Research* (J. Fogh and B. Giovanella, eds.), Academic Press Inc., New York, pp. 385–409.

Stiles, C. D., Capone, G. T., Scher, C. D., Antoniades, H. N., Van Wyk, J. J., and Pledger, W. J., 1979, Dual control of cell growth by somatomedins and platelet-derived growth factor, *Proc. Natl. Acad. Sci. U.S.A.* **76:**1279–1283.

Stoker, M., and MacPherson, I., 1961, Studies on transformation of hamster cells by polyoma virus *in vitro*, *Virology* **14:**359–370.

Stoker, M. G. P., and Rubin, H., 1967, Density dependent inhibition of cell growth in culture, *Nature* (*London*) **215:**171–172.

Swierenga, S. H. H., MacManus, J. P., and Whitfield, J. F., 1976, Regulation by calcium of the proliferation of heart cells from young adult rats, *In Vitro* **12:**31–36.

Taub, M. U. B., Chuman, L., Rindler, M. J., Saier, M. H., Jr., and Sato, G., 1981, Alterations in growth requirements of kidney epithelial cells in defined medium associated with malignant transformation, *J. Supramol. Struct. Cell. Biochem.* **15:**63–72.

Temin, H. M., 1971, Stimulation by serum of multiplication of stationary chicken cells, *J. Cell. Physiol.* **78:**161–170.

Temin, H. M., and Rubin, H., 1958, Characteristics of an assay for Rous sarcoma virus and Rous sarcoma cells in tissue culture, *Virology* **14:**359–370.

Teng, M. -H., Bartholomew, J. C., and Bissell, M. J., 1977, Synergism between antimicrotubule agents and growth stimulants in enhancement of cell cycle traverse, *Nature* (*London*) **268:**739–741.

Thomas, G., Martin-Perez, J., Siegmann, M., and Otto, A. M., 1982, The effect of serum, EGF, $PGF_{2\alpha}$, and insulin on S6 phosphorylation and DNA synthesis, *Cell* **30:**235–242.

Thrash, C. R., and Cunningham, D. D., 1973, Stimulation of division of density inhibited fibroblasts by glucocorticoids, *Nature* (*London*) **242:**399–401.

Todaro, G. J., and Green, H., 1963, Quantitative studies of the growth of mouse embryo cells in culture and their development into established lines, *J. Cell Biol.* **17:**299–313.

Todaro, G. J., and Green, H., 1964, An assay for cellular transformation by SV40, *Virology* **23:**117–119.

Todaro, G. J., Lazar, G. K., and Green, H., 1965, The initiation of cell division in a contact-inhibited mammalian cell line, *J. Cell Comp. Physiol.* **66:**325–334.

Tucker, R. W., Pardee, A. B., and Fujiwara, K., 1979a, Centriole ciliation is related to quiescence and DNA synthesis in 3T3 cells, *Cell* **17:**527–535.

Tucker, R. W., Scher, C. D., and Stiles, C. D., 1979b, Centriole deciliation associated with the early response of 3T3 cells to growth factors but not to SV40, *Cell* **18:**1065–1072.

Tupper, J. T., and Zorgniotti, F., 1977, Calcium content and distribution as a function of growth and transformation in the mouse 3T3 cell, *J. Cell Biol.* **75:**12–22.

Ushiro, H., and Cohen, S., 1980, Identification of phosphotyrosine as a product of epidermal growth factor-activated protein kinase in A-431 cell membranes, *J. Biol. Chem.* **255:**8363–8365.

Vaheri, A., Ruoslahti, E., Hovi, T., and Nordling, S., 1973, Stimulation of density-inhibited cell cultures by insulin, *J. Cell. Physiol.* **81:**355–364.

Vogel, A., Ross, R., and Raines, E., 1980, Role of serum components in density-dependent inhibition of growth of cells in culture. Platelet-derived growth factor is the major serum determinant of saturation density, *J. Cell Biol.* **85:**377–385.

Walthall, B. J., and Ham, R. G., 1981, Multiplication of human diploid fibroblasts in a synthetic medium supplemented with EGF, insulin, and dexamethasone, *Exp. Cell Res.* **134:**303–311.

Whittenberger, B., and Glaser, L., 1977, Inhibition of DNA synthesis in cultures of 3T3 cells by isolated surface membranes, *Proc. Natl. Acad. Sci. U.S.A.* **74:**2251–2255.

Whittenberger, B., Raben, D., Lieberman, M. A., and Glaser, L., 1978, Inhibition of growth of 3T3 cells by extract of surface membranes, *Proc. Natl. Acad. Sci. U.S.A.* **75:**5457–5461.

Wolfe, R. A., Wu, R., and Sato, G. H., 1980, Epidermal growth factor-induced down-regulation of receptors does not occur in HeLa cells grown in defined medium, *Proc. Natl. Acad. Sci. U.S.A.* **77:**2735–2739.

Yamada, M., and Olden, K., 1978, Fibronectins—Adhesive glycoproteins of cell surface and blood, *Nature (London)* **275:**179–184.

Yarden, Y., Schreiber, A. B., and Schlessinger, J., 1982, A nonmitogenic analogue of epidermal growth factor induces early responses mediated by epidermal growth factor, *J. Cell Biol.* **92:**687–693.

Yen, A., Warrington, R. C., and Pardee, A. B., 1978, Serum-stimulated 3T3 cells undertake a histidinol-sensitive process which G_1 cells do not, *Exp. Cell Res.* **114:**458–462.

Young, D. V., Cox, F. W., III, Chipman, S., and Hartman, S. C., 1979, The growth stimulation of SV3T3 cells by transferrin and its dependence on biotin, *Exp. Cell Res.* **118:**410–414.

Zetter, B. R., Sun, T. T., Chen, L. B., and Buchanan, J. M., 1977, Thrombin potentiates the mitogenic response of cultured fibroblasts to serum and other growth promoting agents, *J. Cell. Physiol.* **92:**233–240.

CHAPTER 3

Growth and Differentiation of Preadipocyte Cell Lines in Serum-Free Medium

GINETTE SERRERO-DAVÉ

Introduction

Obesity arises from an abnormal growth and differentiation of adipose cells. Previous works have shown that obesity could be due to either an increase in adipose cell number (hyperplasia) or an increase in the triglyceride content of adipocytes (hypertrophy) (Faust *et al.*, 1979; Bjorntorp, 1979). The origin of adipocytes is still unknown. They derive from mesodermal tissue and are thought to result from the reprogramming of committed fibroblastlike cells, called preadipocytes, populating the adipose tissue (Green, 1979). The reprogramming takes place during embryonic life, but some evidence indicates that the formation of new adipocytes can also occur in adults (Faust *et al.*, 1978). Preadipocyte cells constitute a pool of cells in the stromal-vascular fraction of the already developed adipose tissue. Bjorntorp and co-workers (1978) succeeded in cultivating cells isolated from the stromal-vascular fraction and showed that the cells could undergo proliferation and adipose conversion in culture. But the identification of the replicating preadipocyte or differentiated adipocyte has yet to be accomplished (Hausman *et al.*, 1980).

Only by studying the normal development of adipose tissue can one expect to understand such a pathological state as obesity. In order

GINETTE SERRERO-DAVÉ • Cancer Center, University of California, San Diego, La Jolla, California 92093. *Present address*: W. Alton Jones Cell Science Center, Lake Placid, New York 12946.

to determine the mechanisms controlling the normal development of adipose tissue, it is necessary to explore phenomena related to the commitment of mesenchymal cells to become preadipocytes and to the conversion of preadipocyte cells into functionally differentiated adipocytes. Various hypotheses have been advanced regarding the control of the differentiation process (Brunner, 1977). According to one of these, it is possible to assume (1) that environmental factors will control the determination and differentiation of preadipose cells and, therefore, (2) that any qualitative and quantitative changes of these extracellular factors will create abnormal conditions leading to pathological states. Hence, it is of importance to identify the nature of the extracellular factors that influence normal growth and differentiation of adipocytes as a prerequisite of understanding how the onset of obesity can occur.

Earlier *in vivo* studies performed on intact and hypophysectomized rats (Hollenberg and Vost, 1968) and *in vitro* experiments performed on freshly isolated adipocytes (Murakawa and Raben, 1968; Kazdova and Vrana, 1970) indicated that adipose differentiation was regulated by hormones and, in particular, by growth hormone and insulin. But *in vivo* experiments could not determine whether the two hormones were augmenting cellular proliferation in the same or in separate cell populations because of the heterogeneity in cell composition of the adipose tissue. In addition, *in vitro* experiments done with freshly isolated adipocytes gave a limited amount of information and made any detailed biochemical studies hard to perform because of the short survival time of adipocytes in culture. The establishment in culture of cell lines able to accumulate triglycerides provided the opportunity to work with stable and reproducible systems and opened new ways to investigate the process of adipose conversion. As the differentiated cells resulting from the *in vitro* conversion of these cell lines presented the same characteristics as mature *in vivo* adipocytes, these cell culture systems were considered good models to study adipocyte growth and differentiation *in vitro*. In all cases, the preadipose cells retain a fibroblastic appearance and proliferate until they reach confluency. At the confluent stage, they undergo adipose conversion in a sequence of events that will be described below. In most cases, the growth and differentiation took place when the cells were cultivated in the presence of fetal calf serum. Because serum contains many factors of known and unknown identities, it appeared necessary to use a new approach to investigate the nature of the extracellular factors influencing growth and differentiation of adipose cells, using an *in vitro* cell culture system as a model. Such an approach could be the recent development of hormone-supplemented media for long-term growth of cells in the absence of serum, permitting a detailed analysis of extracellular factors required for cell differentiation as well as growth (Barnes and Sato, 1980). We decide to apply this approach to the study of the growth and differentiation of adipocytelike cells *in vitro*. We attempted to define a hormone-supplemented medium that would support the growth of a chosen cell line and then examine conditions allowing the cells to differentiate. Such an approach should provide answers to various questions, for example:

1. What is the nature of the factors controlling growth and differentiation of adipocytelike cells *in vitro*?
2. Does fetal calf serum contain an adipogenic factor? What is its mechanism of action?
3. Do lipogenic and lipolytic hormones, known to act on mature adipocytes, have a role in the adipose conversion of preadipocytes?

We chose to study a triglyceride-producing, teratoma-derived cell line called 1246 in serum-free medium (Darmon *et al.*, 1981). Before describing the results obtained with this cell line, information about the various models of adipocytelike cells and about the biochemical events accompanying the differentiation program will be provided.

In Vitro Models of Adipocytelike Cells

Several cell lines, able to accumulate triglycerides, have been isolated from various species, either from total embryos, from newborn or adult adipose tissues, from stromal-vascular fractions, or from the differentiated part of adipose tissue (for a review, see Ailhaud, 1982). These cell lines are as follows:

1. The 3T3-L1 and 3T3-F442 A cell lines were cloned from Swiss mouse embryo fibroblast 3T3 cultures (Green and Kehinde, 1974, 1976).
2. The Ob17, HG-Fu cell lines were established from the epididymal fat pad of adult Ob/Ob C57B1 mouse and its lean counterpart (Negrel *et al.*, 1978).
3. The N10 cell line was isolated from the epididymal fat pad of Swiss mouse (Ailhaud *et al.*, 1979).
4. The PFC6 cell line was established from the stromal-vascular fraction of mouse epididymal fat pad (Ailhaud, 1982).
5. ST13 was obtained from a spontaneous mammary carcinoma of ddN mouse (Hirogun *et al.*, 1980).

6. Forest *et al.* (unpublished data) have established a cell line from newborn human adipose tissue.
7. The 1246 cell line was clonally obtained from a myogenic cell line derived from C3H mouse teratoma (Darmon *et al.*, 1981).

In addition, various preadipose cell strains have been obtained from adipose tissue and bone marrow of human and rat (Ailhaud, 1982).

Differentiation Program

During exponential growth, these cells remain in fibroblastic appearance. When they reach confluency, they undergo adipose conversion that corresponds to the following sequence of events (Green, 1979). A series of morphological changes are apparent. The confluent cells develop a reduced adherence to the substratum and adopt a spherical shape. These changes are correlated with differentiation of enzymatic activities controlling lipid metabolism. There is an increase in the activities of enzymes involved in fatty acid synthesis including: ATP-citrate lyase, acetyl-coA carboxylase, fatty-acid synthetase, malic enzyme, and pyruvate carboxylase (Green and Kehinde, 1975; Mackall *et al.*, 1976; Mackall and Lane, 1977; Negrel *et al.*, 1978; Verrando *et al.*, 1981). In addition, the enzymes required for fatty acid esterification and acylation of glycerol such as glycerophosphate acyl transferase, glycerophosphate dehydrogenase, and diglyceride acyl transferase (Kuri-Harcuch and Green, 1977; Grimaldi *et al.*, 1978; Coleman *et al.*, 1978; Negrel *et al.*, 1978; Wise and Green, 1979; Verrando *et al.*, 1981) show increased activity. The activity of lipoprotein lipase, triglyceride lipase, diglyceride lipase, monoglyceride lipase, cholesterol ester hydrolase, fatty acyl coA ligase (Rothblat and deMartinis, 1977; Eckel *et al.*, 1977; Wise and Green, 1978; Grimaldi *et al.*, 1978; Spooner *et al.*, 1979; Kawamura *et al.*, 1981; Forest *et al.*, 1981; Murphy *et al.*, 1981) are also regulated during differentiation.

The induction level of these enzymes is variable being sixfold for diglyceride acyl transferase, 80- to 180-fold for lipoprotein lipase, and 1000-fold for glycerophosphate dehydrogenase. The kinetics of induction are also variable; glycerophosphate dehydrogenase activity is induced maximally after eight days and acyl-coA ligase after 15 days.

The differentiation of preadipose cells is accompanied by an increase in the number of insulin receptors (Reed *et al.*, 1977), by an increased sensitivity to lipolytic hormones, and by the appearance of response to ACTH (Butcher and Baird, 1968; Rubin *et al.*, 1977; Negrel *et al.*, 1978; Serrero and Khoo, 1982); appearance of a limited mitogenic response is also seen during differentiation. This mitogenic response has been shown to occur in 3T3-F442A cells (Pairault and Green, 1979) and in Ob17 cells (Djian *et al.*, 1983). It is of limited duration and concerns only the cells susceptible to adipose conversion. Proliferation of the preadipose cells responding to the adipogenic factor leads to a selective increase in their progeny, all of which will form adipose cells. The mitogenic response is a form of clonal selection (Green, 1979).

The cells that undergo adipose conversion accumulate triglycerides in the form of lipid droplets that fuse together. This accumulation is affected by many factors such as β-adrenergic compounds, ACTH, and glucagon, which stimulate lipolysis (Kawamura *et al.*, 1981; Negrel *et al.*, 1978; Serrero and Khoo, 1982).

Factors Influencing Adipose Conversion *in Vitro*

Regarding the factors that influence differentiation of adipocytes *in vitro*, it is necessary to distinguish between two sets of factors: the regulators and the accelerators of differentiation. A regulator will be defined as any factor, in the absence of which the development of the adipose conversion program will not take place. The accelerators are factors that are able to increase the amplitude of an event or events accompanying adipose conversion, but unable to act on it in the absence of a regulator of differentiation. The study of the various established cell lines described below has permitted the identification of numerous factors capable of influencing adipose conversion (Table I). But the fact that cells were cultivated in the presence of serum made it difficult to distinguish between accelerators and regulators of differentiation among these various factors. One possible way to identify the regulators and the accelerators was to eliminate serum from the culture medium of the cells undergoing differentiation, replacing it with hormones. The study of the various established cell lines described above has provided evidence that numerous factors stimulate or inhibit the process of differentiation of cells cultivated in the presence of fetal calf serum (Table I) and can be considered accelerators and inhibitors of differentiation.

Various works have shown that fetal calf serum contains an adipogenic factor that may be considered a regulator of differentiation. Kuri-Harcuch and Green (1978) have demonstrated that 3T3-F442A cells grow in a medium supplemented with cat serum, but do not undergo adipose conversion unless fetal calf serum is added to the

Table I. Accelerators and Inhibitors of Differentiation of Adipocytelike Cells in Culture

Accelerators
Insulin (Green and Kehinde, 1974; Negrel *et al.*, 1978)
Dexamethasone (Rubin *et al.*, 1978)
3-Isobutyl, 1-methylxanthine (Russel and Ho, 1976; Rubin *et al.*, 1978)
Prostaglandins PGF2α, PGE_1 (Russel and Ho, 1976)
Indomethacin (Williams and Polakis, 1977)
Clofenapate (Verrando *et al.*, 1981)
Hemin (Chen and London, 1981)
Triiodothyronine (T3) (Gharbi-Chihi *et al.*, 1981)
Inhibitors
Retinoic acid (Murray and Russell, 1980)
5-Bromodeoxyuridine (Green and Kehinde, 1975)
Interferon (Keay and Grossberg, 1980)
α-Difluoromethyl ornithine (Bethell and Pegg, 1982)
Aminotriazole (Chen and London, 1981)

medium. Hayashi *et al.* (1981) showed that beef pituitary extracts contained an adipogenic activity for the 3T3-F442A cells that can replace that of fetal calf serum. These pituitary extracts contain also antiadipogenic activities that might be attributable to the presence of fibroblast growth factor (FGF) in the extracts.

Grimaldi *et al.* (1982) demonstrated that an adipose-converting fraction, ACF, could be isolated from fetal calf serum by passing the serum through an ion-exchange resin. The treated fetal calf serum could support Ob17 cell growth but not its differentiation. The ACF activity, eluted from the resin, was able to induce Ob17 cell differentiation when added to the culture medium containing treated fetal calf serum.

By culturing 3T3-L1 cells in serum-free medium, we demonstrated that the cells needed the presence of an adipogenic factor in fetal calf serum to undergo adipose differentiation (Serrero *et al.*, 1979). It was shown that 3T3-L1 cells could proliferate in DME-F12 medium containing insulin, transferrin, FGF, and a fraction prepared from rat submaxillary extract. Under these conditions, the cells grew until they reached confluency in a fashion similar to cells cultivated in DME-F12 medium containing 10% fetal calf serum. But in contrast to the cells maintained in serum-supplemented medium, the cells maintained in serum-free medium did not undergo differentiation unless 1% fetal calf serum was added to the medium, indicating that fetal calf serum contained a factor that controlled the onset of the differentiation program in 3T3-L1 cells. Interestingly, other adipocyte-like cell lines, such as Ob17, HG, and N10, could also grow in the same serum-free medium as the one developed for the growth of 3T3-L1 cells (Gaillard *et al.*, unpublished data; Serrero, unpublished observations). As with 3T3-L1 cells, Ob17 cells cultured in serum-free medium will differentiate only if fetal calf serum is added to the culture medium.

By culturing 3T3-L1 cells in serum-free medium, it was possible to demonstrate not only the presence of adipogenic activity(ies) in fetal calf serum, but also of novel growth factors in submaxillary gland extract. Two factors have been purified in the submaxillary gland extract (McClure *et al.*, 1981). Only one of them (mol. wt., 40,000) stimulates the growth of 3T3-L1 cells in serum-free medium (Serrero, unpublished observations).

We applied serum-free culture techniques to the study of a teratoma-derived fibroadipogenic cell line. Teratocarcinoma of the mouse is one of the most attractive systems for studying cell determination and cell differentiation *in vitro* (Stevens, 1967). The malignant embryonal carcinoma cells (EC cells) of the teratocarcinomas are multipotent. When injected into mice, they develop into tumors that contain not only stem cells but also various differentiated tissues. Embryonal carcinoma cells can lose their malignancy and participate in normal embryonic development when grafted into mouse blastocysts (Papaioannou *et al.*, 1975). Under appropriate conditions, EC cells can differentiate *in vitro* into derivatives of the three germ layers (Jakob *et al.*, 1973; Nicolas *et al.*, 1975). Many differentiated cell lines have been isolated either from these *in vitro* cultures or from teratocarcinoma tumors (Nicolas *et al.*, 1981). They correspond to embryonic and extraembryonic tissues and represent either very primitive multipotent cell lines (EC cells) or terminally differentiated ones. Because of their differentiation properties, teratoma-derived cell lines provide excellent models to understand phenomena related to phenotypic expression and also to modulation of phenotypic expression and dedifferentiation, and to study the control of differentiation by the extracellular environment by culturing them in hormonally defined media.

Results and Discussion

Isolation of the Teratoma-Derived Adipogenic Cell Line, 1246

1246 is a triglyceride-accumulating cell line that has been clonally isolated from a myogenic cell line C17-S1-D-T984. T984 was established

Table II. Biochemical Phenotype of 1168, 1246, and 1246 B2 Cell Lines[a]

Markers	1168	1246	1246 B2
α-Bungarotoxin bound (fmole/mg protein)	200	0	0
Creatine phosphokinase activity (mU/mg protein)	400	31	20
Triglycerides (mg/mg protein)	0.21	2.1	0.17

[a] Cells were cultured in DME supplemented with 15% fetal calf serum. Binding of α-bungarotoxin and creatine phosphokinase activity were assayed five days after confluency. Triglycerides were measured 15 days after confluency. Techniques are described in Darmon *et al.* (1981).

from a contractile zone from a culture of a tumor obtained in a mouse after injection of the embryonal carcinoma cell line, C17-S1 (McBurney, 1976; Darmon *et al.*, 1981). Mass cultures of T984 cells grow and differentiate with high efficiency into striated muscular cells (Jakob *et al.*, 1978). However, at clonal densities, T984 form myogenic, adipogenic and fibroblastic cells. It was possible to select clones representative of the three phenotypes. They were 1168, a myogenic cell line; 1246-B2, a fibroblastic cell line; and 1246, a fibroadipogenic cell line. 1246 was considered as fibroadipogenic because when subcloned, it always gave rise to fibroblastic and adipogenic clones. Table II gives the biochemical markers characterizing the phenotypes of 1168, 1246, and 1246-B2. An interesting feature of 1246 cells is that when plated at high denisty (10^7 cells/100 mm dish) or when treated with 5-azacytidine at low density, the cells can be induced to form patches of myotubes (Darmon *et al.*, 1981). Because of their potentiality, we chose the adipogenic cell line as a model to study adipose conversion *in vitro*.

Growth of 1246 Cells in Serum-Free Medium

The simplest medium allowing the best growth of 1246 cells consisted of DME-F12 containing insulin (10 μg/ml), transferrin (10 μg/ml), FGF (100 ng/ml), and fibronectin (5 μg/ml). This medium, called 4F medium, supported a growth rate of 1246 cells similar to that seen in DME-F12 supplemented with 10% FCS, as shown in Fig. 1. Insulin was added at 10 μg/ml but a dose response curve (Fig. 2) showed that 1246 cells responded to concentrations of insulin as low as 1 ng/ml. Figure 2 shows that in the absence of insulin the cells do not grow after they undergo one doubling. Insulin appears to be the most critical hormone in the 4F medium. The cells maintained in medium lacking insulin did not die. If insulin was added to the cells maintained in its absence, they started growing again until they reached a density

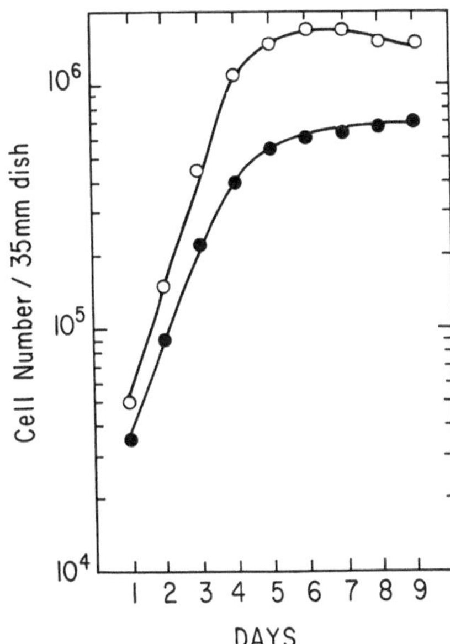

Figure 1. Growth curve of the clonal line 1246 in DME-F12 medium containing 10% FCS (○—○) or supplemented with insulin (10 μg/ml), fibronectin (5μg/ml), transferrin (10 μg/ml), and FGF (100 ng/ml)(●—●).

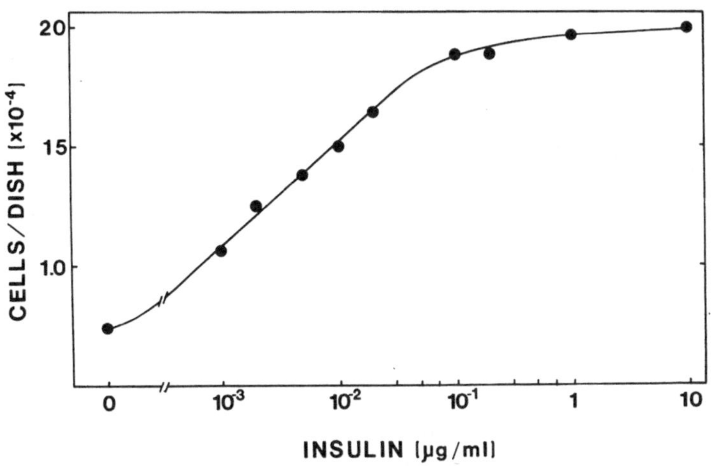

Figure 2. Growth of 1246 cells as a function of insulin concentration. Concentrations of transferrin, fibronectin, and FGF were held constant. Cell number was counted on day 4.

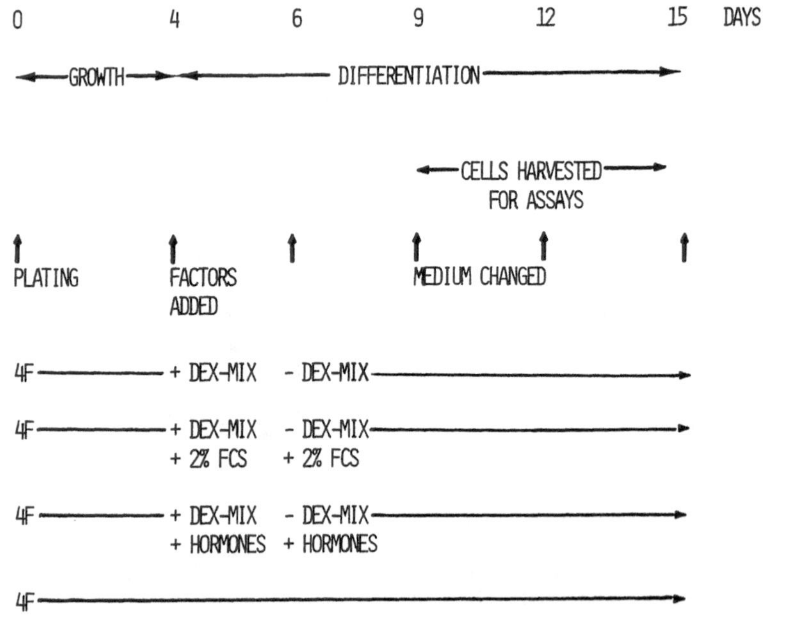

Figure 3. Assay system for measuring differentiation in 1246 cells.

equivalent to the one reached by the cells maintained constantly in the presence of insulin (results not shown).

Differentiation of 1246 Cells

As described previously, the *in vitro* adipose conversion of preadipose cells (3T3-L1, 3T3-F442A, Ob17. . .) takes place according to a sequence of events including morphological changes, enzyme induction, increased sensitivity to hormones, and appearance of a limited mitogenic response (Green, 1979). The questions we proposed to solve were as follows: Under which conditions will 1246 cells, cultivated in serum-free medium, undergo adipose conversion? Are 1246 cells dependent on the presence of a serum adipogenic factor to undergo adipose conversion similar to 3T3-L1, 3T3-F442A, and Ob17 cells? To explore this problem, we defined an assay system described in Fig. 3. We followed the development of various differentiation markers in 1246 cells treated with dexamethasone (2×10^{-7} M) and 3-isobutyl, 1-methylxanthine (2×10^{-9} M) in the absence and in the presence of

Table III. Development of Enzymatic Activities Controlling Lipid Metabolism in 1246 Cells

Activity	4F Day 4	4F + dex-mix Day 9	4F + dex-mix + 2% FCS Day 9
[^{14}C]Acetate incorporated			
Total	270,000	378,000	430,000
Into triglycerides	7,500	120,000	180,000
Glycerophosphate dehydrogenase[a]	N.D.[b]	108	210
Triglyceride lipase	3.6	60.4	60.0
Diglyceride lipase	8.7	198.0	330.0
Monoglyceride lipase	3.1	238.0	685.0
Cholesterol esterase	0.6	44.0	69.2
Lipoprotein lipase	6.2	575.0	625.0
Triglyceride accumulated[c]	N.D.	1.4	4.2

[a] Glycerophosphate dehydrogenase was measured on day 15 after inoculation.
[b] N.D. = not detectable.
[c] Triglycerides were measured on day 15 after inoculation and are expressed in nmole of triglycerides/mg of protein. [^{14}C]Acetate incorporation is expressed in cpm/10^6 cells. All the enzymatic activities are expressed in mU/mg protein. 1 mU of enzymatic activity corresponds to the hydrolysis of nmole of substrate per minute.
(The techniques used for preparing cell extracts and measuring activities have been described previously by Serrero and Sato, 1982; Serrero and Khoo, 1982).

2% fetal calf serum. The markers were morphological changes, triglyceride accumulation, and various enzymatic activities controlling lipolysis and lipogenesis (Serrero and Khoo, 1982; Serrero and Sato, 1982). Table III gives a summary of the different activities measured in cells treated with dexamethasone and 3-isobutyl, 1-methylxanthine (dex-mix), in the presence or absence of 2% fetal calf serum. In the absence of 2% fetal calf serum, the cells treated with dex-mix in serum-free medium developed activities controlling lipid metabolism and acquired characteristics of differentiated adipocytes. Among the activities measured were found activities controlling *de novo* synthesis of triglycerides measured by [^{14}C]acetate incorporation into triglycerides, lipase activities, and glycerophosphate dehydrogenase activity. Fig. 4 shows that the triglyceride lipase activity induced in 1246 cells, treated with dex-mix in serum-free medium, is inhibited by a specific inhibitor of cAMP-dependent protein kinase and stimulated by the addition of an excess of exogenous cAMP-dependent protein kinase indicating that 1246 cells contain an enzyme having the same characteristics as hormone-sensitive lipase, a specific marker of adipocyte-like cells. As shown in Table 3,

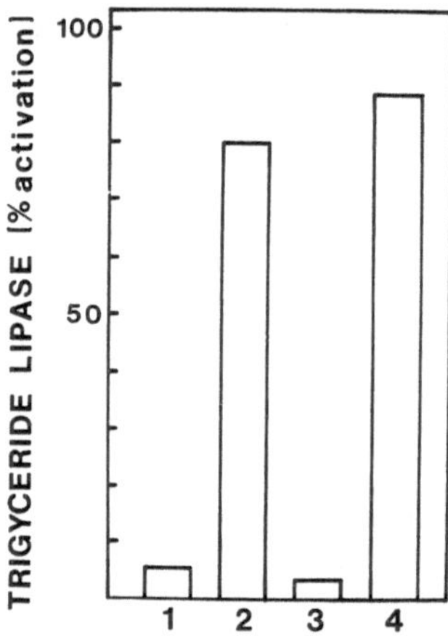

Figure 4. Effect of cAMP-dependent protein kinase inhibitor (PKI) and cAMP-dependent protein kinase (PK) on neutral triglyceride lipase activity of differentiated 1246 cells. 1. Mg-ATP; 2. Mg-ATP + cAMP; 3. Mg-ATP + cAMP + PKI (2 μg/ml). 4. Mg-ATP + cAMP + PKI + PK (10μg/ml). Activity was assayed as described by Serrero and Khoo (1982). The values are given as percentages of activation of triglyceride lipase activity measured in the absence of Mg-ATP and cAMP.

the addition of 2% fetal calf serum to the 4F culture medium at the time of treatment with dex-mix resulted in a greater increase (twofold to threefold) in the various specific activities measured than in the absence of serum, indicating that the addition of serum had an additional effect on the development of adipose conversion program in 1246 cells. However, in contrast to what was observed for 3T3-L1 and 3T3-F442A cells (Kuri-Harcuch and Green, 1978; Serrero *et al.*, 1979), the addition of fetal calf serum was not a requirement for the onset of the adipose differentiation of 1246 cells.

The results obtained in Table III show that fetal calf serum contains other factors that have the ability to further stimulate the development of enzymatic activities controlling lipid metabolism already induced in serum-free medium. As it was possible to have 1246 cells grow and differentiate in serum-free medium, this system was considered adequate to identify the additional factors that could substitute for serum in its ability to stimulate 1246 cell differentiation. We attempted to replace residual serum by factors added at the time of dex-mix treatment and measure glycerophosphate dehydrogenase activity in comparison to that found in cells treated with dex-mix in the presence of serum.

Three factors were found to be active in increasing glycerophosphate dehydrogenase activity (Table IV). These were triiodothyronine (T3), found to accelerate the adipose conversion of Ob17 cells in the presence of serum (Gharbi-Chihi *et al.*, 1981); ethanolamine ($EtNH_2$), shown to replace fetal calf serum in stimulating the growth of a rat mammary carcinoma cell line (68-24 cell line) (Kano-Sueoka and Errick, 1981); and human growth hormone (hGH), known to bind to adipose

Table IV. Development of Glycerophosphate Dehydrogenase (G3PDH) Activity in 1246 Cells: Replacement of Residual Serum by Various Factors

Condition	G3PDH mU/mg protein
Dex-mix	108
Dex-mix + 2% FCS	202
Dex-mix + T3	141
Dex-mix + $EtNH_2$	134
Dex-mix + h GH	224
Dex-mix + GH + T3 + $EtHN_2$	290

Cells were inoculated in DME-F12 supplemented with fibronectin, insulin, transferrin, and FGF. Dex-mix was added at day 4 alone or with 2% fetal calf serum (FCS), triiodothyronine (T3) 10^{-11} M, ethanolamine ($EtNH_2$) 5×10^{-6} M, or human growth hormone (hGH) 50 ng/ml. Dex-mix was removed on day 6 but other factors were maintained until day 15, the time at which cells were harvested and G3PDH assay performed.

cells and to influence their carbohydrate and lipid metabolism (Goodman, 1968; Gavin *et al.*, 1982; Lewis *et al.*, 1980).

Addition of growth hormone resulted in a twofold higher glycerophosphate dehydrogenase specific activity than the one measured in cells treated only with dex-mix. The specific activity then reached (202 mU/mg) was equivalent to the one measured in cells treated with dex-mix in the presence of 2% fetal calf serum. Triiodothyronine (T3) and ethanolamine had some ability to stimulate glycerophosphate dehydrogenase induction but their effect was only 40% and 34%, respectively, higher than in the presence of dex-mix alone. In this system, serum can thus be replaced by known factors, the most potent being growth hormone.

Morikawa *et al.* (1982) had found that human growth hormone could replace fetal calf serum in stimulating the adipose conversion of 3T3-F442A cells maintained in a medium containing insulin, transferrin, EGF, biotin, sodium ascorbate, 1% calf serum, and 1.5% cat serum. They concluded that growth hormone was the adipogenic factor and was responsible for the acitivity found in fetal calf serum that controlled the differentiation of 3T3-F442A cells reported by Kuri-Harcuch and Green (1978). In the case of 1246 cells, it was shown that growth hormone was able to stimulate the increase of glycerophosphate dehydrogenase activity in serum-free medium in the presence of insulin, transferrin, FGF, dexamethasone, and 3-isobutyl, 1-methylxanthine. The growth hormone preparation consists of a complex of proteins

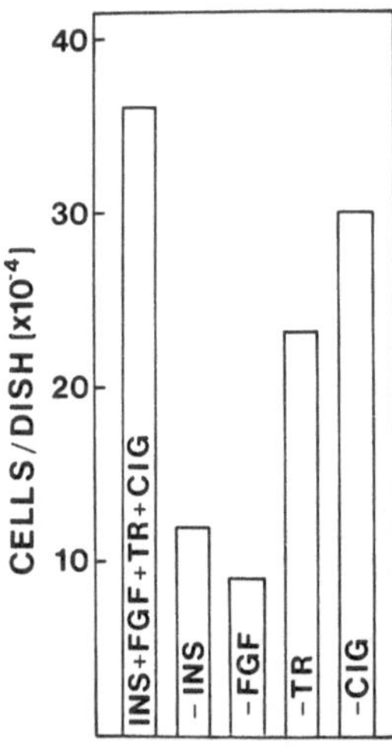

Figure 5. Effect of removing individual factors from 4F medium on the growth of 1246. Cells were counted on day 4. (INS + FGF + TR + CIG) 4F medium; (−INS) 4F medium lacking insulin; (−FGF) 4F medium lacking fibroblast growth factor; (−TR) 4F medium lacking transferrin; (−CIG) 4F medium lacking fibronectin.

that have been isolated and purified (Lewis *et al.*, 1980). The 1246 cell line represents a good system to assay these different fractions of growth hormone for their ability to stimulate adipose differentiation (Serrero, unpublished results).

Regulator versus Accelerator of Differentiation

1246 cells that behave like adipocytelike cells can grow and undergo adipose differentiation in a serum-free medium of relatively simple composition. Among the various factors that have an ability to stimulate adipose differentiation, one has to distinguish between the regulators of differentiation and the accelerators of differentiation (see p. 57). Cultivating 1246 cells in serum-free medium appears to be a good system to determine which factor is going to be important for their differentiation of adipocytelike cells. To investigate this problem, cells were cultivated in 4F medium lacking one factor. Growth and differentiation were measured by counting cell number and assaying glycerophosphate dehydrogenase activity (Fig. 5, Table V). Omission of insulin or FGF reduced the growth threefold and fourfold, respectively. Omission of transferrin and fibronectin reduced cell growth by 37% and 20%, respectively. As shown in Fig. 2, insulin concentrations as low as 1 ng/ml stimulated 1246 cell growth.

Table V showed that when insulin or transferrin was omitted, G3PDH-specific activity in 1246 cells remained at undetectable levels.

Table V. Effect of Removal of Individual Factors on Development of G3PDH Activity

Condition	G3PDH Activity (mU/mg protein)
4F	283
Insulin	0
Transferrin	0
FGF	280
Fibronectin	240

Cells were cultivated in the presence or absence of individual factors from day 0. At day 4, dex-mix was added, removed on day 6. G3PDH activity was measured at day 15.

In both cases, no triglyceride accumulation could be observed. However, it was possible to observe cells accumulating triglycerides when they had been maintained more than 15 days in the absence of transferrin. In the absence of FGF and fibronectin, the specific activities measured were equivalent to the ones found in cells maintained in 4F medium (Table V). When cells were cultured in the presence of insulin concentrations between 1 ng/ml and 10 μg/ml, no induction of G3PDH specific activity was observed for insulin concentrations lower than 10 ng/ml. At higher concentrations, the specific activities increased and reached a plateau at concentrations above 1 μg/ml (Fig. 6). The insulin dose response curve for the increase of G3PDH activity is shifted by one order of magnitude when compared with the insulin dose response curve for growth (Figs. 2 and 6). From these results, it appears that the various factors constituting the serum-free medium able to support growth and differentiation of 1246 cells can be classified in three categories: (1) factors influencing only 1246 cell growth (FGF); (2) factors influencing only 1246 cell differentiation (transferrin, dexamethasone, 3-isobutyl 1-methylxanthine, triiodothyronine, growth hormone); and (3) factors influencing both cell growth and differentiation (insulin).

As insulin influences both growth and differentiation of 1246 cells in serum-free medium, two questions can be raised: (1) Is insulin the regulating factor for 1246 cell differentiation? and (2) Is the effect of insulin on differentiation related to its effect on growth; that is, does the omission of insulin impair cell differentiation because it impairs cell growth?

In the case of 3T3-F442A cells, Green and co-workers consider that insulin enhances the differentiation process (Green and Kehinde, 1975) and that growth hormone is responsible for the adipose conversion

Table VI. Influence of Insulin on the Development of G3PDH Activity in 1246 Cells

Condition	G3PDH (mU/mg protein)
1.	
−Ins	0
−Ins + hGH	0
−Ins + 3A	0
−Ins + ins	114
2.	
+Ins	75
+Ins + hGH	201
+Ins + 3A	125
+Ins − ins[a]	13.4

1. Cells were inoculated in 4F medium lacking insulin. At day 4, dex-mix was added alone and with 50 ng/ml of human growth hormone (hGH), 20% dilution of conditioned medium from 1246-3A cells (3A), or 10 μg/ml insulin (ins). Dex-mix was removed at day 6. The other factors were maintained in the culture medium until day 15, at which time cells were collected and enzymatic activity measured.
2. Cells were inoculated in the presence of insulin (4F medium). At day 4, cells were treated with dex-mix for 48 hours under the same conditions as above.
[a] Insulin was removed from the culture medium at day 4. Cells were maintained in the absence of insulin until day 15.

(Morikawa *et al.*, 1982). In the case of 1246 cells, we conclude the contrary. In the absence of insulin, 1246 cells cannot differentiate even if growth hormone is present (Table VI). Under these conditions, glycerophosphate dehydrogenase activity is not induced. Dexamethasone, 3-isobutyl 1-methylxanthine, triiodothyronine, and ethanolamine are also unable to influence the process of differentiation of 1246 cells if the medium is deprived of insulin. If 1246 cells are cultured for four days in the presence of insulin and if insulin is removed at the time of treatment with dex-mix, glycerophosphate dehydrogenase specific activity is fourfold lower than the one found in cells maintained in the presence of insulin (Table VI). Conversely, when 1246 cells are cultured in the absence of insulin, but are treated with insulin at the time of the addition of dex-mix, an increase of glycerophosphate dehydrogenase occurs. Moreover, the growth of 1246 cells at an insulin concentration of 10 ng/ml is 75% of the growth in the presence of 10μg/ml of insulin, whereas the glycerophosphate dehydrogenase specific activity at 10 ng/ml of insulin is only 15% of the one at 10μg/ml of insulin (Fig. 2 and 6).

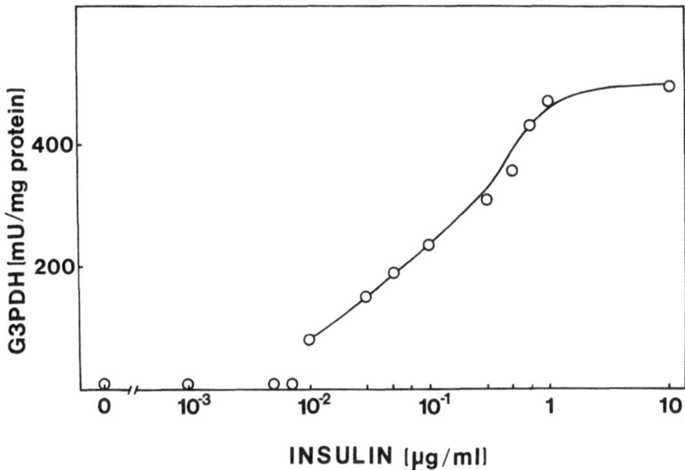

Figure 6. Development of glycerophosphate dehydrogenase activity as a function of insulin concentration. Activity was assayed on day 15.

It appears that (1) insulin plays the role of a regulator of differentiation, (2) growth hormone is an accelerator of differentiation, and (3) the controls by insulin of growth and differentiation are not directly related. Another line of evidence regarding the latter conclusion has been provided by the results obtained with the isolation of a variant of 1246 cells independent of insulin for its growth in serum-free medium.

Insulin-Independent Variant 1246-3A

When 1246 cells are cultivated in the absence of insulin, they do not grow after one doubling. But if these cells are maintained more than 15 days in the absence of insulin, it is possible to observe groups of cells that began to grow, indicating that they had become independent of insulin for their growth. These cells have been isolated and cloned. One of the clones, called 1246-3A, has been studied. Figure 7 shows that the growth curves for 1246-3A cells cultivated in the absence and presence of insulin are identical and that 1246-3A cells grow with a shorter generation time and reach higher cell densities than the parent cell line, 1246. By assaying medium conditioned by 1246-3A cells for growth-promoting activity on 1246 cells maintained in medium lacking insulin, it was demonstrated that 1246-3A cells produce a factor able to replace insulin in promoting growth of 1246 cells in serum-free

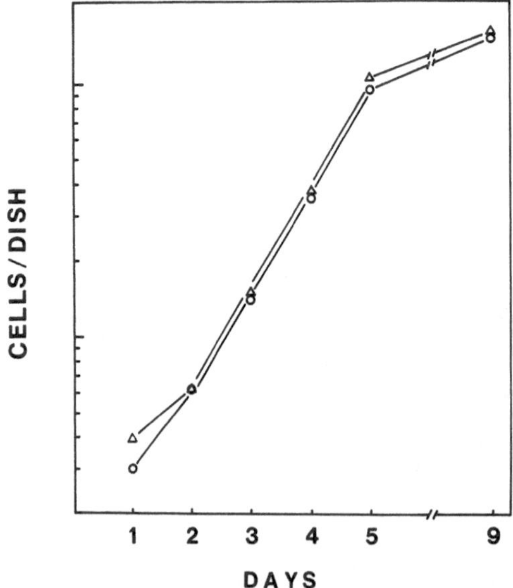

Figure 7. Growth curve for 1246-3A cell line in 4F medium (cv) and 4F medium lacking insulin (○).

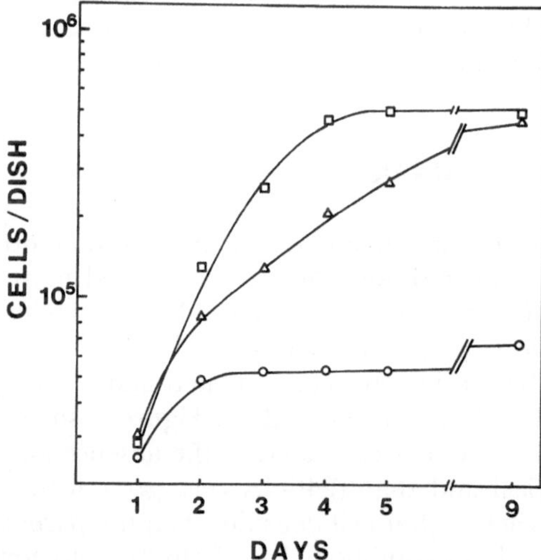

Figure 8. Growth curve for 1246 cell line in 4F medium (cv), 4F medium lacking insulin (○) and 4F medium lacking insulin but containing 20% of DME-F12 medium conditioned by 1246-3A cell line (□).

medium (Fig. 8). 1246 cells cultivated in the presence of a 20% dilution of 1246-3A-conditioned medium reach the same cell density as the ones cultivated in the presence of 10 μg/ml of insulin. By assaying glycerophosphate dehydrogenase activity, we demonstrated that the factor produced by 1246-3A cells was unable to stimulate increase of enzymatic activity in cells maintained in the absence of insulin. When added to cells cultivated in the presence of insulin, 1246-3A-conditioned medium did not induce any change in the expression of glycerophosphate dehydrogenase activity. These results indicate that 1246-3A-conditioned medium did not contain activity stimulating or inhibiting 1246 cell differentiation. It appears, then, that the effects of insulin on growth and on differentiation of 1246 cells can be separated.

The characterization of the activity produced by 1246-3A cells is under way. By radioimmunoassay, it has been shown to be different from insulin. The activity remained unchanged when conditioned medium was acidified to pH 2.0, alkalinized to pH 11.0, and submitted to a temperature of 85°C. The activity was eluted out of a Sephadex G100 column with an apparent molecular weight between 20,000 and 30,000 daltons. Its purification is under way.

The study of the growth and differentiation of 1246 cells in serum-free medium could provide results concerning the role of insulin on growth and differentiation of adipocytelike cells and could furnish a model to study the effect of growth hormone on adipose differentiation. The 1246 cell line represents the first example of a preadipocytelike cell able to grow and differentiate in serum-free medium. The results could show that the ability of 1246 cells to differentiate does not appear to depend on an external signal present in fetal calf serum. One can consider either that insulin is the regulating factor for 1246 cell differentiation or that 1246 cells produce their own adipogenic factor that is modulated by insulin. These hypotheses are under study.

ACKNOWLEDGMENTS. GS thanks Professor Gordon H. Sato for helpful discussion and for the use of his laboratory, and Ms. M. A. Zurbach for processing the manuscript. This work was supported by the following grants: NCI CA 19731, NIH GM 17702, and NSF PCM 82-02797.

References

Ailhaud, G., 1982, Adipose cell differentiation in culture, *Mol. Cell. Biochem.* **49:**17–31.

Ailhaud, G., Grimaldi, P., Negrel, R., Serrero, G., and Verrando, P., 1979, Multiplication and differentiation of preadipocyte cell lines isolated from mouse epididymal fat pad, in: *Obesity: Cellular and Molecular Aspects* (G. Ailhaud, ed.), Colloques de l'inserm, pp. 51–64.

Barnes, D., and Sato, G., 1980, Methods for growth of cultured cells in serum-free medium, *Anal. Biochem.* **102:**255–270.

Bethell, D. R., and Pegg, A. E., 1982, Polyamines are needed for the differentiation of 3T3-L1 fibroblasts into adipose cells, *Biochem. Biophys. Res. Commun.* **102:**272–278.

Bjorntorp, P., 1979, Development of adipocytes from precursor cells obtained from epididymal fat pads of rats, in: *Obesity: Cellular and Molecular Aspects* (G. Ailhaud, ed.), Colloques de l'inserm, pp. 89–102.

Bjorntorp, P., Karlsson, M., Pertoft, H., Petterson, P., Sjostrom, L., and Smith, U., 1978, Isolation and characterization of cells from rat adipose tissue developing into adipocytes, *J. Lipid Res.* 19:316–324.

Brunner, G., 1977, Membrane impression and gene expression: Towards a theory of cytodifferentiation, *Differentiation* **8:**123–132.

Butcher, R. W., and Baird, C. E., 1968, Effects of prostaglandins on adenosine, 3′, 5′ monophosphate levels in fat and other tissues, *J. Biol. Chem.* **243:**1713–1717.

Chen, J. J., and London, I. M., 1981, Hemin enhances the differentiation of mouse 3T3 cells to adipocytes, *Cell* **26:**117–122.

Coleman, R. A., Reed, B. C., Mackall, J. C., Student, A. K., Lane, M. D., and Bell, R. M., 1978, Selective changes in microsomal enzymes of triacylglycerol, phosphatidylcholine and phosphatidylethanolamine biosynthesis during differentiation of 3T3-L1 preadipocytes, *J. Biol. Chem.* **253:**7256–7261.

Darmon, M., Serrero, G., Rizzino, A., and Sato, G., 1981, Isolation of myoblastic, fibroadipogenic and fibroblastic clonal cell lines from a common precursor and study of their requirements for growth and differentiation, *Exp. Cell. Res.* **132:**313–327.

Djian, P., Grimaldi, P., Negrel, R., and Ailhaud, G., 1982, Adipose conversion of OB17 preadipocytes: Relationships between cell division and fat cell cluster formation, *Exp. Cell. Res.* **142:**273–281.

Eckel, R. H., Fujimoto, W. Y., and Brunzell, J. D., 1977, Development of lipoprotein lipase in cultured 3T3-L1 cells, *Biochem. Biophys. Res. Commun.* **78:**288–293.

Faust, I. M., Johnson, P. R., Stern, J. S., and Hirsch, J., 1979, Diet induced adipocyte number increases in adult rats: A method to provide a new model of obesity, *Am. J. Physiol.* **235:**E279–E276.

Forest, C., Negrel, R., and Ailhaud, G., 1981, Development of lipolytic response to isoproterenol during adipose conversion of Ob17 preadipocyte cells, *Biochem. Biophys. Res. Commun.* **102:**577–587.

Garin, J. R. III, Saltman, R. J., and Tollefsen, S. E., 1982, Growth hormone receptors in isolated rat adipocytes, *Endocrinology* **110:**637–643.

Gharbi-Chihi, J., Grimaldi, P., Torresani, J., and Ailhaud, G., 1981, Triiodothyronine and adipose conversion of Ob17 preadipocytes: Binding to high affinity sites and effect on fatty acid synthesizing and esterifying enzymes, *J. Recept. Res.* **2:**153–173.

Goodman, H. M., 1968, Growth hormone and the metabolism of carbohydrate and lipid in adipose tissue, *Ann. N.Y. Acad. Sci.* **148:**418–440.

Green, H., 1979, Adipose conversion: A program of differentiation, in: *Obesity: Cellular and Molecular Aspects* (G. Ailhaud, ed.), Collogues de l'inserm, pp. 15–24.

Green, H., and Kehinde, O., 1974, Sublines of mouse 3T3 cells that accumulate lipid, *Cell* **1:**113–116.

Green, H., and Kehinde, O., 1975, An established preadipose cell line and its differentiation in culture, II. Factors affecting the adipose conversion, *Cell* **5:**19–27.

Green, H., and Kehinde, O., 1976, Spontaneous heritable changes leading to increased adipose conversion in 3T3 cells, *Cell* **7:**105–113.

Grimaldi, P., Negrel, R., and Ailhaud, G., 1978, Induction of the triglyceride pathway enzymes and of lipolytic enzymes during differentiation in a preadipocyte cell line, *Eur. J. Biochem.* **84:**369–376.

Grimaldi, P., Djian, P., Negrel, R., and Ailhaud, G., 1982, Differentiation of Ob17 preadipocytes to adipocyte: Requirement of adipose conversion factor(s) for fat cell cluster formation, *EMBO J.* **6:**687–692.

Hausman, G. J., Campion, D., and Martin, R. J., 1980, Search for the adipocyte precursor cell and factors that promote its differentiation, *J. Lipid Res.* **21:**657–670.

Hayashi, I., Nixon, T., Morikawa, M., and Green, H., 1981, Adipogenic and antiadipogenic factors in the pituitary and other organs, *Proc. Natl. Acad. Sci. U.S.A.* **78:**3969–3972.

Hirogun, A., Sato, M., and Mitsui, H., 1980, Establishment of a clonal cell line that differentiates into adipose cells *in vitro, In Vitro* **16:**685–693.

Hollenberg, C. H., and Vost, A., 1968, Regulation of DNA synthesis in fat cells and stromal elements from rat adipose tissue, *J. Clin. Invest.* **47:**2485–2498.

Jakob, H., Boon, T., Gaillard, J., Nicolas, J. F., and Jacob, F., 1973, Teratocarcinome de la souris: Isolement, culture et propríetés de cellules a 'potentialites multiples, *Ann. Microbiol.* **124B:**269.

Jakob, H., Buckingham, M., Cohen, A., Dupont, L., Fiszman, M., and Jacob, F., 1978, A skeletal muscle cell line isolated from a mouse teratocarcinoma undergoes apparently normal terminal differentiation *in vitro, Exp. Cell. Res.* **114:**403–410.

Kano-Sueoka, T., and Errick, J. E., 1981, Effect of phosphoethanolamine and ethanolamine on growth of mammary carcinoma cells in culture, *Exp. Cell. Res.* **136:**137–145.

Kawamura, M., Jensen, D., Wancewicz, E. V., Joy, L. L., Khoo, J. C., and Steinberg, D., 1981, Hormone sensitive lipase in differentiated 3T3-L1 cells and its activator by cAMP-dependent protein kinase, *Proc. Natl. Acad. Sci. U.S.A.* **78:**732–736.

Kazdova, L., and Vrana, A., 1970, Insulin and adipose tissue cellularity, *Horm. Metab. Res.* **2:**117–123.

Keay, S., and Grossberg, S. E., 1980, Interferon inhibits the conversion of 3T3-L1 mouse fibroblasts into adipocytes, *Proc. Natl. Acad. Sci. U.S.A.* **77:**4099–4103.

Kuri-Harcuch, W., and Green, H., 1977, Increasing activity of enzymes on pathway of triacylglycerol synthesis during adipose conversion of 3T3 cells, *J. Biol. Chem.* **252:**2158–2160.

Kuri-Harcuch, W., and Green, H., 1978, Adipose conversion of 3T3 cells depends on a serum factor, *Proc. Natl. Acad. Sci. U.S.A.* **75:**6107–6109.

Lewis, U. J., Singh, R. N. P., Tutwiler, G. F., Sigel, M. B., Vanderlaan, E. F., and Vanderlaan, W. P., 1980, Human growth hormone: A complex of proteins, *Recent Prog. Horm. Res.* **36:**477–508.

Mackall, J. C., and Lane, M. D., 1977, Role of pyruvate carboxylase in fatty acid synthesis: Alterations during preadipocyte differentiation, *Biochem. Biophys. Res. Commun.* **79:**720–725.

Mackall, J. C., Student, A., Polakis, S. E., and Lane, M. D., 1976, Induction of lipogenesis during differentiation in a "preadipocyte" cell line, *J. Biol. Chem.* **251:**6462–6464.

McBurney, M., 1976, Clonal lines of teratocarcinoma cells *in vitro*: Differentiation and cytogenetic characteristics, *J. Cell. Physiol.* **89:**441.

McClure, D. B., Ohasa, S., and Sato, G., 1981, Factors in the rat submaxillary gland that stimulate growth of cultured glioma cells: Identification and partial characterization, *J. Cell. Physiol.* **87:**195–207.

Morikawa, M., Nixon, T., and Green, H., 1982, Growth hormone and the adipose conversion of 3T3 cells, *Cell* **31:**783–789.

Murakawa, S., and Raben, M. S., 1968, Effect of growth hormone and placental lactogen on DNA synthesis in rat costal cartilage and adipose tissue, *Endocrinology* **87:**645–650.

Murphy, M. G., Negrel, R., and Ailhaud, G., 1981, Lipoprotein lipase and monoacylglycerol lipase activaties during maturation of Ob17 preadipocytes, *Biochem. Biophys. Acta* **664:**240–245.

Murray, T., and Russell, T. R., 1980, Inhibition of adipose conversion in 3T3-L1 cells by retinoic acid, *J. Supramol. Struct. Cell. Biochem.* **14:**255–266.

Negrel, R., Grimaldi, P., and Ailhaud, G., 1978, Establishment of preadipocyte clonal cell line from epididymal fat pad of Ob10b mouse that responds to insulin and lipolytic hormones, *Proc. Natl. Acad. Sci. U.S.A.* **75:**6054–6058.

Nicholas, J. F., Dubois, P., Jakob, H., Gaillard, J., and Jacob, F., 1975, Teratocarcinome de la souris: Differenciation en culture d'une lignée de cellules primitives a' potentialités multiple, *Ann. Microbiol.* **126A:**3.

Nicolas, J. F., Jakob, H., and Jacob, F., 1981, Teratocarcinoma derived cell lines and their use in the study of differentiation, in: *Functionally Differentiated Cell Lines* (G. Sato, ed.), Alan R. Liss, New York, pp. 185.

Pairault, J., and Green, H., 1979, A study of the adipose conversion of suspended 3T3 cells using glycerophosphate dehydrogenase as differentiation marker, *Proc. Natl. Acad. Sci. U.S.A.* **76:**5138–5142.

Papaioannou, V., McBurney, M., and Gardner, R., and Evans, M., 1975, Fate of teratocarcinoma cells injected into early mouse embryos, *Nature* (*London*) **258:**70.

Reed, B. C., Kaufmann, S. H., Mackall, J. C., Student, A. R., and Lane, M. D., 1977, Alterations in insulin binding accompanying differentiation of 3T3-L1 preadipocytes, *Proc. Natl. Acad. Sci. U.S.A.* **74:**4876–4880.

Rothblat, G. H., and deMartinis, F. D., 1977, Differentiation of 3T3-L1 fibroblasts to adipocytes: The effect of indomethacin, prostaglandin E_1 and cyclic AMP on the process of differentiation, *Biochem. Biophys. Res. Commun.* **78:**45.

Rubin, C. S., Lai, E., and Rosen, O. M., 1977, Acquisition of increased hormone sensitivity during adipocyte development, *J. Biol. Chem.* **252:**3554–3557.

Rubin, C. S., Hirsch, A., Fung, C., and Rosen, O. M., 1978, Development of hormone receptors and hormonal responsiveness *in vitro*: Insulin receptors and insulin sensitivity in the preadipocyte and adipocyte forms of 3T3-L1 cells, *J. Biol. Chem.* **253:**7570–7578.

Russell, T. R., and Ho, R., 1976, Conversion of 3T3 fibroblasts into adipose cells: Triggering of differentiation by prostaglandin F2α and 1-methyl, 3-isobutylxanthine, *Proc. Natl. Acad. Sci. U.S.A.* **73:**4516–4520.

Serrero, G., and Khoo, J. C., 1982, An *in vitro* model to study adipose differentiation in serum-free medium, *Anal. Biochem.* **120:**351–359.

Serrero, G., and Sato, G., 1982, Growth and differentiation of a teratoma-derived fibroadipogenic cell line in serum-free medium, in: *Growth of Cells in Hormonally Defined Media*, Volume 9 (G. H. Sato, A. B. Pardee, and D. A. Sirbasku, eds.), Cold Spring Harbor Conferences on Cell Proliferation, Cold Spring Harbor Laboratory, New York, pp. 943–955.

Serrero, G., McClure, D., and Sato, G., 1979, Growth of mouse 3T3 fibroblasts in serum-free hormone-supplemented media, in: *Hormones and Cell Culture*, Volume 6 (G. H. Sato and R. Ross, eds.), Cold Spring Harbor Conference on Cell Proliferation, Cold Spring Harbor Laboratory, New York, pp. 523–530.

Spooner, P. M., Chernick, S. S., Garrison, M. M., and Scow, R. O., 1979, Development of lipoprotein lipase activity and accumulation of triacylglycerol in differentiating 3T3-L1 adipocytes, *J. Biol. Chem.* **254:**1305–1311.

Stevens, L., 1967, The biology of teratomas, *Adv. Morphogenesis* **6:**1.

Verrando, P., Negrel, R., Grimaldi, P., Murphy, M., and Ailhaud, G., 1981, Differentiation of Ob17 preadipocytes to adipocytes: Triggering effects of clofenapate and indomethacin, *Biochem. Biophys. Acta* **663:**255–265.

Williams, I. H., and Polakis, S. E., 1977, Differentiation of 3T3-L1 fibroblasts to adipocytes: The effect of indomethacin, prostaglandin E_1, and cyclic AMP on the process of differentiation, *Biochem. Biophys. Res. Commun.* **77:**175–186.

Wise, L., and Green, H., 1978, Studies of lipoprotein lipase during the adipose conversion of 3T3 cells, *Cell* **13:**233–242.

Wise, L. S., and Green, H., 1979, Participation of one isozyme of cytosolic glycerophosphate dehydrogenase in the adipose conversion of 3T3 cells, *J. Biol. Chem.* **254:**273–275.

CHAPTER 4

Growth and Differentiation of Human Myelomonocytic Leukemia Cell Lines in Serum-Free Medium

THEODORE R. BREITMAN, BEVERLY R. KEENE, and HIROMICHI HEMMI

Introduction

In recent years the development of human myelomonocytic cell lines has provided useful models for studying regulation of both cell proliferation and differentiation. This can be very important for studies on the treatment of leukemia because the finding that some of these cell lines can be induced to terminally differentiate by a wide variety of compounds has suggested an alternative approach to the therapy of certain types of leukemias.

The growth advantage that myeloid leukemia cells have *in vivo* over normal cells is not because of a more rapid growth rate, but rather because of an apparent inability to mature to functional terminally differentiated end cells. It is possible that some leukemia cells do not mature either because they have a decreased ability to respond to exogenous differentiative factors or because the production of specific gene products obligatory for differentiation is altered. The availability of tissue-culture cell lines has made it possible to study the regulation

THEODORE R. BREITMAN, BEVERLY R. KEENE, and HIROMICHI HEMMI • Laboratory of Tumor Cell Biology, National Cancer Institute, National Institutes of Health, Bethesda, Maryland 20205.

of proliferation and differentiation of specific hematopoietic cell types and the effect on these cells of known mediators and modulators. These studies have had an impact on the approach to the treatment of some leukemias. Cancer treatment has traditionally consisted of a two-pronged attack consisting of surgical removal where applicable and treatment with cytotoxic agents including chemotherapy, ionizing radiation, and immunotherapy. The employment of cytotoxic agents has been the prime treatment for leukemia, but it is associated with many toxic side effects. Now with the finding that some leukemic tissue-culture cell lines can be induced to differentiate with physiological substances, it is feasible that treatment may be approached not with the aim of killing the leukemia population, but rather to induce them to mature to a more normal and functional nongrowing cell type. With this approach, the hope is that treatment will be more selective with diminished toxic side effects.

Human Myelomonocytic Cell Lines

Some properties of human myeloid leukemia cell lines are summarized in Table I. The first clearly human nonlymphoid leukemia cell line, K562, was established from the pleural effusion of a patient with chronic myelogenous leukemia (CML) in blast crisis (Lozzio and Lozzio, 1975). These cells are induced to differentiate along the erythroid pathway with the production of hemoglobin when treated with butyric acid or hemin (Anderson *et al.*, 1979; Rutherford *et al.*, 1979; Hoffman *et al.*, 1979). More recently, K562 has been reported to be a multipotential cell line differentiating in long-term culture without added inducers into recognizable progenitors of the granulocytic, monocytic, and erythrocytic series (Lozzio *et al.*, 1981).

The KG-1 cell line, isolated from a patient with acute myelogenous leukemia (AML), is composed primarily of myeloblasts and promyelocytes and differentiates into nondividing macrophagelike cells by exposure to phorbol diesters (Koeffler and Golde, 1980). These cells develop many of the characteristics of macrophages including adhesion, phagocytosis, lysozyme secretion, nitroblue tetrazolium reduction, Fc receptors, and nonspecific esterase activity (Koeffler *et al.*, 1981).

U-937 is a human histiocytic lymphoma cell line with monoblastlike characteristics (Sundstrom and Nilsson, 1976). These cells differentiate into morphologically mature macrophagelike cells with increased activity in antibody-dependent cytotoxicity assays after treatment with phorbol diesters (Nilsson *et al.*, 1980) or the conditioned medium from mixed-

Table I. Some Properties of Human Myeloid Leukemia Cell Lines[a]

Characteristics	HL-60	U-937	THP-1	KG-1	K-562
Source	APL	Histocytic lymphoma	AMoL	AML	CML in blast crisis
Predominant cell	Promyelocyte	Monoblastoid	Monocyte	Myeloblast	Blast cell
Mean doubling-time, h	36	50	48	40–50	12
Karyotype	Aneuploid	Aneuploid	Diploid	Hypodiploid	Ph^1
Lymphocyte makers	None	None	None	None	None
Colony formation in agar					
Spontaneous	Yes	No	No	No	Yes
Induced by CSA	Yes	No	Yes	Yes	No
Differentiation in culture					
Spontaneous	Yes	No	Yes	No	Yes
Induced by DMSO	Yes	No	No	No	No
Induced by hemin	No			No	Yes
Induced by RA	Yes	Yes	Yes	No	No
Induced by DIA	Yes	Yes	Yes	Yes	
Induced cell	G or M	M	M	M	Erythroid, G, or M

[a] Abbreviations: APL, acute promyelocytic leukemia; AMol, acute monocytic leukemia; AML, acute myelogenous leukemia; CML, chronic myelogenous leukemia; Ph^1, Philadelphia chromosome; G, granulocyte; M, monocyte/macrophage; CSA, colony stimulating activity; DMSO, dimethylsulfoxide; RA, all-*trans*-β-retinoic acid; DIA, T-lymphocyte-derived differentiation inducing activity.

lymphocyte cultures (Koren *et al.*, 1979). More recently, U-937 was shown to differentiate into monocytelike cells by incubation with retinoic acid (Olsson and Breitman, 1982). These induced cells are phagocytic, reduce nitroblue tetrazolium, and show an increased hexose monophosphate shunt activity, consistent with monocytelike cells.

THP-1 is a human acute monocytic leukemia cell line established by Tsuchiya *et al.* (1980). The monocytic nature of this cell line was based on morphology, nonspecific esterase activity, lysozyme production, and phagocytosis of latex beads and sensitized sheep erythrocytes. When these cells are treated with phorbol diester, they become adherent and there are increases in phagocytosis, nonspecific esterase activity, and hexose monophosphate shunt activity (Tsuchiya *et al.*, 1982). These results have been interpreted to indicate that THP-1 cells can be converted into mature cells with functions of macrophages.

HL-60 is the prima donna of the human myeloid leukemia cell lines. It was the first human cell line with distinct myeloid features to be developed (Collins *et al.*, 1977). This cell line, isolated from the blood of a patient with acute promyelocytic leukemia, proliferates continuously in suspension culture, and consists predominantly of promyelocytes. HL-60 is induced to terminally differentiate to cells having many of the morphological features of mature granulocytes by exposure to a wide variety of compounds, including dimethylsulfoxide, dimethylformamide, hypoxanthine, butyric acid, actinomycin D, and retinoic acid (Collins *et al.*, 1978; Breitman *et al.*, 1980b; Honma *et al.*, 1980). Moreover, these induced HL-60 cells have many of the functional characteristics of normal human peripheral blood granulocytes including phagocytosis, lysosomal enzyme release, complement receptors, chemotaxis, hexose monophosphate shunt activity, superoxide anion generation, and the ability to reduce nitroblue tetrazolium (Collins *et al.*, 1978, 1979; Newburger *et al.*, 1979). HL-60 lacks leukocyte alkaline phosphatase (Gallagher *et al.*, 1979), and although differentiating HL-60 cells do not develop secondary (specific) granules and the specific protein markers for these granules, lactoferrin and B_{12}-binding protein (Newburger *et al.*, 1979; Fontana *et al.*, 1980; Olsson and Olofsson, 1981), they still provide a unique system for studying growth and differentiation of human myeloid cells *in vitro*.

Growth of HL-60 in Defined Medium

HL-60 can be grown in serum-free nutrient medium supplemented only with 5 μg insulin/ml and 5 μg transferrin/ml (Breitman *et al.*,

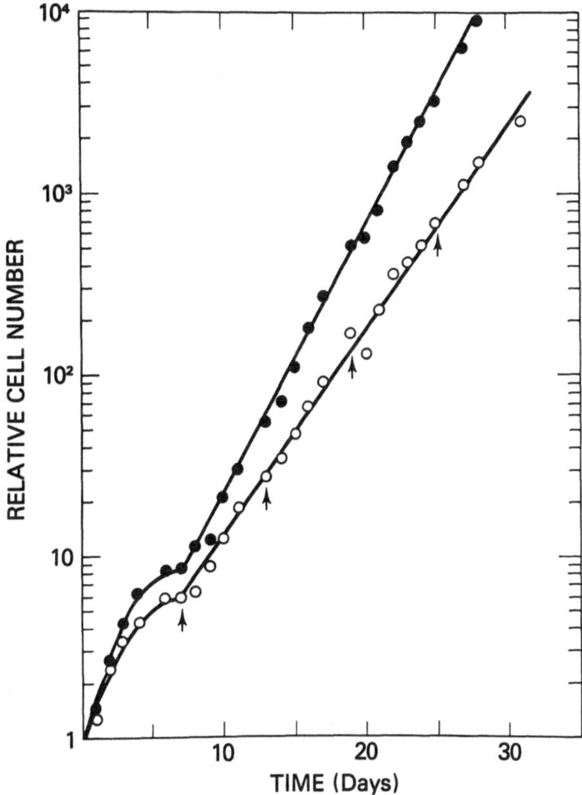

Figure 1. Long-term growth of HL-60 cells in defined medium. Cells were cultured either in nutrient medium supplemented with 10% FBS (●) or in nutrient medium supplemented with 5 μg insulin/ml and 5 μg transferrin/ml (○). At each subculture (arrows), cells from defined medium were harvested, washed, and inoculated into the two media.

1980a). The nutrient medium can be either a 1 : 1 mixture of Dulbecco's modified Eagle medium and Ham's F12 medium containing 14.3 mM $NaHCO_3$ and 15 mM HEPES at pH 7.2 (Mather and Sato, 1979) or RPMI 1640. In this insulin- and transferrin-supplemented medium, referred to as defined medium, long-term growth of HL-60 continues at a rate approximately 80% of that occurring in medium supplemented with serum (Fig. 1). The saturation density is lower in defined medium (1.5×10^6 cells/ml) than in serum-supplemented medium (3×10^6 cells/ml). Growth of HL-60 cells has continued in defined medium for over 30 passages, including at least 60 population doublings. This translates into a 10^{18}-fold increase in cell number in the absence of serum, making it unlikely that residual serum components are contributing to the growth of the cells.

There is an absolute requirement for both insulin and transferrin to maintain continuous proliferation of HL-60 cells (Breitman *et al.*, 1980a). These requirements are more pronounced at low cell concentrations where a lag before the onset of growth is also observed. In nutrient medium supplemented with 10% fetal bovine serum (FBS), there is essentially no lag in the growth of HL-60 after cultures are seeded with initial cell concentrations as low as 3000 cells/ml (Breitman and Keene, 1982). However, in defined medium this lag is seen even at initial cell concentrations of 2.5×10^5/ml and can be decreased but not eliminated by the addition of low concentrations of bovine serum albumin (BSA) (Fig. 2). The cause of the lag in growth in defined medium appears to involve not only a delay before the onset of growth, but a decrease in the number of cells in suspension because of the attachment of cells to the tissue culture vessel surface (Breitman and Keene, 1982). The finding that, in the presence of insulin and transferrin, the addition of BSA decreases the lag before the onset of growth indicated that a major function of FBS is to provide a protein-rich environment. Indirect support for this view is that the addition of 1% FBS to defined medium supports growth of HL-60 such that after four days it is approximately 90% of the growth obtained in medium with 10% FBS (Fig. 3). Medium with 1% FBS has a total protein concentration of approximately 400 μg/ml, of which 240 μg/ml is serum albumin. Therefore, comparable increases in growth of HL-60 in defined medium occur at similar protein concentrations of either BSA or FBS. The extreme sensitivity of the growth of HL-60 to microgram quantities of various proteins should be taken into consideration if the growth of these cells in defined medium is used for analysis of possible growth factors.

Although growth of HL-60 in suspension culture is supported by defined medium, colony growth of HL-60 in semisolid medium appears to be absolutely dependent on serum (Breitman *et al.*, 1980a). HL-60 cells form colonies in semisolid medium (methylcellulose) supplemented with FBS and their cloning efficiency increases with the addition of colony-stimulating activity (CSA) (Gallo *et al.*, 1979; Ruscetti *et al.*, 1981). However, when HL-60 cells are suspended in a serum-free semisolid medium supplemented with insulin and transferrin, no colonies form even at high cell densities or in the presence of CSA (Table II). The addition of BSA to the defined medium has no effect on the number of colonies formed. Thus, it appears clear that factor(s) in serum are

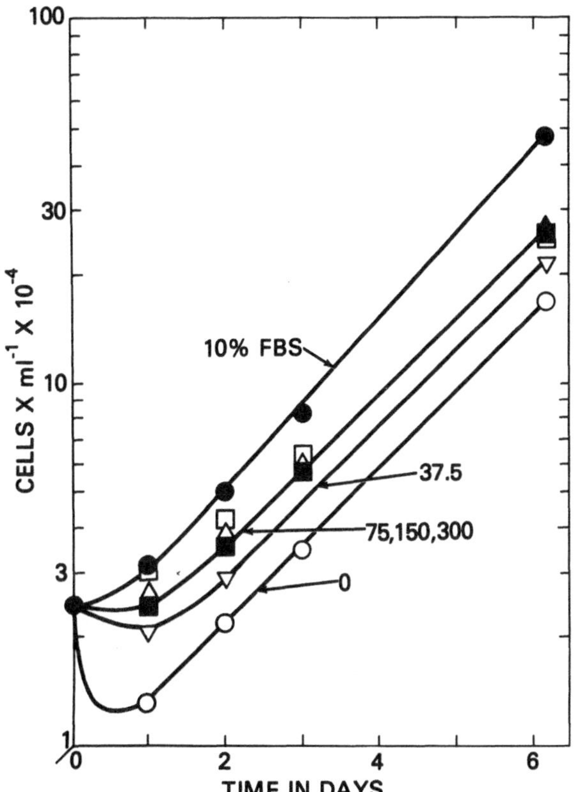

Figure 2. Prevention by BSA of the lag in growth of HL-60 cells in defined medium. BSA concentrations (μg/ml): (○) zero, (▽) 37.5, (■) 75, (△) 150, (□) 300; (●) 10% FBS.

required to support colony formation of HL-60. The identification of this activity would be a fruitful area of future investigation. The ability of a defined medium to support the growth of single cells may be considered the penultimate test for the efficacy of a growth medium. By this criterion serum-free medium supplemented only with insulin and transferrin does not support the growth of individual cells of HL-60 into colonies. This may be a manifestation of what was observed in suspension culture where defined medium did not support maximal growth of HL-60 without an initial lag; the extent of the lag being inversely related to the initial cell concentration.

In spite of these deficiencies, defined medium has been very useful in investigating differentiation of HL-60, as discussed in the next

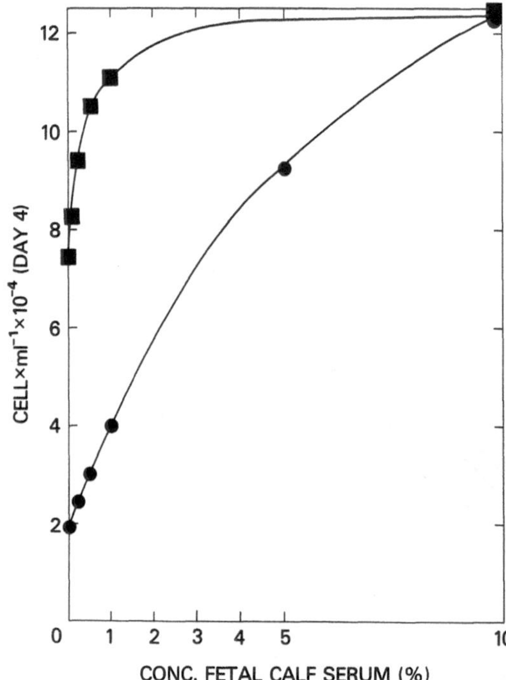

Figure 3. Growth of HL-60 in nutrient medium (●) and in defined medium (■) as a function of the concentration of FBS. The initial cell concentration was 2.0 × 10^4/ml.

Table II. HL-60 Colony Formation in Serum-Free and Serum-Supplemented Semisolid Medium (Methylcellulose)

No. of cells plated	No. of colonies	
	Serum-supplemented	Serum-free
500 (Without CSA)	18[a]	0
500 (With CSA)[b]	45	0
5,000 (Without CSA)	182	0
5,000 (With CSA)	412	0
50,000 (Without CSA)	>1,000	0
50,000 (With CSA)	>1,000	0

[a] Numbers given represent the averages of duplicate experiments.
[b] Conditioned medium containing CSA was added at a 10% (v/v) concentration.

section, where studies have been carried out in the absence of undefined serum inhibitors or enhancers of differentiation providing a useful means of assessing HL-60 response to physiologic differentiation inducing compounds having relevance to the control mechanisms of normal granulopoiesis.

Differentiation of HL-60 in Defined Medium

Until recently, terminal differentiation of HL-60 could be induced either by nonphysiological compounds, for example, dimethyl sulfoxide, or by physiological chemicals at markedly greater than physiological concentrations, for example, hypoxanthine. Recently, all trans-β-retinoic acid (RA) was found to be the most potent inducer of granulocytic differentiation of HL-60 (Breitman *et al.*, 1980b; Homma *et al.*, 1980). This compound induces differentiation at 10^{-6} to 10^{-3} the concentration of other inducers, induces relatively more extensive morphological differentiation than other inducers (Breitman *et al.*, 1980b), and probably more importantly induces at concentrations that are physiological (DeRuyter *et al.*, 1979).

The extent of RA-induced differentiation of HL-60, as was the case with dimethyl sulfoxide-induced differentiation is the same in defined medium as in serum-supplemented medium (Fig. 4). These results are in contrast to those observed with murine F9 embryonal carcinoma cells where the RA requirements for induction of differentiation are markedly decreased in the absence of serum (Rizzino and Crowley, 1980). Because of the high capacity that serum albumin has for binding RA (Ong and Chytil, 1975), the latter findings are not surprising. However, serum albumin at concentrations as high as 4.5 mg/ml has essentially no effect on RA-induced differentiation of HL-60 in defined medium (Fig. 5). At the highest concentration of albumin used in this experiment, the molar ratio of albumin to RA was 220 (67 μM/0.3 μM). Thus, even though a value for the equilibrium constant for the binding of RA to albumin has not been reported, it would be expected that under these conditions the concentration of free RA would be decreased. The absence of any major effect by FBS or BSA on RA-induced differentiation raises the possibility that bound RA is as active as the free-form and suggests that the "triggering" for differentiation occurs at the plasma membrane.

Notwithstanding the above speculation, the exact mechanism by which RA enhances differentiation of HL-60 cells or affects other cells is unknown. A proposed mechanism is that RA exerts an effect in the

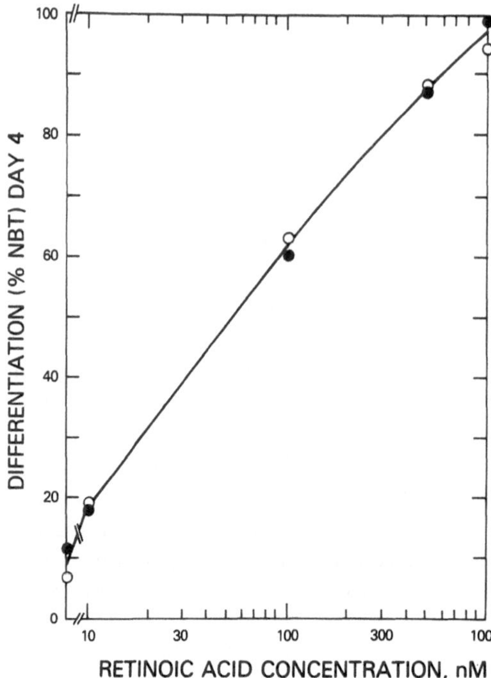

Figure 4. RA-induced differentiation of HL-60 in defined medium (○) and in serum-supplemented medium (●).

nucleus after binding to a highly specific cytoplasmic receptor (cellular RA-binding protein) (Ong and Chytil, 1975). However, we have been unable to demonstrate such a binding protein in HL-60 cells either with the sucrose-density-gradient sedimentation assay (Breitman *et al.*, 1981) or with a much more sensitive polyacrylamide gel electrophoresis technique (A. M. Jetten and T. R. Breitman, unpublished data). These results are in conflict with the conclusion of Takenaga *et al.* (1981), who report that HL-60 has RA-binding protein, but are in agreement with Douer and Koeffler (1982), who also could not detect this protein.

The possibility that RA acts on HL-60 after its conversion to 4-hydroxy- and 4-keto-retinoic acid (Roberts and Frolik, 1979) was investigated indirectly by testing the ability of these two metabolites to induce differentiation (Breitman, 1982). Both compounds were approximately one tenth as effective as RA. In the same study other retinoids were tested and the results indicated that the most effective retinoid inducers of HL-60 differentiation possess a carboxylic acid function at the C-15 terminal carbon (RA and 13-*cis*-RA). This activity was retained, although somewhat diminished, in spite of alterations in

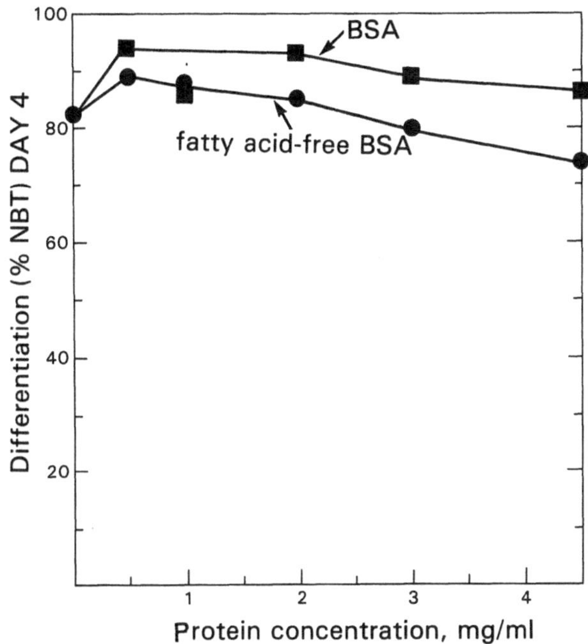

Figure 5. Inability of bovine serum albumin (BSA) to inhibit induction of differentiation of HL-60 by 0.3 μM RA in defined medium.

the ring as in the 4-hydroxy and 4-keto substituted derivatives and α-RA. Substitutions at the C-15 position resulted in essentially a complete loss of activity (retinal, retinol, retinyl acetate). These activity–structure relationships emphasize the specificity of the RA effect on HL-60 and makes more likely the possibility that this phenomenon, observed *in vitro*, is an expression of a true physiological process.

A very speculative hypothesis for the mechanism of action of RA in some cell types, presently under investigation in this laboratory, is that RA is metabolized in a series of reactions analogous to the activation and transport of fatty acids in mitochondria (Fig. 6). Thus, an acyl-CoA synthetase would catalyze the formation of a thioester bond between the carboxyl group of RA and the thiol group of coenzyme A: RA + ATP + CoA-SH ↔ Retinoyl-CoA + AMP + PPi. In the next reaction, similar to the formation of fatty acyl-carnitine, an oxygen ester is formed between the high-energy retinoyl-CoA and an hydroxyl group of an acceptor: Retinoyl-CoA + R-OH → Retinoyl-OR + CoA-SH. In the metabolism of fatty acids, a high energy fatty acyl-carnitine

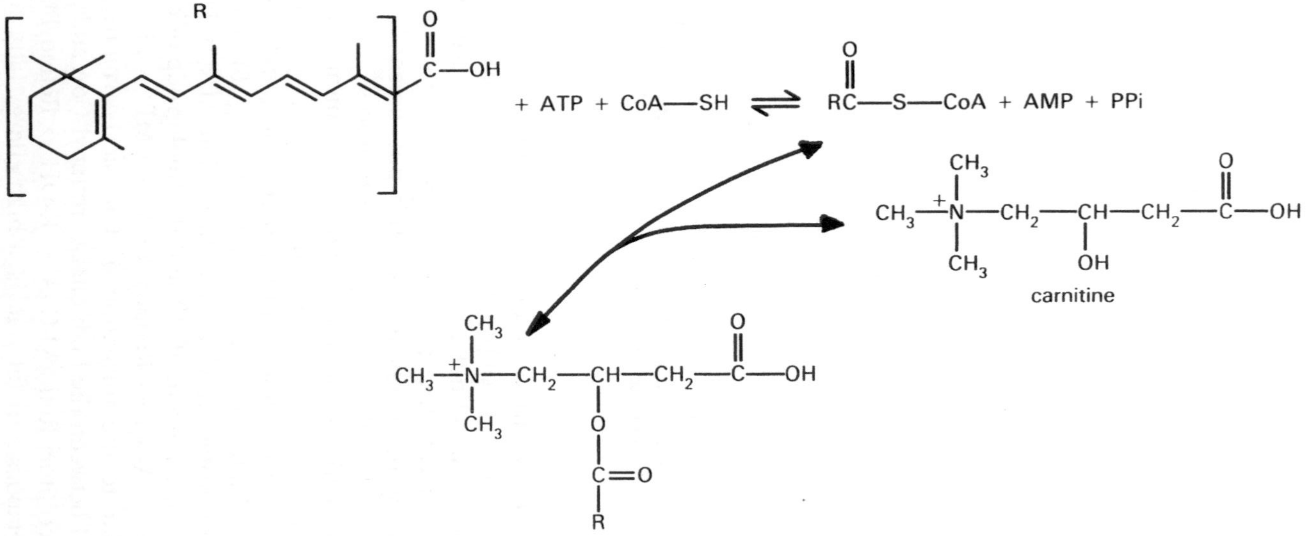

+ ATP + CoA—SH ⇌ RC(=O)—S—CoA + AMP + PPi

carnitine

Figure 6. Possible metabolism of RA based on fatty acid activation and formation of fatty acyl-S-CoA. With RA, the receptor of the retinoyl moiety may be a hydroxyl group of a macromolecule. R = decarboxylated portion of RA.

ester is formed, but it is possible that with RA the acceptor is an hydroxyl group of a macromolecule, for example, the threonine, serine, or tyrosine moieties of a protein. This ester would be of low energy resulting in essentially a one-way reaction for its formation and the formation of a stable covalent bond.

An extension of this hypothesis is that retinoylation and another modification (phosphorylation, methylation, acetylation, and others) occurs at the same site. When this other modification dominates, the cell continues to proliferate and does not differentiate. In a transformed or neoplastic cell, the balance is shifted even further in the direction of the other modification. This could be because of an increase in the cells' capacity to carry out the other modification or because the information for this modification, carried by an oncogene or an infective virus, is activated or amplified. Under these conditions a higher-than-normal dose of RA may shift the balance back towards an increase in retinoylation, effectively blocking the competing modification, and allowing the cell to differentiate. With some cells, the higher-than-normal dose of RA may result in a retinoylation that competes with an essential modification, thus leading to cell death. This model, or one similar to it, can explain why treatment of cells with RA has chemoprevention, differentiation inducing, and cytotoxic activities.

Involvement of Intracellular Cyclic Adenosine 3′:5′-Monophosphate (cAMP) in Differentiation of HL-60

Agents such as prostaglandin E (PGE), cholera toxin, and dibutyryl cAMP that increase the intracellular cAMP level markedly potentiate RA-induced differentiation of HL-60 (Breitman, 1982; Breitman and Keene, 1982; Olsson *et al.*, 1982). Prostaglandin E_1, E_2, A_1, and A_2 promote differentiation of HL-60 in a concentration-dependent manner (Figs. 7 and 8). Compared with RA, at least 60-fold greater concentrations of PGE are required to promote differentiation to the same extent. The most striking finding is that PGE_1 and PGE_2 are very effective inducers in combination with RA. That this combination is synergistic and not additive is demonstrated clearly by isobolograms (Fig. 9). Prostaglandin E_2 inhibits proliferation of HL-60 without having any effect on viability (Fig. 10). Retinoic acid is present in normal human peripheral blood at 10 nM (DeRuyter *et al.*, 1979) and PGE_2 is routinely found in human peripheral blood at 1 nM (Jaffe *et al.*, 1973) and can increase to 400 nM in localized areas of trauma, inflammation, or

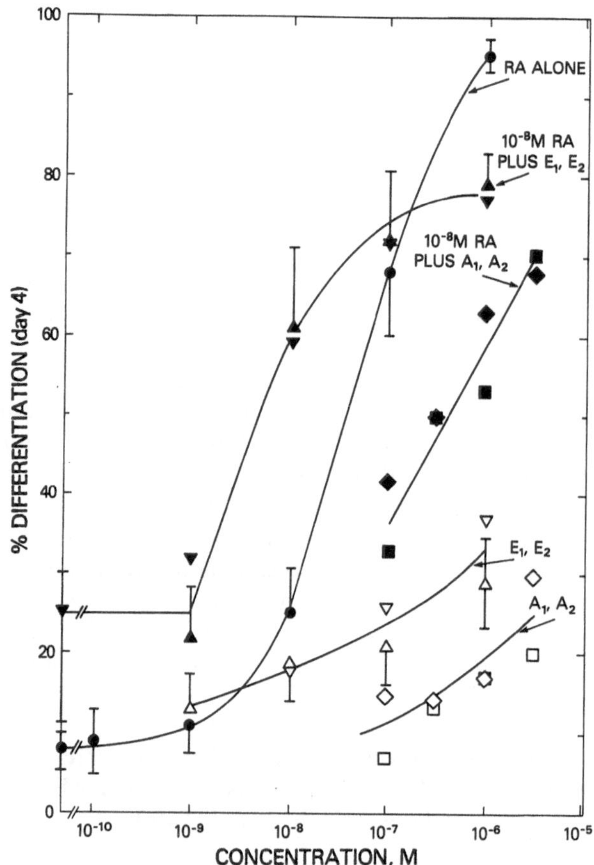

Figure 7. Differentiation of HL-60 induced by various concentrations of PGE and PGA in the presence and absence of 10 nM RA. Differentiation is expressed as the percentage of cells reducing nitroblue tetrazolium.

infection (Berenbaum *et al.*, 1976). Thus, it is possible that the effects on differentiation with 10 nM RA and various concentrations of PGE (Figs. 7 and 9) reflect *in vitro* what can occur *in vivo* where concentrations of RA can be expected to be fairly steady and concentrations of PGE vary widely.

The other PGs, either alone or in combination with RA, are much less active than PGE in inducing differentiation of HL-60 (Table III), indicating a great degree of biological specificity by HL-60 in distinguishing between these structurally similar compounds. This specificity is related to substitutions in the cyclopentane ring, as there are no

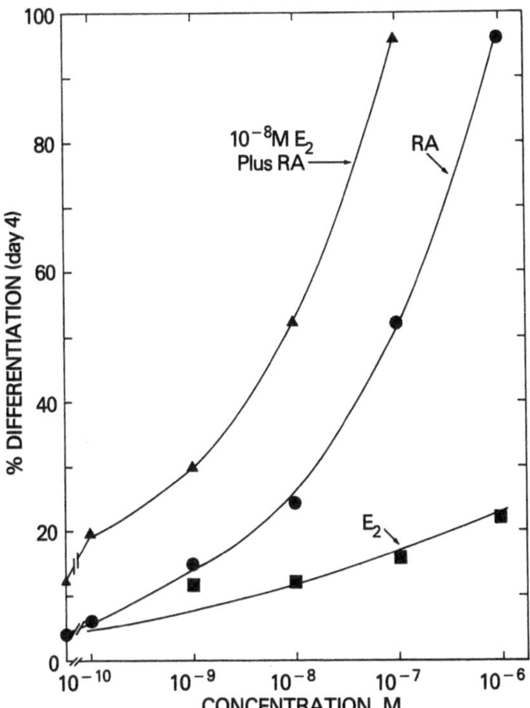

Figure 8. Differentiation of HL-60 induced by various concentrations of RA or PGE_2 and by RA in the presence of 10 nM PGE_2. Differentiation is expressed as the percentage of cells reducing nitroblue tetrazolium.

significant differences in biological activity between the PG-1s and PG-2s in each class.

PGE acts on many cell types by activating membrane adenylate cyclase and increasing intracellular levels of cyclic AMP (Gilman and Nirenberg, 1971; Kuehl and Humes, 1972; Hittelman and Butcher, 1973). A similar mechanism is probably operating in HL-60 cells, as PGE_2 promotes a marked increase in cAMP (Fig. 11). $PGF_{2\alpha}$ is much less potent than PGE_2 in increasing the cAMP level of HL-60 (Breitman and Keene, unpublished experiments) indicating that for the PGs there is a positive correlation between the extent of the increase in cAMP and the potentiation of RA-induced differentiation. Retinoic acid has no effect on the intracellular cAMP level either alone or in combination with PGE_2. The combination of cholera toxin, [which also increases HL-60 intracellular cAMP (Breitman and Keene, unpublished experi-

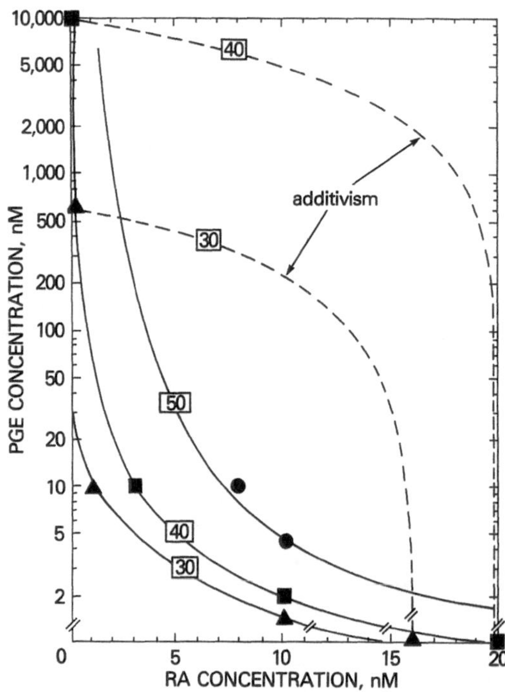

Figure 9. Isobolograms showing synergy for induction of differentiation of HL-60 cells by combinations of PGE and PGA. The experimental points were obtained from the data shown in Figs. 7 and 8 and other experiments, and are the concentrations of PGE and RA in combination inducing 30% (▲), 40% (■), and 50% (●) differentiation as indicated by the boxed values on each curve. The dashed lines connecting the values of PGE and RA alone would have been obtained if the two compounds in combination were additive.

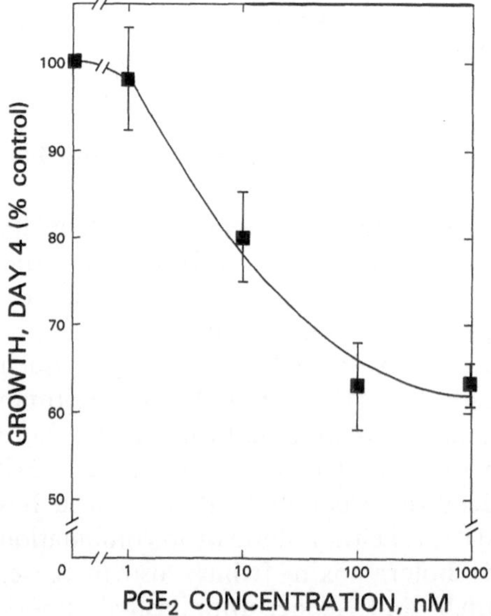

Figure 10. Inhibition by PGE_2 of growth of HL-60 cells. Cells (2.5×10^5/ml) were grown for 4 days in defined medium containing the indicated concentration of PGE_2. At day 4, control cells had grown to a concentration of 8×10^5·cells/ml (mean of 5 experiments). Bars: S.E. of 5 experiments. Viability, measured by trypan blue exclusion, was >90% under all conditions.

Table III. Differentiation of HL-60 Cells by Combinations of RA, PGF, PGB, and PGA.

		Differentiation, Day 4 (% NBT), PG Concentration, μM			
PG	RA, 10 nM	0	1	3	10
$F_{1\alpha}$	−	8	4	7	11
	+	25	24	25	53
$F_{2\alpha}$	−		12	16	25
	+		26	27	47
B_1	−		12	20	33
	+		26	37	82
B_2	−		9	11	23
	+		34	52	76
A_1	−		19	25	
	+		62	68	
A_2	−		18	25	
	+		53	70	

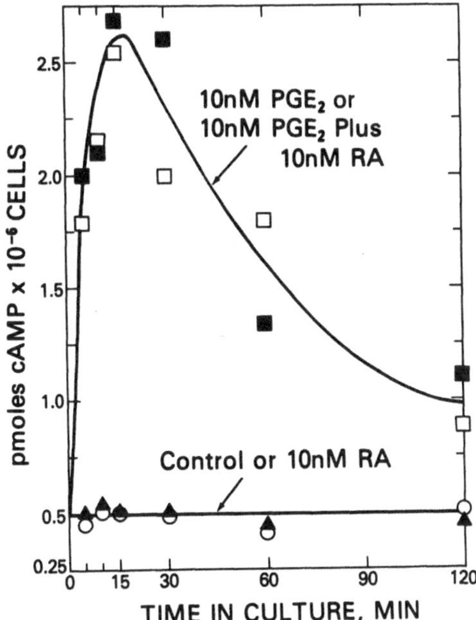

Figure 11. Increase in intracellular cAMP of HL-60 cells by PGE_2. Cells (2.5×10^5 cells/ml) were incubated in defined medium alone (○) or in defined medium containing 10 nM RA (▲), 10 nM PGE_2 (□), or 10 nM PGE_2 + 10 nM RA (■).

Table IV. Modulation of RA-Induced Differentiation of HL-60 by Dibutyryl cAMP (dbcAMP)

Condition	Differentiation, Day 4 (% NBT)
Control	8 ± 3
RA, 10 nM	25 ± 6
dbcAMP,	
10 μM	12
100 μM	33 ± 5
dbcAMP, 100 μM plus 10 nM RA	92 ± 7

ments)], or dibutyryl cAMP with RA give increases in differentiation of HL-60 that were synergistic (Table IV and Breitman, 1982). These results and the report of Levine and Ohuchi (1978) that in MDCK cells RA enhances the deacylation of cellular lipids with a consequent increase in PG production suggested the possibility that RA induction of HL-60 is mediated by endogenous synthesis of PG. However, indomethacin and aspirin, inhibitors of the PG synthetic enzyme cyclo-oxygenase, has no inhibitory effect on RA-induced differentiation of HL-60 (Table V and Breitman, 1982). The concentration of aspirin and the lowest concentration of indomethacin used here are at least fivefold greater than concentrations reported to suppress completely PG synthesis in a wide variety of cell types and cell lines including rat neutrophils, murine macrophages and human monocytes, MDCK cells, HeLa cells, and

Table V. Effects of Aspirin and Indomethacin on RA-Induced Differentiation of HL-60 Cells

Condition	Differentiation, Day 4 (% NBT)
Control	8 ± 3
RA, 0.1 μM	68 ± 8
1 μM	95 ± 2
Aspirin, 0.56 mM	7
plus RA, 1 μM	95
Indomethacin, 2.8 μM	10 ± 3
plus RA, 0.1 μM	76
plus RA, 1 μM	90
Indomethacin, 14 μM	9
plus RA, 0.1 μM	88
Indomethacin, 28 μM	10
plus RA, 0.1 μM	94

human fibroblasts. In addition, Bonser *et al.* (1981) have shown that indomethacin does not block the increases in chemotactic formyl peptide receptor binding and hexose monophosphate shunt activity in HL-60 induced to differentiate by exposure to dimethyl sulfoxide. Thus, it appears unlikely that endogenously synthesized PGs play a role in the induction of differentiation of HL-60. It is more likely that the results we have obtained on HL-60 cells *in vitro* reflect events occurring during normal granulocytic differentiation *in vivo*. It is well known that PGE and cAMP are potent inhibitors of both normal bone marrow granulocyte/macrophage colony-forming units and of myeloid leukemia cell lines in semisolid medium (Kurland *et al.*, 1978; Williams, 1979; Koeffler and Golde, 1980; Taetle and Koessler, 1980). However, these studies have not adequately provided a clear distinction between decreased colony numbers resulting from terminal differentiation and those resulting from cytotoxicity. In the microenvironment of the bone marrow, the interplay between RA, derived from the diet, and PGE and colony-stimulating activity derived from monocytes may be one of the important controls of granulocytic maturation.

Differentiation Effects of Agents That Increase Intracellular cAMP on HL-60 Primed with RA

The finding that differentiation of HL-60 by RA is potentiated by agents known to increase intracellular cAMP prompted a study on the effects of sequential treatment with these agents (Olsson *et al.*, 1982). It was found that HL-60 could be primed for differentiation by treatment for approximately one day with 10 nM RA followed by exposure to a cAMP-inducing agent (cholera toxin or PGE_2). The reverse sequence was ineffective. Thus, HL-60 could be primed by incubation for less than 20 h with 10 nM RA to respond by differentiation to the addition of 10 nM PGE_2 or 1 nM cholera toxin, whereas 10 nM RA alone was inactive. Priming of HL-60 did not depend on the normal rate of protein synthesis, as it occurred even better in the presence of cycloheximide 1 μg/ml, a concentration that inhibited growth completely and protein synthesis by 86%. However, the resulting maturation was inhibited by cycloheximide. These results suggest that a decrease in synthesis of some protein(s) favors RA-induced differentiation. The nature of these proteins is unknown as are the direct consequences of inhibition of their synthesis. One possibility is that differentiation in HL-60 is inhibited by a polypeptide. Inhibition of protein synthesis by cycloheximide could diminish the production of this inhibitor and facilitate the modulating activities on differentiation by cAMP-inducing agents. Another untested hypothesis is that RA induces the production of a mRNA whose translation initiates the phenotypic changes characteristic of differentiated cells; cycloheximide could increase the production and the half-life of this mRNA as has been reported for human fibroblast interferon mRNA (Cavalieri *et al.*, 1977).

It is possible that the effects of RA and cAMP-inducing agents on HL-60 may be explained on the basis of a cAMP-dependent protein kinase activity. The phosphorylation of specific proteins has been implicated as playing a central role in a large number of biological processes and most, if not all, effects of cAMP are thought to be mediated by cAMP-dependent protein kinases (Greengard, 1978). In murine melanoma cells, RA increases cAMP-dependent protein kinase activity (Ludwig *et al.*, 1980) with no change in intracellular cAMP levels (Lotan *et al.*, 1978; Ludwig *et al.*, 1980). In HL-60, RA may also induce or increase the synthesis of a specific protein kinase. The activity of this enzyme would be dependent on the intracellular level of cAMP, which in vitro is increased by adding PGE, cholera toxin, or dibutyryl cAMP to HL-60 cultures. Alternatively, RA may induce the production of a product that is phosphorylated in the process of differentiation. Recent findings (Olsson *et al.*, 1982) have shown that exogenous ATP or dATP have a differentiation-inducing effect on HL-60 either in the presence of a low concentration of RA or on cells primed with RA. Because nucleotides are not transported into the intracellular space and because cAMP-dependent protein kinases are present on the outer surface of the plasma membrane of some cells (Mastro and Rozengurt, 1976), these results give indirect support for the involvement of phosphorylation reactions on the plasma membrane in the differentiation of HL-60.

Growth and Differentiation of Other Myelomonocytic Cell Lines in Defined Medium

K-562 and KG-1 proliferate in defined medium. As with HL-60, the growth rate of these two cells lines is approximately 80% of that obtained in serum-containing medium (Gallo *et al.*, 1982; Cioe *et al.*, 1981).

Both hemin and butyric acid induce hemoglobin synthesis in K-562 independent of the presence of serum in the growth medium (Cioe *et al.*, 1981). However, the extent of both spontaneous differentiation and induced differentiation is higher in serum-containing medium. In fact, shifting cultures of K-562 from defined medium to serum-containing medium causes an induction of hemoglobin synthesis, suggesting that components in serum could induce differentiation. Thus, the use of K-562 cells growing in defined medium could be used as indicator cells for the identification of these components.

Koeffler *et al.* (1981) have reported that the phorbol diester-induced differentiation of KG-1 and HL-60 to monocyte (macrophage)-like cells occurs as well in either serum-containing alpha medium or in serum-free alpha medium. The variables examined were morphology, adhesion, pseudopodia formation, reduction of nitroblue tetrazolium, and nonspecific esterase activity. Though it is not reported by the above authors, neither HL-60 nor KG-1 would be expected to grow in serum-free alpha medium that does not contain insulin and transferrin. Thus, the phorbol diester-induced differentiation of these cell lines probably occurs in the absence of conditions required for growth, a finding similar to that reported by Ferrero *et al.* (1982) for maturation of HL-60 along the granulocytic pathway. In addition, these studies can be interpreted as indicating that insulin and transferrin are not required for this induced differentiation of KG-1 and HL-60, at least over the four-day experimental period. In a related study, Welsh *et al.* (1982) reported that approximately 20-fold higher concentrations of phorbol diester are required in defined medium than in serum-containing medium to induce similar levels of macrophagelike differentiation of HL-60. This requirement for a higher dose of phorbol diester in defined medium was accompanied by a substantial loss of cellular polyunsaturated fatty acids and a large increase in the palmitoleic acid content of cells grown in the absence of serum. As serum is a rich source of fatty acids, the use of a serum-free medium should aid in studies on the possible linkage of phospholipid metabolism in the induction process.

The monoblastlike and monocytelike cell lines, U-937 and THP-1, grow and are induced to mature by RA in defined medium. In this medium, RA-treated cells and, to a lesser extent, untreated cells adhere to the plastic surface. If, however, 600 μg/ml of bovine serum albumin is added, there is essentially no adherence, and the extent of differentiation induced by RA is identical to that obtained in serum-containing medium (Olsson and Breitman, 1982; Hemmi and Breitman, unpublished results).

References

Anderson, L. C., Jokinen, M., and Gahmberg, C. G., 1979, Induction of erythroid differentiation in the human cell line K-562, *Nature* (*London*) **278:**364–366.

Berenbaum, M. C., Cope, W. A., and Bundick, R. V., 1976, Synergistic effect of cortisol and prostaglandin E_2 on the PHA response, *Clin. Exp. Immunol.* **26:**534–541.

Bonser, R. W., Siegel, M. I., McConnell, R. T., and Cuatrecasas, P., 1981, The appearance of phospholipase and cyclo-oxygenase activities in the human promyelocytic leukemia cell line HL60 during dimethyl sulfoxide-induced differentiation, *Biochem. Biophys. Res. Commun.* **98:**614–620.

Breitman, T. R., 1982, Induction of terminal differentiation of HL-60 and fresh leukemic cells by retinoic acid, in: *Expression of Differentiated Functions in Cancer Cells* (R. P. Revoltella, Giuseppe M. Pontieri, Claudio Basilico, Giovanni Rovera, Robert C. Gallo, and John H. Subak-Sharpe, eds.), New York, pp. 257–273.

Breitman, T. R., and Keene, B. R., 1982, Growth and differentiation of human promyelocytic cell line HL-60 in a defined medium, in: *Growth of Cells in Hormonally Defined Media*, Volume 9 (Gordon H. Sato, Arthur B. Pardee, and David A. Sirbasku, eds.), Cold Spring Harbor Conferences on Cell Proliferation, Cold Spring Harbor Laboratory, New York, pp. 691–702.

Breitman, T. R., Collins, S. J., and Keene, B. R., 1980a, Replacement of serum by insulin and transferrin supports growth and differentiation of the human promyelocytic cell line, HL-60, *Exp. Cell Res.* **126:**494–498.

Breitman, T. R., Selonick, S. E., and Collins, S. J., 1980b, Induction of differentiation of the human promyelocytic leukemia cell line (HL-60) by retinoic acid, *Proc. Natl. Acad. Sci. U.S.A.* **77:**2936–2940.

Breitman, T. R., Collins, S. J., and Keene, B. R., 1981, Terminal differentiation of human promyelocytic leukemic cells in primary culture in response to retinoic acid, *Blood* **57:**1000–1004.

Cavalieri, R. L., Havell, E. A., Vilcek, J., and Pestka, S., 1977, Induction and decay of human fibroblast interferon mRNA, *Proc. Natl. Acad. Sci. U.S.A.* **74:**4415–4419.

Cioe, L., McNab, A., Hubbell, H. R., Meo, P., Curtis, P., and Rovera, G., 1981, Differential expression of the globin genes in human leukemia K562(S) cells induced to differentiate by hemin or butyric acid, *Cancer Res.* **41:**237–243.

Collins, S. J., Gallo, R. C., and Gallagher, R. E., 1977, Continuous growth and differentiation of human myeloid leukemic cells in suspension culture, *Nature* (*London*) **270:**347–349.

Collins, S. J., Ruscetti, F. W., Gallagher, R. E., and Gallo, R. C., 1978, Terminal differentiation of human promyelocytic leukemia cells induced by dimethyl sulfoxide and other polar compounds, *Proc. Natl. Acad. Sci. U.S.A.* **75:**2458–2462.

Collins, S. J., Ruscetti, F. W., Gallagher, R. E., and Gallo, R. C., 1979, Normal functional characteristics of cultured human promyelocytic leukemia cells (HL60) after induction of differentiation by dimethyl sulfoxide, *J. Exp. Med.* **149:**969–974.

DeRuyter, M. G., Lambert, W. E., and DeLeenheer, A. P., 1979, Retinoic acid: An endogenous compound of human blood. Unequivocal demonstration of endogenous retinoic acid in normal physiological conditions, *Anal. Biochem.* **98:**402–409.

Douer, D., and Koeffler, H. P., 1982, Retinoic acid. Inhibition of the clonal growth of human myeloid leukemia cells, *J. Clin. Invest,* **69:**277–283.

Ferrero, D., Tarella, C., Gallo, E., Ruscetti, F. W., and Breitman, T. R., 1982, Terminal differentiation of the human promyelocytic leukemia cell line, HL-60, in the absence of cell proliferation, *Cancer Res.* **42:**4421–4426.

Fontana, J. A., Wright, D. G., Schiffman, E., Corcoran, B. A., and Deisseroth, A. B., 1980, Development of chemotactic responsiveness in myeloid precursor cells: Studies with a human leukemia cell line, *Proc. Natl. Acad. Sci. U.S.A.* **77:**3664–3668.

Gallagher, R., Collins, S., Trujillo, J., McCredie, K., Ahearn, M., Tsai, S., Metzgar, R., Aulakh, G., Ting, R., Ruscetti, F., and Gallo, R., 1979, Characterization of the continuous, differentiating myeloid cell line (HL-60) from a patient with acute promyelocytic leukemia, *Blood* **54:**713–733.

Gallo, R. C., Ruscetti, F. W., Collins, S., and Gallagher, R., 1979, Human myeloid leukemia cells: Studies on oncornaviral related information and *in vitro* growth and differentiation, in: *Hematopoietic Cell Differentiation* (D. W. Golde, M. J. Cline, D. Metcalf, and C. F. Fox, eds.), Academic Press, New York, pp. 335–354.

Gallo, R. C., Breitman, T. R., and Ruscetti, F. W., 1982, Proliferation and differentiation of human myeloid leukemia cell lines *in vitro*, in: *Maturation Factors and Cancer* (M. A. S. Moore, ed.), Raven Press, New York, pp. 255–271.

Gilman, A. G., and Nirenberg, M., 1971. Regulation of Adenosine 3′,5′-cyclic monophosphate metabolism in cultured neuroblastoma cells, *Nature* (*London*) **234:**356–358.

Greengard, P., 1978, Phosphorylated proteins as physiological effectors, *Science* **199:**146–151.

Hittelman, K. J., and Butcher, R. W., 1973, Cyclic AMP and the mechanism of action of the prostaglandins, In: *The Prostaglandins* (M. F. Cuthbert, ed.), J. B. Lippincott, Philadelphia, pp. 151–165.

Hoffman, R., Murnane, M. J., Benz, E. J., Jr., Prohaska, R., Floyd, V., Dainiak, N., Forget, B. G., and Furthmayr, H., 1979, Induction of erythropoietic colonies in a human chronic myelogenous leukemia cell line, *Blood* **54:**1182–1187.

Honma, Y., Takenaga, K., Kasukabe, T., and Hozumi, M., 1980, Induction of differentiation of cultured human promyelocytic leukemia cells by retinoids, *Biochem. Biophys. Res. Commun.* **95:**507–512.

Jaffe, B. M., Behrman, H. R., and Parker, C. W., 1973, Radioimmunoassay measurement of prostaglandins E, A, and F in human plasma, *J. Clin. Invest.* **52:**398–405.

Koeffler, H. P., and Golde, D. W., 1980, Humoral modulation of human acute myelogenous leukemia cell growth *in vitro, Cancer Res.* **40:**1858–1862.

Koeffler, H. P., Bar-Eli, M., and Territo, M. C., 1981, Phorbol ester effect on differentiation of human myeloid leukemia cell lines blocked at different stages of maturation, *Cancer Res.* **41:**919–926.

Koren, H. S., Anderson, S. J., and Larrich, J. W., 1979, *In vitro* activation of a human macrophage-like cell line, *Nature* (*London*) **279:**328–331.

Kuehl, F. A., Jr., and Humes, J. L., 1972, Direct evidence for a prostaglandin receptor and its application to prostaglandin measurements, *Proc. Natl. Acad. Sci. U.S.A.* **69:**480–484.

Kurland, J. I., Hadden, J. W., and Moore, M. A. S., 1978, Role of cyclic nucleotides in the proliferation of committed granulocyte-macrophage progenitor cells, *Cancer Res.* **37:**4534–4538.

Levine, L., and Ohuchi, K., 1978, Retinoids as well as tumour promoters enhance deacylation of cellular lipids and prostaglandin production in MDCK cells, *Nature* (*London*) **276:**274–275.

Lotan, R., Giotta, G., Nork, E., and Nicolson, G. L., 1978, Characterization of the inhibitory effects of retinoids on the *in vitro* growth of two malignant murine melanomas, *J. Natl. Cancer Inst.* **60:**1035–1041.

Lozzio, B. B., Lozzio, C. B., Bamberger, E. G., and Feliu, A. S., 1981, A multipotential leukemia cell line (K-562) of human origin (41106), *Proc. Soc. Exp. Biol. Med.* **166:**546–550.

Lozzio, C. B., and Lozzio, B. B., 1975, Human chronic myelogenous leukemia cell line with positive Philadelphia chromosome, *Blood* **45:**321–334.

Ludwig, K. W., Lowey, B., and Niles, R. M., 1980, Retinoic acid increases cyclic AMP-dependent protein kinase activity in murine melanoma cells, *J. Biol. Chem.* **255:**5999–6002.

Mastro, A. M., and Rozengurt, E., 1976, Endogenous protein kinase in outer plasma membrane of cultured 3T3 cells, *J. Biol. Chem.* **251:**7899–7906.

Mather, J. P., and Sato, G. H., 1979, The growth of mouse melanoma cells in hormone-supplemented, serum-free medium, *Exp. Cell Res.* **120:**191–200.

Nilsson, K., Andersson, L. C., Gahmberg, C. G., and Forsbeck, K., 1980, Differentiation *in vitro* of human leukemia and lymphoma cell lines, in: *International Symposium on New Trends in Human Immunology and Cancer Immunotherapy* (B. Serrou and C. Rosenfeld, eds.), Doin, Paris, pp. 271–292.

Newburger, P. E., Chovaniec, M. E., Greenberger, J. S., and Cohen, H. J., 1979, Functional changes in human leukemia cell line HL-60: A model for myeloid differentiation, *J. Cell Biol.* **82:**315–322.

Olsson, I. L., and Breitman, T. R., 1982, Induction of differentiation of the human histiocytic lymphoma cell line U-937 by retinoic acid and cyclic adenosine 3′:5′-monophosphate-inducing agents, *Cancer Res.* **42:**3924–3927.

Olsson, I., and Olofsson, T., 1981, Induction of differentiation in a human promyelocytic leukemic cell line (HL-60). Production of granule proteins, *Exp. Cell Res.* **131:**225–230.

Olsson, I. L., Breitman, T. R., and Gallo, R. C., 1982, Priming of human myeloid leukemic cell lines HL-60 and U-937 with retinoic acid for differentiation effects of cyclic adenosine 3′:5′-monophosphate-inducing agents and a T-lymphocyte-derived differentiation factor, *Cancer Res.* **42:**3928–3933.

Ong, D. E., and Chytil, F., 1975, Retinoic acid-binding protein in rat tissue. Partial purification and comparison to rat tissue retinol-binding protein, *J. Biol. Chem.* **250:**6113–6117.

Rizzino, A., and Crowley, C., 1980, Growth and differentiation of embryonal carcinoma cell line F_9 in defined media, *Proc. Natl. Acad. Sci. U.S.A.* **77:**457–461.

Roberts, A. B., and Frolik, C. A., 1979, Recent advances in the *in vivo* and *in vitro* metabolism of retinoic acid, *Fed. Proc.* **38:**2524–2527.

Ruscetti, F. W., Collins, S. J., Woods, A. M., and Gallo, R. C., 1981, Clonal analysis of the response of human myeloid leukemia cell lines to colony-stimulating activity, *Blood* **58:**285–292.

Rutherford, T. R., Clegg, J. B., and Weatherall, D. J., 1979, K-562 human leukemic cells synthesize embryonic hemoglobin in response to hemin, *Nature (London)* **280:**164–166.

Sundstrom, C., and Nilsson, K., 1976, Establishment and characterization of a human histocytic lymphoma cell line (U937), *Int. J. Cancer* **17:**565–577.

Taetle, R., and Koessler, A., 1980, Effects of cyclic nucleotides and prostaglandins on normal and abnormal human myeloid progenitor proliferation, *Cancer Res.* **40:**1223–1229.

Takenaga, K., Honma, Y., and Hozumi, M., 1981, Cellular retinoid-binding proteins in cultured human and mouse myeloid leukemia cells, *Cancer Lett.* **13:**1–6.

Tsuchiya, S., Yamabe, M., Yamaguchi, Y., Kobayashi, Y., Konno, T., and Tada, K., 1980, Establishment and characterization of a human acute monocytic leukemia cell line (THP-1), *Int. J. Cancer* **26:**171–176.

Tsuchiya, S., Kobayashi, Y., Goto, Y., Okumura, H., Nakae, S., Konno, T., and Tada, K., 1982, Induction of maturation in cultured human monocytic leukemia cells by a phorbol diester, *Cancer Res.* **42:**1530–1536.

Welsh, C. J., Sager, A. M., Littlefield, L. C., and Cabot, M. C., 1982, Modification of lipid acyl groups by serum deprivation does not affect phorbol ester induced differentiation of human leukemia cells, *Cancer Lett.* **16:**145–154.

Williams, N., 1979, Preferential inhibition of murine macrophage colony formation by prostaglandin E, *Blood* **53:**1089–1094.

CHAPTER 5

In Vitro Immunization and Growth of Hybridomas in Serum-Free Medium

STEPHEN D. WOLPE

Introduction

The experiments of Köhler and Milstein (1975, 1976) leading to the first practical method of producing cell lines secreting specific monoclonal antibodies are now classic ones and have been the subject of numerous reviews (Melchers *et al.*, 1978; Kennett *et al.*, 1980; Goding, 1980; Fazekas de St. Groth and Scheidegger, 1980; Galfré and Milstein, 1981). The basic principles involved are illustrated in Fig. 1. Lymphoid cells, usually obtained from the spleens of animals previously immunized with a specific antigen, are fused with a myeloma cell line in the presence of polyethylene glycol (PEG). This fusion gives rise to three different types of cells: those derived solely from the original spleen cells, those derived solely from the original myeloma cell line, and those that are hybrids between the two. It is the latter hybrids, or hybridoma cells, that are of interest and that must be separated from the other cell types present in the fusion mixture. This is usually accomplished through the use of selective media, most commonly the hypoxanthine, aminopterin, thymidine (HAT) system devised by Littlefield (1964) (Fig. 2).

The HAT system uses the folic acid antagonist aminopterin to block the main biosynthetic pathways leading to the de novo production of purines and pyrimidines by inhibiting the enzyme dihydrofolate

STEPHEN D. WOLPE • The Population Council, New York, New York 10021.

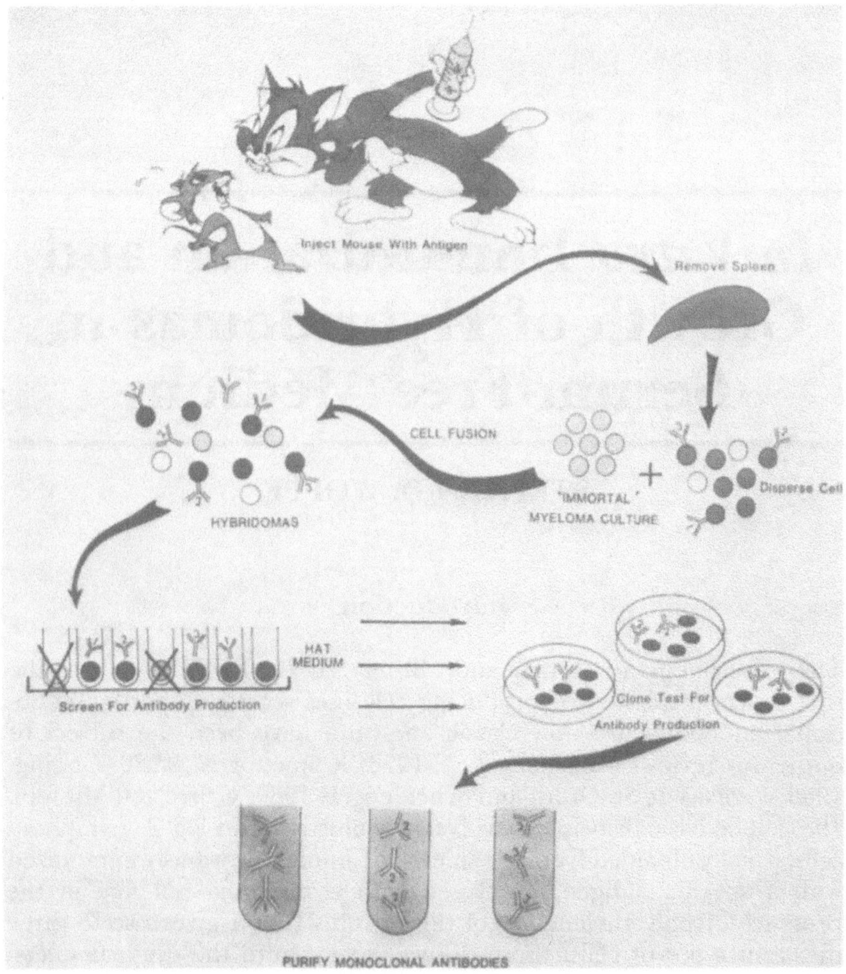

Figure 1. Schematic representation of monoclonal antibody production.

reductase. Aminopterin, however, does not affect the salvage pathways in which nucleotides are recycled for use. These pathways depend on the enzymes thymidine kinase (TK) for pyrimidines and hypoxanthine guanine phosphoribosyl transferase (HPGRT) for purines; a cell that is treated with aminopterin can therefore still synthesize DNA via the salvage pathways provided that thymidine and hypoxanthine are supplied and the enzymes TK and HPGRT are functional. A cell lacking one or both of these enzymes will thus be selected against relative to cells with functional TK and HPGRT in a medium containing HAT.

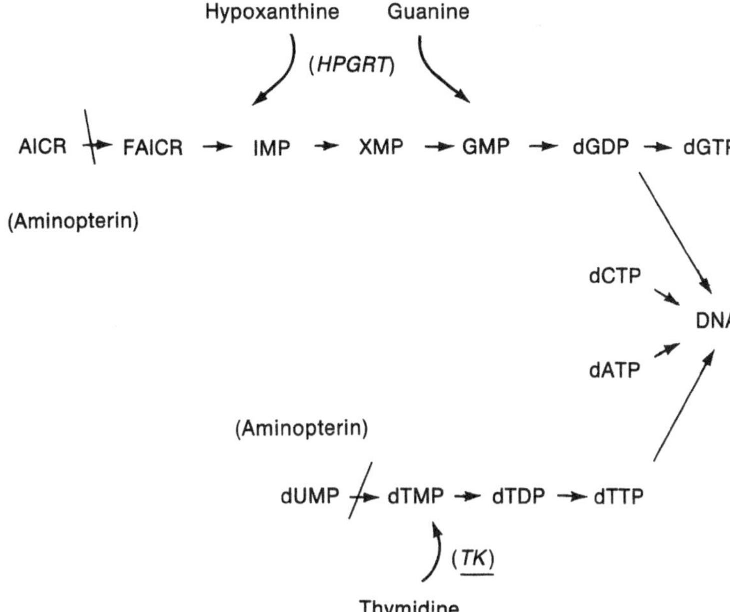

Figure 2. HAT selection (see text). Tetrahydrofolic acid is a coenzyme that is necessary for the conversion of 5-aminoimidazole-4-carboximide ribonucleotide (AICR) to 5-formaidoimidazole-4-carboximide ribonucleotide (FAICR) in purine biosynthesis and for the conversion of deoxyuridine monophosphate (dUMP) to deoxythymidine monophosphate (dTMP) in pyrimidine biosynthesis. Aminopterin, a competitive inhibitor of folic acid, prevents the formation of tetrahydrofolic acid by binding to folic acid reductase and therefore blocks purine and pyrimidine biosynthesis at the steps indicated. Cells poisoned by aminopterin, however, can continue to produce these nucleotides via scavenger pathways that utilize the enzymes hypoxanthine-guanine phosphoribosyl transferase (HGPRT) for purines and thymidine kinase (TK) for pyrimidines. Myeloma cells that have been selected for deficiency in one of these enzymes will die in the presence of aminopterin, but can be rescued if they fuse with a lymphocyte that can supply the missing enzyme. The full HAT selection technique utilizes one cell type deficient in HPGRT and another deficient in TK; for hybridoma formation, only the myeloma cells must be selected against as normal lymphocytes do not proliferate in culture. Abbreviations: the monophosphates of inosine (IMP), xanthine (XMP), guanine (GMP); the deoxymonophosphates of uridine (dUMP), thymidine (dTMP); the deoxydiphosphates of guanine (dGDP) and thymidine (dTMP); the deoxytriphosphates of guanine (dGTP), cytidine (dCTP), adenine (dATP), and thymidine (dTTP).

The full HAT selective system uses one cell type lacking in TK and another lacking in HPGRT; in HAT medium only hybrids derived from a parent of each cell type will survive. For hybridoma production, the so-called half-selective system is generally used. Normal spleen cells do not proliferate in culture under the conditions used and so it is only necessary to select against the myeloma cell line. Myeloma cell lines deficient in HPGRT have been selected through the use of toxic drugs such as thioguanine or 8-azaguanine; as these purine analogs are incorporated into DNA via HPGRT, only HPGRT$^-$ cells survive the selection. Such variants are relatively easy to obtain since HPGRT is an X-linked gene and, due to X-inactivation, only one allele is normally expressed in a cell. Similar selective strategies for TK$^-$ mutants using the toxic pyrimidine analogue 5′-bromodeoxyuridine (Budr) are more difficult since it is necessary to obtain a cell in which two rare mutations have occurred simultaneously. Still, myeloma cell lines using both types of mutations (as well as other selective systems) are available for hybridoma production (Table I).

The most commonly used myeloma cell lines were all derived from MOPC-21, a BALB/c myeloma that secretes IgG_1 antibodies (heavy and light chains) (Potter, 1972). This myeloma cell line was adapted to tissue culture (Horibata and Harris, 1970) and renamed "P3K" in accordance with the nomenclature then in use at the Salk Institute. An 8-azaguanine subline of P3K was developed by Milstein and co-workers (P3-x63Ag8 or "P3") and used for making the first hybridomas (Köhler and Milstein, 1975, 1976).

Because the cell line used by Köhler and Milstein still produced the original MOPC-21 antibody, the hybridomas derived from it consisted of a random assortment of heavy and light chains from the myeloma and from the spleen cell that fused with it. In the case of an IgG-producing spleen cell, for example, there are 16 possible combinations of antibodies that could be produced resulting from the intermixing of the two heavy and two light chains of each of the IgG molecules synthesized by the two parental cell types. As the avidity of an IgG antibody depends in part on the bivalent interaction of its two combining sites with antigen (Karush, 1978), those antibodies with a nonfunctional combining site (resulting from the presence of one or more myeloma chains) bind antigen less efficiently than do those whose immunoglobulin chains are derived solely from the spleen. This leads, for example, to much lower signals in radiobinding assays (Stähli *et al.*, 1980).

In order to eliminate this problem, it is necessary to obtain variants of the myeloma cell line that do not synthesize immunoglobulin. The

Table I. Cell Lines Available for Fusion

Name (Abbreviation)	Derived from	Immunoglobulin expression	Model # of chromosomes	Selective medium (or drug resistance)	Reference
Mouse					
P3-X63 Ag 8 (Ag8)	P3K	IgG_1	65	HAT	Köhler and Milstein, 1975
NS1-Ag4/1 (NS1)	P3K	K light chains	65	HAT	Köhler and Milstein, 1976
P3-X63 $Ag8U_1$ (P3U1)	Ag8	K light chains		HAT	Yelton *et al.*, 1978
X63-Ag8 653 (653)	Ag8	None	58	HAT	Kearney *et al.*, 1979
SP2/0 Ag14 (SP2)	(Ag8 × Balb/c)	None	72	HAT	Kearney *et al.*, 1978
FO	Clone of SP2	None	72	HAT	Fazekas de St. Groth and Scheidegger, 1980
S194/5x x 0.Bu.1	BALB/c Myeloma	None		BuDr	Trowbridge, 1978
MPC11-45.6, TG1.7	BALB/c Myeloma	IgG2b	62	HAT	Margulies, *et al.*, 1976
FOX-NY	NS-1		62	HAT or AAT	Taggart and Samloff, 1983
Rat					
Y3-Agl.2.3. (Y3)	R210.RCUY3	Rat K light chain		HAT	Galfré *et al.*, 1979
YB2/0	(Y3 × A0 rat) hybrid	None		HAT	Kilmartin *et al.*, 1982
Human					
U-266AR_1	U-266 myeloma	IgE	44	HAT	Olsson and Kaplan, 1980
GM1500 6TG2	GM1500	IgG_2		HAT	Croce *et al.*, 1980
KR-4	Ouabain-resistant variant of GM15006TG2	IgG_2		Ouabain	Kozdor *et al.*, 1982
LICR-Low-HMy2		IgG_1	60	HAT	Edwards *et al.*, 1982
UM-1-6TGr	UM-1	IgM		HAT	Bloom and Nakamura, 1974
UM-21-5	UM-21	K light chains		Citrullinemic	Bloom and Nakamura, 1974

first step in this direction was the isolation of a variant that did not produce heavy chain and hence was a "nonsecretor" of immunoglobulin (Köhler and Milstein, 1975). This variant, which was called P3-NS1-Ag4-1 (or "NS1"), still produced light chains that resulted in the production of heterogenous antibodies when it was used for hybridomas. Subsequently, cell lines that were true nonproducers, devoid of both heavy and light chains, were isolated. The two in most common use are SP2/O-Ag14 ("SP2") (Shulman *et al.*, 1978) and P3x63-Ag8.653 ("653") (Kearney *et al.*, 1979). Sp2 cells are, in reality, hybridoma cells that do not produce immunoglobulin and that were reselected for resistance to 8-azaguanine. We (Wolpe and Mather, unpublished observations) and others (Galfré and Milstein, 1981; Reading, 1982) find the 653 cell line to be superior to SP2 in hybridoma formation. This may be due in part to its lower chromosome number (Table 1) and possibly due to the inherent genetic instability of fusing spleen cells with a hybridoma cell line as opposed to a myeloma cell line.

As the procedures for making hybridomas are well standardized it is likely that, if an animal is capable of mounting an immune response to an antigen, that response can be immortalized. Technical improvements would appear, therefore, to lie in optimizing the fusion protocol (Fazekas de St. Groth and Scheidegger, 1980), increasing the response to weakly immunogenic antigens (Luben and Mohler, 1980a,b; Fox *et al.*, 1981; Stähli *et al.*, 1980), and improving the yields of monoclonal antibodies. This review will discuss the contributions that the recent advent of serum-free media for lymphocytes makes to these problems.

In Vitro Immunization

One of the drawbacks of the hybridoma method is the low yield of antigen-specific hybrids. Theoretically, the percentage of positive clones obtained by hybridization should reflect the percentage of positive spleen cells in the original spleen suspension. This can be estimated through the use of the Jerne plaque assay (Jerne and Nordin, 1963; Jerne *et al.*, 1974) or its variations. This assay depends on the fact that serum complement components (C′) will lyse sheep red blood cells (SRBC) that have antibody affixed to their surface. Spleen cells from an animal that has previously been immunized against SRBC are mixed with a suspension of sheep red blood cells and immobilized with them in a thin layer of agar. Those spleen cells that are actively secreting antibody to the immunogen will be surrounded by a zone of SRBC to which that antibody is bound. When C′ is added to these cells they will lyse, leaving a clear area or plaque in the center of which is the antibody-secreting spleen cell. When C′ alone is added only those cells secreting IgM (which activates complement much more effectively than IgG) form plaques ("direct plaques"). In the presence of a secondary antibody (e.g., rabbit antimouse IgG), those cells secreting IgG will also form plaques ("indirect plaques"). This assay is exceedingly sensitive and less than 1 antigen-specific cell in 10^6 can be detected. Further, the assay can be modified to detect other antigens provided that they can be coupled to SRBC or other suitable red blood cells.

Using this assay it has been observed that there is approximately a tenfold increase in the proportion of the antigen-specific hybridomas obtained after fusion in comparison with what would be expected from the frequency of antigen-specific spleen cells in the fusion mixture (Köhler and Milstein, 1976; Köhler and Shulman, 1978; Andersson and Melchers, 1978; Shulman *et al.*, 1978). It has therefore been suggested that cells that have recently been exposed to antigen and are thus rapidly proliferating preferentially form fusion products with the myeloma cells. It is for this reason that immunizations are generally conducted in such a manner as to allow the primary (mostly IgM) response to subside and to boost shortly (3 to 4 days) prior to fusion. The aim is to activate the clones of cells responding to the specific immunogen and to fuse at the peak of their proliferative activity.

It is this amplification of the proportion of positive cells that makes the hybridoma method a useful one. On the average, only about 1 out of 10^5 spleen cells yield viable hybrids (Milstein *et al.*, 1977) and this fraction does not increase by more than a factor of two or three even under optimized conditions (see, for example, Table XIV in Fazekas de St. Groth and Scheidegger, 1980). For SRBC, the frequency of plaque-forming cells is on the order of 1% (Andersson and Melchers, 1978). These cells, of course, are complex antigens with many determinants and 10^{-2} probably represents an upper limit of the number of antigen-specific cells that can be expected in an immunized spleen, which in actuality may range from 10^{-4} to 10^{-2}. When multiplied by the frequency of spleen cells that form viable hybrids, one can therefore expect specific hybridomas to occur on the order of about 10^{-8} to 10^{-6} spleen cells fused, provided one includes the amplification factor of 10 mentioned above. Since mouse spleens contain approximately 10^8 cells, this results in potentially 1 to 100 positive hybridomas per spleen. The actual number obtained will vary depending on the immunogenicity of the antigen, the number of splenic precusors capable of reacting to it, and the efficiency of the immunization and fusion protocols.

From these considerations it can be seen that one method of increasing the efficiency of hybridoma production would be to raise the percentage of antigen-specific cells in the original fusion mixture. One way to improve this would be to reduce the antigenic load on the responding animal by keeping it as free of environmental antigens (e.g., pathogens) as possible and by making the *last* injection of antigen as pure as is practical. (As earlier injections only serve to prime the animal they need not be as pure.) Another method is to adoptively transfer cells from an immunized animal to an X-irradiated recipient and fuse with the spleen cells of the latter after *in vivo* boosting (Fox *et al.*, 1981). A more generally applicable method, and one with several advantages is to immunize *in vitro* prior to fusion (Hengartner *et al.*, 1978; Luben and Mohler 1980a,b; Luben *et al.*, 1982; Fox *et al.*, 1981; Reading, 1982).

Immunization *in vitro* has a number of advantages over the *in vivo* method usually employed: (1) the immunization period is shortened to 4 or 5 days; (2) substantially smaller amounts of antigen are required; (3) greater responses to many antigens are observed (due in part to a reduction in the activity of tolerance and suppression mechanisms *in vitro*); (4) there is a greater percentage of responding cells; and (5) the influence of various factors on the efficiency of immunization (e.g., cell populations, lymphokines, and monokines) can more easily be tested.

Present methods for *in vitro* immunization trace back to two main culture systems developed in the late 1960s. The suspension culture method was developed by Mishell and Dutton (1966, 1967) and was the first successful means of inducing an immune response totally *in vitro*. The critical factors thought to be necessary for this to occur were the use of low oxygen tension, high cell densities, gentle agitation, daily feeding of the cultures, and the use of selected lots of fetal calf serum and SRBC (which were used as immunogen). Subsequently, Click *et al.* (1972a,b) investigated the culture conditions required for optimal production of plaque-forming cells. These authors found that the inclusion of thiols, particularly β-mercaptoethanol (2me), dramatically increased the *in vitro* response. Addition of nucleic acid precursors, glutamine and nonessential amino acids, along with adjusting the Ca^{2+} levels and including 2me, eliminated the necessity for agitation and daily feeding of the cultures and also allowed for the generation of response in an atmosphere of 5% CO_2 and air. These authors found that the inclusion of glutamine in the medium was essential for the formation of plaque-forming cells, whereas others have found that it has little effect (Schreier and Nordin, 1977); the necessity for rocking and feeding of the cultures has also been debated (Schreier and Nordin, 1977).

The other main culture system used for *in vitro* immunization is the diffusion culture method developed by Marbrook (1967). Here cells grow in a small chamber on a dialysis membrane that exchanges with a large reservoir of medium. This method also eliminates the necessity for constant rocking and feeding of cultures and allows for the use of a 10% CO_2–air atmosphere.

In 1978, Hengartner *et al.* reported the first use of *in vitro* immunization in hybridoma formation. These authors reported that they obtained a frequency of antigen-specific hybridomas almost one order of magnitude greater after *in vitro* immunization with SRBC than that normally obtained *in vivo.* They attributed this increase to both the selective survival of antigen-stimulated cells during immunization in culture and to an escape from *in vivo* regulatory mechanisms. Similar increases in the percent of positive clones have been noted by other authors using these techniques (Luben and Mohler, 1980a,b; Luben *et al.*, 1982; Fox *et al.*, 1981; Reading, 1982). In addition, evidence has been presented that tolerance and suppression mechanisms are reduced under the conditions of *in vitro* immunization (Luben and Mohler, 1980b) and that it is possible to produce responses to antigens that are normally not very immunogenic *in vivo* (Reading, 1982). Using this technique it has been possible to produce monoclonal antibodies using picomole levels of antigen (Luben *et al.*, 1982).

One disadvantage of the *in vitro* immunization method for the production of monoclonal antibodies is that it often leads to a greater proportion of IgM clones than is obtained normally (Luben and Mohler, 1980a). This is because the immunization schedule *in vivo* is usually aimed at fusing during the secondary response, whereas that obtained *in vitro* is generally a primary response. Secondary responses can, however, be obtained under completely *in vitro* conditions, particularly in Marbrook cultures (North and Maizels, 1977). Resting B cells evidently require 6 to 10 days in culture to mature into IgG-secreting cells; Mishell-Dutton cultures, even with repeated media changes, cannot support cell viability for this length of time. Marbrook cultures, on the other hand, with their large reservoir of media do not require refeeding and thus promote cell viability without disturbing cell–cell contacts that may be important for differentiation. In fact, feeding Marbrook cultures during immunization is actually counterproductive, perhaps due in part to the pH and temperature shocks that may accompany feeding (North and Maizels, 1977).

In addition to the requirement for longer periods of time to obtain a secondary response *in vitro*, the transition from IgM- to IgG-secreting cells in culture may be affected by T-cell suppression (see below). Although suppressive mechanisms appear to be reduced under *in vitro* conditions (Luben and Mohler, 1980b) T-cell suppression can be demonstrated for at least some antigens (Schreier and Lefkovits, 1978). Suppressor cells are more radiosensitive than B cells or helper T cells (Eardley and Gershon, 1976; Fauci, 1979) and are sensitive to hydrocortisone. Addition of 10^{-8} M hydrocortisone to spleen cells immunized against SRBC in Mishell-Dutton cultures increased the percentage of indirect (IgG) plaques from 15% to 50% with a resultant increase in the percentage of IgG-secreting hybridomas obtained from these immunizations (Hengartner *et al.*, 1978).

It is apparent from the discussion in this section that *in vitro* immunization is a useful procedure for amplifying the specific immune response to an antigen prior to fusion, particularly if that antigen is only weakly immunogenic and/or is available in only small quantities. However, it appears that optimal reactions only occur in the presence of high levels of serum (Schreier and Nordin, 1977) that can cause polyclonal activation (Schreier, 1981; Mosier, 1981) most likely due, at least in part, to the antigenic properties of the serum itself. Syngeneic mouse serum does not support an *in vitro* response (Mishell and Dutton, 1967). It is therefore likely that a serum-free culture system would further enhance the advantages of *in vitro* immunization. Although attempts in this direction have been made (Burger, 1977; Iscove and Melchers, 1978; Mosier, 1981), none of the reported serum-free media for lymphocytes give as good a response as that obtained in the presence of serum (Mosier, 1981).

A further benefit of serum-free media in hybridoma production is the ease of purifying monoclonal antibodies from supernatants that have been cultured in the absence of serum proteins. As *in vitro* immunization requires the interactions of several cell types and is dependent on multiple factors (see Section - Cellular Interactions and Soluble Factors), it is likely that the requirements for optimal growth of transformed clones of cells such as hybridomas will be less stringent than those required for successful *in vitro* immunization. Nevertheless, what has been learned in one system may still be applicable to the other and so they will be considered together in the following sections.

The Role of Serum

Schreier and Nordin (1977) and Schreier (1981) have investigated the role of serum in *in vitro* immunization and found it to be a complex one. Previously it had been noted that only certain lots of fetal calf serum were supportive of *in vitro* responses, whereas others were deficient or even suppressive (Mishell and Dutton, 1967). The difference is that, whereas a deficient serum simply does not support an optimal response, a suppressive serum will actually inhibit the response obtained under optimal conditions. At least part of the difference between a supportive batch of fetal calf serum and a deficient one is that the former one often is found to have been contaminated with bacteria. The bacterium involved appears to be a gliding bacterium and the stimulatory product in nonprotein and distinct from lipopolysaccharide (LPS) (Shiigi *et al.*, 1977). However, it is unlikely that any one factor is involved as fractionation of supportive serum (e.g., on a Sephadex G-200 column) reveals that a stimulatory effect of every fraction can be found when it is added to deficient serum (Schreier and Nordin, 1977; Schreier, 1981). Also, many factors have been found that can increase the response obtained with defective serum (see Sections-Cellular Interactions and Soluble Factors, and Thiols). For example, changing the basal medium from Eagle's minimal essential medium to RPMI 1640, adding 2me, and raising the serum concentration from 5% to 20% increased the percentage of supportive batches of fetal calf serum from 15% to up to 75% of randomly tested lots. Still, highly supportive batches of FCS were rare (Schreier and Lefkovits, 1978). One function of serum in the mitogenic response of lymphocytes is to facilitate the G_1-S transition (Kristensen *et al.*, 1981). Possible factors that may be involved in creating a supportive environment are discussed in the following sections.

Thiols

Broome and Jeng (1973) tested the effect of various thiols on the growth of the mouse lymphoma cell line L1210, a line whose optimal growth is dependent on the presence of high levels of cysteine in the medium. Out of 30 thiols tested, nine were found to be active; the most stimulatory were β-thioglycerol and 2me. Both were readily oxidized in culture and the half-life of 2me was found to be 5.9 h. In common with the other active thiols found, these are aliphatic compounds with a primary unsubstituted sulfhydryl group. Their mode of action is still a matter of debate.

Chen and Hirsch (1972) have reported that 2me can substitute for the requirement of adherent cells (see Section-Cellular Interaction and Soluble Factors) in the primary *in vitro* immune response to SRBC. This report is subject to the general criticism inherent in all such studies that the removal of the adherent population may not have been total. β-Mercaptoethanol could exert a synergistic effect with the remaining cells, amplifying a suboptimal response and therefore appearing to substitute for adherent cell function. It is therefore possible that the role of thiols in enhancing lymphoma cell growth and in the primary *in vitro* immune response are the same.

Several theories have been proposed for the function of thiols in these systems; often they are presented as conflicting but there is no a priori reason for thinking that they should be mutually exclusive. The first point of contention is whether thiols act directly on the cells involved or whether they act on serum factors that in turn mediate the stimulatory effects. Although some authors find that thiols are effective in serum-free media (Soulillou *et al.*, 1975; Goodman and Weigle, 1977, 1981; Mosier, 1981), others find that serum is required (Hewlett *et al.*, 1981; Nordin, 1978; Sidman and Unanue, 1978; Hoffeld and Oppenheim, 1980a,b). The discrepancies may be due in part to artifacts in the various systems used. As none of the serum-free media that have been described to date yield optimal *in vitro* responses, it is difficult to evaluate the lack of stimulation by 2me in the absence of serum. Conversely, as lymphocytes spend a good deal of time in both vascular and lymphatic compartments, it is possible that they carry with them serum components adsorbed to their cell surface that can be activated in the presence of 2me even in serum-free conditions. Probably the only definitive experiment that could answer the question of whether 2me has direct effect on cells would be to have cell lines available that can be grown in serum-free medium and that, when cultured under serum-free conditions with antigen, can yield an immune response. Such a system is unavailable at this time.

Another complicating factor is that 2me may have multiple effects. Goodman and Weigle (1981) have shown that the mitogenic effects of thiols (i.e., the ability of thiols per se to induce proliferation) can be dissociated from the adjuvant effects of these compounds (i.e., their ability to enhance the response to another mitogen; e.g., lipopolysaccharide). Thus both 2me and a disulfide derivative (2-hydroxyethyldisulfide) were found to be mitogenic but only 2me exhibited adjuvant effects as well. Both these effects were found to be dependent on more than just the ability of these compounds to act as reducing agents; both a free sulfhydryl group and a hydroxyl group are necessary to retain both activities.

Although it is still unclear as to whether thiols can have a direct effect on cells, there is substantial evidence that certain serum components can interact with thiols to yield stimulatory products. Two laboratories (Hewlett *et al.*, 1981; Sidman and Unanue, 1978) have isolated a protein with a molecular weight similar to albumin that when treated with 2me and added back to macrophage-depleted cultures can support *in vitro* immunizations. The factor binds tightly to a CM Affi-Blue gel and is resistant to heat inactivation at 56 °C for 30 minutes. Whether or not the factor is identical to albumin is under investigation.

Hoffeld and Oppenheim (1980a,b) have presented evidence that 2me also exerts an effect on the primary antibody response by its effect on glutathione. Of a variety of sera examined, only fetal bovine serum (FBS) contains glutathione (Bump and Reed, 1977); sera from many species other than fetal bovine are deficient in their ability to support an *in vitro* response (Schreier and Nordin, 1977). Further, the levels of reduced glutathione in FBS lots correlate with how supportive they are (Hoffeld and Oppenheim, 1980a). Addition of reduced glutathione to deficient batches of FBS made them supportive (Hoffeld and Oppenheim, 1980b) and addition of several structurally unrelated compounds, all of which blocked various steps in the lipid peroxidation process, could substitute for 2me in stimulating LPS mitogenesis and *in vitro* immunization (Hoffeld, 1981). Thus one of the main functions of reduced glutathione in these systems may be the prevention of peroxide damage to cells, although glutathione is also involved in many biochemical reactions (Meister, 1982). In this connection it is of interest to note the recent observation by Darfler and Insel (1983a) that the high levels of BSA or casein needed for lymphocyte growth in some serum-free media could be eliminated by the substitution of catalase, an enzyme that degrades hydrogen peroxide. Zmuda and Friedenson (1983) have shown that 2me induces large increases in the intracellular levels of glutathione in cells undergoing a mitogenic response induced by Con A. Glutathione may therefore be involved in the adjuvant effects of thiols; data linking it to the mitogenic effects, however, are not yet persuasive.

Finally, Ohmori and Yamamoto (1983a,b) have recently shown that one function of 2me is to stimulate the uptake of cysteine via the formation of a mixed disulfide between 2me and cysteine. Mixed disulfides were also found to form between 2me and serum components; these were suggested to serve as a repository for oxidized 2me that later could be reduced by lymphocytes to liberate free 2me. This would then be able to interact with cysteine, facilitating its uptake.

Cellular Interactions and Soluble Factors

One of the advantages of *in vitro* immunization is its ability to dissect out various cellular constituents of the immune system and examine their role in the induction of immune responses. As was mentioned briefly in the last section, there is a requirement for adherent cells in the culture system in order to obtain a primary response. The cell types involved have been assumed to be macrophages, although recently there has been the identification of another cell, the dendritic cell, that may play a vital role (Steinman, 1981). In addition, T cells (i.e., lymphocytes that have undergone differentiation in the thymus) are necessary in the case of most antigens in order for B cells (presumptive antibody-secreting cells) to proliferate and mature into antibody secretors. These cellular interactions are beyond the scope of this chapter and have been extensively reviewed (Katz, 1977; Bach, 1978; Benacerraf and Unanue, 1979; Eisen, 1980; Golub, 1981). Although direct cell–cell contact may be of importance, at least some of the interactions involved are mediated by soluble factors. These are termed "lymphokines" if their source is activated lymphocytes and "monokines" if derived from macrophages. The study of the production and biological effects of soluble mediators of immunity is an extremely active field at the present time (Hudden and Steward, 1981; Cohen *et al.*, 1979; Rumjanek *et al.*, 1982; Pick, 1981).

Because the original definitions of monokines and lymphokines were based on their activities in various assay systems, the same molecules were called by different names in different laboratories leading to much confusion in the field. This was resolved at the Second International Lymphokine Workshop for two molecules that were termed "interleukins" (for "between leukocytes") 1 and 2 (Aarden *et al.*, 1979). Interleukin 1 is a monokine and induces a subset of T cells to produce interleukin 2 (IL2, also called "T-cell growth factor" [TCGF]). Although resting T cells do not have receptors for IL2, those that have been stimulated by antigen or mitogen do, and they respond to the presence of IL2 by proliferating. It has thus been possible to clone T cells by stimulating them with either antigen or mitogen and then maintaining them by providing exogenous IL2.

The successful use of growth factors in the cloning of T cells has stimulated the search for similar factors that may be active with B cells. The present status of this research has recently been reviewed. Several distinct factors have been described that are involved to some extent in controlling B-cell proliferation. As with T cells, it appears that the first step is activation by antigen or mitogen. Subsequent to this step is a proliferation phase followed by maturation in which the B cell differentiates into an immunoglobulin-secreting plasma cell (Schimpl and Wecker, 1975). Some of the factors that appear to be involved in these various stages include various B-cell growth factors (Howard *et al.*,

1982; Leanderson *et al.*, 1982; Maizel *et al.*, 1982; Pike *et al.*, 1982) and T-cell-replacing factor (TRF) (Schimpl and Wecker, 1975). The growth factors may be involved in the proliferation phase and TRF in the maturation phase; recently it has been shown that at least two distinct factors are involved in these steps (Leanderson *et al.*, 1982). It is possible that both IL2 and B-cell growth factors share a common 14kd subunit, and that the cellular specificities of these factors are determined by the association of this subunit with a larger peptide (Fresno *et al.*, 1982).

These factors are of particular interest because they are antigen nonspecific and cause polyclonal growth of activated lymphocytes. There is evidence, however, that antigen-specific factors are also involved in the initiation of immune response (Tada and Okumura, 1979). These factors bind both antigen and anti-Ia antibody, the particular Ia specificity depending on whether the factor enhances or suppresses the immune response. In addition to this variability, a number of both suppressor and helper factors have been described with widely ranging molecular weights and genetic restrictions in their modes of action. A unifying hypothesis that explains the roles of these various factors in the immune response has not yet been achieved.

In addition to factors produced by cells that classically are considered to function in the immune response, there have been reports of other cell types that may secrete soluble mediators. Astaldi *et al.* (1980, 1981) have reported the presence of a growth factor for hybridomas and myeloma cells that is present in the culture supernatants of human umbilical cord endothelial cells. We (Wolpe and Mather, unpublished observations) have confirmed these results using supernatants from the clonal rat testicular endothelial cell line TR-1 (Mather *et al.*, 1982). A shown in Table II, the growth-promoting activity present in TR-1 supernatants was not observed in those obtained from other clonal cell lines such as those of Sertoli, Leydig, or myoid cell origin (see Mather *et al.*, 1982, for a description of these cell lines).

In vitro antibody responses appear to be influenced by neuroendocrine hormones (Johnson *et al.*, 1982). Adrenocorticotropic hormone (ACTH) and α-endorphin strongly inhibited the plaque-forming response to SRBC in Mishell-Dutton cultures; Leu- and Met-enkephalin were intermediately suppressive and β- and γ-endorphins were only minimally inhibitory. In contrast, β-endorphin, but not α-endorphin or Met-enkephalin potentiated the T-cell response to mitogens (but not the B-cell response) (Gilman *et al.*, 1982). Although the competitive opiate inhibitor naloxone could block the suppressive effects of α-endorphin on the formation of plaque-forming cells (Johnson *et al.*,

Table II. Growth Stimulation by Supernatants from the TR1 Endothelial Cell Line

Cell line	Derived from	SF medium alone	SF medium (TM4)	SF medium (TRM)	SF medium (TR1)
SP2	Myeloma	8.0×10^5	7.3×10^5	9.1×10^5	1.9×10^6
653	Myeloma	3.2×10^5	3.5×10^5	3.1×10^5	8.8×10^5
C48	(653 × mouse)	8.2×10^5	6.9×10^5	6.8×10^5	1.7×10^6
C294	(653 × mouse)	4.1×10^5	2.8×10^5	3.2×10^5	9.1×10^5
C654	(653 × mouse)	4.2×10^5	3.1×10^5	3.7×10^5	8.9×10^5

Cells were grown in 15% fetal bovine serum, washed twice, and plated in 60 mm dishes in the medium of Murakami *et al.* (1982); 0.5 ml of conditioned (serum free) supernatant was added to each dish containing 4.5 ml medium and 10^4 cells. Cells were counted on day 4. All counts were done in triplicate. TM4, mouse Sertoli cell line; TRM, Rat myoid-like cell line; TR1, Rat endothelial cell line. See Mather *et al.* (1982) for derivation of the cell lines used for supernatants.

1982), it could not inhibit the potentiation of T-cell responses to mitogens by β-endorphin (Gilman *et al.*, 1982).

It is not clear which of these various factors are of particular importance in the response to antigen *in vitro*. For this reason, many *in vitro* immunization schemes use a thymocyte-conditioned medium that serves as a source of lymphokines (Andersson *et al.*, 1977; Luben and Mohler, 1980a,b; Luben *et al.*, 1982; Reading, 1982). Clearly, then, it is too early to have a "completely defined" medium which supports most *in vitro* immunizations. However, there have been some advances made in eliminating the serum requirement in this procedure. A supportive serum-free medium would be an important first step in delineating which factors are of particular importance in the *in vitro* response to antigen.

Serum-Free Media

Iscove and Melchers (1978) have shown that mitogen-induced B cells can be grown in a medium supplemented with albumin, transferrin and soybean lipid. The addition of thymocytes as feeder cells (Andersson and Melchers, 1977) allowed for clonal growth of B cells at limiting dilutions under serum-free conditions. Cells grown in this medium were capable of underoing the switch from IgM to IgG secretion.

Murakami *et al.* (1981) developed a serum-free medium that could support the growth and immunoglobulin secretion of a plasmacytoma cell line that was an ouabain- and thioguanine-resistant derivative of MPC-11 (MPC11 TG 1.7 Ova 3). These authors showed that transferrin, luteinizing hormone (LH), luteinizing hormone releasing-hormone

(LHRH), prostaglandin E_1 (PGE_1), epidermal growth factor (EGF), glucagon, selenium, prostaglandin $F_{2\alpha}$ ($PGF_{2\alpha}$), and nerve growth factor (NGF) were all growth stimulatory for these cells. Transferrin was the most important component, whereas some of the others showed significant stimulation only in long-term experiments. Suprisingly, insulin was not found to be stimulatory for this cell type.

This work has recently been extended to include the serum-free growth of hybridoma cells (Murakami *et al.*, 1981, 1982). The medium was initially derived through experiments with MPC11-BL, a hybridoma between the subline of MCP11 used in the previously mentioned study and a BALB/c spleen cell. It is of particular interest that this hybridoma exhibited different growth requirements than did the parental plasmacytoma. In addition to an absolute requirement for insulin (not required by a parental cell) and transferrin, the hybridoma was also strictly dependent upon the presence of ethanolamine or phosphoethanolamine in the medium. Selenium was found to strongly stimulate cell growth but was not an absolute requirement. In addition, relaxin, prolactin, ACTH and growth hormone were stimulatory.

Because medium supplemented with insulin, transferrin, ethanolamine, and selenium (ITES) was found to be sufficient to grow the MPC11-BL hybridoma, this combination was used in growth experiments with hydridomas from a number of parental cells. Variable results, with growth rates from 18% to 144% of that seen with serum, were obtained with cells derived from MPC-11 and SP2/0. Hybridomas derived from NS1 did not grow well at all in ITES medium. This has recently been explained as a requirement by NS1 cells for low-density lipoprotein (LDL) and either oleic or linoleic acids (Kawamoto *et al.*, 1982). A subline of NS1 that does not require LDL for growth and exhibits a higher fusion frequency than the original NS1 line has been derived (Kawamoto *et al.*, 1983). Although hybridoma cells derived from some cell lines can grow in ITES medium alone in monolayer cultures, they exhibit a requirement for phospholipids, particularly phosphatidylethanolamine and/or phosphatidylglycerol, when grown in spinner cultures (Murakami *et al.*, 1982). Phospholipids derived from soybean were more effective than those derived from egg yolk, and this was attributed to the greater percentage of linoleic acid in the former. This is in keeping with the observation that (human) lymphocytes require both a saturated and unsaturated free fatty acid for optimal growth *in vitro* (Spieker-Polet and Polet, 1981). However, it is clear that the requirement for lipids is not universal as some hybridoma lines can grow in medium supplemented with only transferrin, insulin, and nonessential amino acids (Chang *et al.*, 1980).

Table III. Growth of Myeloma and Hybridoma Cell Lines in Serum-Free Media

Cell line	Derived from	Serum free #1[a] (cell #/% growth)	Serum free #2[b] (cell #/% growth)	15% FBS
SP2	Myeloma	9.8×10^5/61%	8.7×10^5/54%	1.6×10^6
653	Myeloma	4.4×10^5/34%	2.9×10^5/22%	1.3×10^6
P3U1	Myeloma	1.2×10^6/63%	1.4×10^6/74%	1.9×10^6
C294	(653 × mouse)	3.5×10^5/32%	2.0×10^5/18%	1.1×10^6
C48	(653 × mouse)	9.6×10^5/64%	7.9×10^5/53%	1.5×10^6
C654	(653 × mouse)	4.3×10^5/48%	5.2×10^5/58%	9×10^5
53-7.313	(NS1 × rat)	1.4×10^5/18%	9.6×10^4/12%	8×10^5
M1/70.15	(NS1 × rat)	2.0×10^5/17%	3.8×10^5/32%	1.2×10^6

[a] Medium of Murakami *et al.* (1982).
[b] Medium of Darfler and Insel (1983a).
Cells were grown in 15% fetal bovine serum (FBS), washed twice in serum-free medium, and plated in 60 mm dishes (10^4 cells/dish). Cells were counted on day 4. Percent growth refers to percent of growth relative to that in 15% medium. All counts were done in triplicate.

Darfler *et al.* (1980) have devised a medium containing casein, insulin, transferrin, testosterone, and linoleic acid (CITTL) that was designed to support the growth of T-lymphoma cell lines, particularly S49. Subsequently it was shown that a number of other human and murine lymphoid cell lines, as well as hybridomas and normal T cells, could be grown in CITTL (Darfler and Insel, 1983b). Recently the function of casein in this medium has been clarified since it was shown that the addition of catalase to the medium eliminates the need for albumin or casein (Darfler and Insel, 1983a). It was therefore proposed that albumin, casein, and selenium all serve the same purpose in serum-free media, namely to protect against peroxide-mediated cytotoxicity and that, for this function, catalase can function more efficiently. Thus clonal growth of S49 cells was achieved in a medium consisting of catalase, insulin, transferrin, testosterone, selenium, and dilinoleoyl phosphatidylcholine.

We (Wolpe and Mather, unpublished results) have compared some of the serum-free formulations described above for their capacity to promote cell growth of a number of hybridoma cell lines. Some results are shown in Table III. The important point to be derived from this table is that although some hybridomas grow well in some serum-free formulations, others, even if derived from the same parental cell line, do not. We have not been able to achieve growth rates similar to those obtained in the presence of serum with any of these serum-free media under the conditions used. The latter stipulation is important as some authors have noted a lag phase after transfer of cells to serum-free medium. We find that some cell lines require prior adaptation to low serum levels before transfer to serum-free medium. This is accomplished by growing the cells in serum-free medium with additives plus 1% to 2% serum. Vigorously growing cultures can then be transferred to serum-free medium. Adaptation appears to be particularly necessary for the 653 line, although some hybridomas derived from it can be placed directly in serum-free medium. In any case, it seems clear from the data in Table II that not all cell lines grow equally well under serum-free conditions. This is to be expected when one considers the nature of hybridomas that are derived from fusion between two cell types and have a strong tendency to lose chromosomes, particularly during the early stages of clonal growth. If serum-free growth of hybridomas is considered to be of particular importance, it would therefore be advantageous to select for it immediately after fusion. Probably the best strategy would be to use a serum-free medium that supports the growth of the parental cell type to be used and to conduct all stages of hybrid selection in this medium. We are currently attempting to find a serum-free formulation that will support the clonal growth of 653 cells for this purpose.

An alternative to growing hybridomas in serum-free media is to use Marbrook cultures (Kinman and McKearn, 1981). Here cells are grown in serum-free medium in a chamber that communicates with a large reservoir of serum-containing medium through a dialysis membrane. Only low-molecular-weight proteins are found in the supernatants after exhaustive growth of the cells; these proteins are easily separated from the immunoglobulin by gel filtration. Because high cell numbers are achievable in a comparatively small volume of serum-free medium, the antibody concentrations are found to be 10 to 12 times greater than those achieved in conventional cultures. It should be noted, however, that in spite of this increase in antibody concentration to 200 to 400 μg/ml, levels comparable to those seen in ascites fluid (10 to 20 mg/ml) still have not been achieved.

It is of interest that low-molecular-weight (i.e., dialyzable) compounds suffice to maintain immunoglobulin secretion by lymphocytes in the system described above. Recently, Cleveland *et al.* (1983) have shown that immunoglobulin secretion can be maintained in a totally protein-free medium for as long as six months, depending on the cell line. These authors used the basal medium devised by Mosier (1981) (see below) consisting of a 1 : 1 mixture of Ham's F12 (Ham, 1975) and Iscove's modified Dulbecco's medium (Iscove and Melchers, 1978) supplemented with α-thioglycerol and progesterone and modified it by the addition of a complex mixture of trace elements. Actual growth rates compared with serum-containing medium were not examined.

A more complex problem than the serum-free growth of clonal cell lines is that of *in vitro* immunization under serum-free conditions. Part of the complexity is undoubtedly due to the differing nutritional requirements of the interacting cell types. For example, it has been noted that although soybean lipids are stimulatory for mitogen-induced B-cell growth, they are inhibitory for that of T cells (Kristensen *et al.*, 1982). On the other hand, zinc deficiency has been shown to primarily affect T-cell responses *in vivo* (Bach, 1981) (interestingly, these effects can persist even into the second and third generations; Beach *et al.*, 1982). Caution should therefore be exercised in extrapolating the results obtained with serum-free growth of certain cell lines to include that of *in vitro* immunization.

Still, some success has been achieved in the latter system. Burger (1977) has demonstrated that fetuin, a major glycoprotein of FBS, can substitute for serum in the Mishell-Dutton system in medium supplemented with nonessential amino acids. Pronase-digested fetuin showed partial activity. Serum proteins could be completely eliminated by substituting zymosan, an insoluble polysaccharide–protein complex. (Burger *et al.*, 1980).

Although these two systems are less complex than that of serum, they still include poorly defined serum substitutes, Further, neither the fetuin-containing medium described above (Burger *et al.*, 1977) nor the medium of Iscove and Melchers (1978) used to support mitogen-induced B-cell growth were capable of supporting primary *in vitro* responses to the nonmitogenic antigen TNP-Ficoll (Mosier, 1981). This antigen is T-cell independent, and thus the response to it is more simple than those requiring T-cell help. Still, it is a significant advance that a primary *in vitro* response to this antigen could be achieved in a completely defined, serum-free medium (Mosier, 1981). The medium used was a 1 : 1 mixture of Ham's F12 (Ham, 1975) and Iscove's modified Dulbecco's medium (Iscove and Melchers, 1978) supplemented with insulin, transferrin, progesterone, 2me, and trace elements. Although B cells survived for relatively long periods of time in this medium, unstimulated T cells did not. Even so, the presence of small amounts of contaminating T cells resulted in strong suppression of the response to TNP-Ficoll. T-cell suppression may be the reason for the variability of the T-dependent response to sheep red blood cell in this system. Significantly, there was a greatly reduced background in both mitogen-induced proliferation and in primary responses to TNP-Ficoll in this medium compared with that observed in the presence of serum, probably due to the polyclonal activation induced by the latter. This suggests that *in vitro* immunization under serum-free conditions may result in an enhancement of the percentage of specific hybrids formed when used for monoclonal antibody production.

Conclusion

The development of the hybridoma technique has allowed for the immortalization of immune responses. Future advances in hybridoma production are likely to be technical ones such as optimizing the response to poorly immunogenic antigens and simplifying the recovery of monoclonal antibody. Serum-free culture conditions, coupled with the ability to manipulate the immune response *in vitro*, are likely to make significant contributions to these advances. They will be of particular significance in clinical applications where human–human monoclonal antibodies are to be favored. Due to ethical considerations, most of these monoclonal antibodies will be generated through the use of *in vitro* immunization. The ability to eliminate foreign proteins, as well as the probable enhancement of *in vitro* responses once truly supportive serum-free conditions have been defined, will be especially advantageous. Finally, the simplified nature of serum-free conditions will allow for better characterization of those factors that mediate immune reactions.

ACKNOWLEDGMENTS. This work was supported by NIH Grant HD13541 to Dr. Jennie Mather and a postdoctoral fellowship to the author from the Surdna Corp., Inc. I would like to thank Dr. Jennie Mather for introducing me to the methodology of serum-free culture and for her continuing advice and support. I would also like to acknowledge the excellent technical assistance of Alicia L. Byer and Florence Kaczorowski, as well as Linda McKeiver for her help in preparing the manuscript. The illustration in Fig. 1 was drawn by Penny Roberts whom I would also like to thank.

References

Aarden, L., et al., 1979, Revised nomenclature for antigen-nonspecific T cell proliferation and helper factors, *J. Immunol.* **123**:2928–2930.

Andersson, J., and Melchers, F., 1978, The antibody repertoire of hybrid cell lines obtained by fusion of X63-Ag8 myeloma cells with mitogen-activated B-cell blasts, *Curr. Topics Microbiol. Immunol.* **81**:130–140.

Andersson, J., Countinho, A., Lernhardt, W., and Melchers, F., 1977, Clonal growth and maturation to immunoglobulin secretion *in vitro* of every growth-inducible B lymphocyte, *Cell* **10**:27–34.

Astaldi, G., Janssen, M., Lansdorp, P., Willems, C., Zeijlemaker, W., and Oosterhoof, F., 1980, Human endothelial culture supernantant (HECS): A growth factor for hybridomas, *J. Immunol.* **125**:1411–1414.

Astaldi, G., Janssen, M., Lansdorp, P., Zeijlemaker, W., and Willems, C., 1981, Human endothelial culture supernatant (HECS): Evidence for a growth-promoting factor binding to hybridoma and myeloma cells, *J. Immunol.* **126**:1170–1173.

Back, J. F., 1978, *Immunology*, John Wiley, New York.

Back, J. F., 1981, The multi-faceted zinc dependency of the immune system, *Immunol. Today* **2**:225–227.

Beach, R., Gershwin, M. E., and Hurley, L., 1982, Gestational zinc deprivation in mice: Persistence of immunodeficiency for three generations, *Science* **218**:469–471.

Benacerraf, B., and Unanue, E., 1979, *Textbook of Immunology*, Williams and Wilkons, Baltimore.

Bloom, A., Nakamura, F., 1974, Establishment of a tetraploid Immunoglobin-producing cell line from the hybridization of two human lymphocytes, *Proc. Natl. Acad. Sci.* **71**:2689.

Broome, J., and Jeng, M., 1973, Promotion of replication in lymphoid cells by specific thiols and disulphides *in vitro*. Effects on mouse lymphoma cells in comparison with splenic lymphocytes, *J. Exp. Med.* **138**:574.

Bump, E., and Reed, D., 1977, A unique property of fetal bovine serum: High levels of protein-glutathione mixed disulfides, *In Vitro* **13**:115–118.

Burger, M., 1977, A serum-free medium for the Mishell-Dutton system, *Eur. J. Immunol.* **7**:906–908.

Burger, M., Hess, M., and Cottier, H., 1980, The *in vitro* immune response in the absence of serum and its proteins: Stimulating activity of zymosan, *Eur. J. Immunol.* **10**:796–798.

Chang, T., Steplewski, Z., and Koprowski, H., 1980, Production of monoclonal antibodies in serum-free medium, *J. Immunol. Methods* **39**:369–375.

Chen, C., and Hirsch, J., 1972, Restoration of antibody-forming capacity in cultures of nonadherent spleen cells by mercaptoethanol, *Science* **176**:60–61.

Cleveland, W. L., Wood, I., and Erlanger, B., 1983, Routine large-scale production of monoclonal antibodies in a protein-free culture medium, *J. Immunol. Methods* **56**:221–234.

Click, R., Benck, L., and Alter, B., 1972a, Enhancement of antibody synthesis *in vitro* by mercaptoethanol, *Cell Immunol.* **3**:156.

Click, R., Benck, L., and Alter, B., 1972b, Immune responses *in vitro* I. Culture conditions for antibody synthesis, *Cell Immunol.* **3**:264.

Cohen, S., Pick, E., and Oppenhein, J. (eds.), 1979, *Biology of the Lymphokines*, Academic Press, New York.

Crochee, C., Linnenbach, A., Hall, W., Steplewski, Z., Koprowski, H., 1980, Production of human hybridoma secreting antibodies to measos virus, *Nature* **288**:488.

Darfler, F., and Insel, P., 1983a, Clonal growth of lymphoid cells in serum-free media requires elimination of H_2O_2 cytotoxicity, *J. Cell. Physiol.* **115**:31–36.

Darfler, F., and Insel, P., 1983b, Serum-free culture of resting, PHA-stimulated and transformed lymphoid cells, including hybridomas, *Exp. Cell. Res.* **138**:287–295.

Darfler, F., Murakami, H., and Insel, P., 1980, Growth of T-lymphoma cells in serum-free medium: Lack of involvement of the cyclic AMP pathway in long-term cultures, *Proc. Natl. Acad. Sci. U.S.A.* **77**:5993–5997.

Eardley, D., and Gershon, R., 1976, Induction of specific suppressor T cells *in vitro*, *J. Immunol.* **117:**117–313.

Edwards, P., Smith, C., Munro-Neville, A., 1982, Human-human hybridoma System based on a fast-growing mutant of the ARH-77 plasma cell Leukemia-derived line, *Eur. J. Immun.* **12:**641.

Eisen, H., 1980, in: *Immunology*, 2nd ed., Harper and Row, New York.

Fauci, A., 1979, Human B cell function in a polyclonally induced plaque forming cell system. Cell triggering and immunoregulation, *Immunol. Rev.* **45:**116–193.

Fazekas de St. Groth, S. and Scheidegger, D., 1980, Production of monoclonal antibodies: Strategy and tactics, *J. Immun. Methods* **35:**1–21.

Fox, P., Berenstein, E., and Sivaganian, R., 1981, Enhancing the frequency of antigen-specific hybridomas, *Eur. J. Immunol.* **11:**431–434.

Fresno, M., DerSimonian, H., Nabel, G., and Cantor, H., 1982, Proteins synthesized by inducer T cells: Evidence for a mitogenic peptide shared by inducer molecules that stimulate different cell types, *Cell* **30:**707–713.

Galfré, G., Milstein, C., Wright, B., 1979, Rat × Rat hybrid myolomas a monoclonal anti-Sd portion of mouse IgG, *Nature* **277:**131.

Galfré, G., and Milstein, C., 1981, Preparation of monoclonal antibodies: Strategies and procedures, *Methods Enz.* **73:**3–46.

Gilman, S., Schwartz, J., Milner, R., Bloom, F., and Feldman, J., 1982, β-endorphin enhances lymphocyte proliferative responses, *Proc. Natl. Acad. Sci. U.S.A.* **79:**4226–4230.

Goding, J., 1980, Antibody production by hybridomas, *J. Immunol. Methods* **39:**285–308.

Golub, E., 1981, *The Cellular Basis of the Immune Response*, 2nd ed., Sinauer, Stamford, Conn.

Goodman, M., and Weigle, W., 1977, Nonspecific activation of murine lymphocytes. I. Proliferation and polyclonal activation induced by 2-mercaptoethanol and α-thioglycerol, *J. Exp. Med.* **145:**473–489.

Goodman, M., Weigle, W., 1981, Nonspecific activation of murine lymphocytes: VII. Functional correlates of molecular structure of thiol compounds, *J. Immunol.* **126:**20–26.

Ham, R., 1975, Clonal growth of mammalian cells in a chemically defined, synthetic medium, *Proc. Natl. Acad. Sci. U.S.A.* **53:**288–293.

Hengartner, H., Luzzati, A., and Schreier, M., 1978, Fusion of *in vitro* immunized lymphoid cells with X63Ag8, *Curr. Topics Microbiol. Immunol.* **81:**92.

Hewlett, G., Opitz, H. G., and Schlumberger, H., 1981, The role of mercaptoethanol-activated serum factor in primary immune responses *in vitro* and in the growth of lymphoma cells, in: *Lymphokines*, Volume 4, (E. Pick, ed.), Academic Press, New York.

Hoffeld, J. T., 1981, Agents which block membrane lipid peroxidation enhance mouse spleen immune activities *in vitro*: Relationship to the enhancing activity of 2-mercaptoethanol, *Eur. J. Immunol.* **11:**371–376.

Hoffeld, J. T., Oppenheim, J., 1980a, The capacity of fetal calf serum to support a primary antibody response *in vitro* is determined, in part, by its reduced glutathione content, *Cell Immunol.* **53:**325–332.

Hoffeld, J. T., and Oppenheim, J., 1980b, Enhancement of the primary antibody response by 2-mercaptoethanol is mediated by its action of glutathione on the serum, *Eur. J. Immunol.* **10:**391–395.

Horibata, K., and Harris, A., 1970, Mouse myelomas and lymphomas in culture, *Exp. Cell Res.* **60:**61–70.

Howard, M., Farrar, J., Hiffiker, M., Thusou, B., Takatsu, K., Hamaoka, T., and Paul, W., 1982, Identification of a T-cell derived B cell growth factor distinct from interleukin 2, *J. Exp. Med.* **155:**914–923.

Hudden, J., and Steward, W. E., 1981, in: *The Lymphokines: Biochemistry and Biological Activity*, Hunan Press.

Iscove, N., and Melchers, F., 1978, Complete replacement of serum by albumin, transferrin and soybean lipid in cultures of lipopolysaccharrde-reactive B lymphocytes, *J. Exp. Med.* **147**:923–933.

Jerne, N., and Nordin, A., 1963, Plaque formation in agar by single antibody producing cells, *Science* **140**:405.

Jerne, N., Henry, C., Nordin, A., Fuji, H., Koros, A., and Lefkovits, I., 1974, Plaque forming cells: Methodology and theory, *Transplant Rev.* **18**:130–191.

Johnson, H., Smith, E., Torres, B., and Blalock, J. E., 1982, Regulation of the *in vitro* antibody response by neuroendocrine hormones, *Proc. Natl. Acad. Sci. U.S.A.* **79**:4171–4174.

Karush, F., 1978, The affinity of antibody: Range, variability and role of multivalence, in: *Comprehensive Immunology*, Volume 5, (G. Litman and R. Good, eds.) Plenum Press, New York, p. 85.

Katz, D., 1977, *Lymphocyte Differentiation, Recognition and Regulation*, Academic Press, New York.

Kawamoto, T., Sato, J., Le, A., McClure, D., and Sato, G., 1983, Development of a serum-free medium for growth of NS-1 mouse myeloma cells and its application to the isolation of NS-1 hybridomas, *Anal. Biochem.* **130**:445–453.

Kearney, J., Radbauch, A., Liesegancy, B., and Rajewsky, K., 1979, A new mouse myeloma cell line that has lost immunoglobulin expression but permits the construction of antibody-secreting hybrid cell lines, *J. Immunol.* **123**:1548–1550.

Kennett, R., McKearn, T., and Bechtol, K. (eds.), 1980, *Monoclonal Antibodies*, Plenum Press, New York.

Kilmartin, J., Wright, B., Milstein, C., 1982, Rat monoclonal antitublin antibodies derived by using a new nonsecreting rat cell line, *J. Cell Biol.*, **93**:576.

Klinman, D., and McKearn, T., 1981, Dialyzable serum components can support the growth of hybridoma cell lines *in vitro*, *J. Immunol. Methods* **42**:1–9.

Köhler, G., and Milstein, C., 1975, Continuous cultures of fused cells secreting antibody of predefined specificity, *Nature (London)* **256**:495–497.

Köhler, G., and Milstein, C., 1976, Derivation of specific antibody-producing tissue culture and tumor lines by cell fusion, *Eur. J. Immunol.* **6**:511–519.

Köhler, G., and Shulman, M., 1978, *Curr. Topics Microbiol. Immunol.* **81**:143–149.

Kozdor, P., Lagarde, A., Roder, J., 1982, Human hybridomas constructed with antigen-specific Epstein-Bar virus-transformation, *Proc. Natl. Acad. Sci.* **79**:6651.

Kristensen, F., Joncourt, F., and deWeek, A., 1981, The influence of serum on lymphocyte cultures. II. Cell cycle specificity of serum action in spleen cells, *Scand. J. Immunol.* **14**:121–130.

Kristensen, F., Walker, C., Walti, M., and deWeek, A., 1982, Development of serum-free defined culture medium for lymphoblast transformation tests of mouse spleen and thymus cells, *Scand. J. Immunol.* **16**:209–216.

Leanderson, T., Lundgren, E., Ruuth, E., Rorg, H., Persson, H., and Coutinho, A., 1982, B-cell growth factor: Distinction from T-cell growth factor, *Proc. Natl. Acad. Sci. U.S.A.* **79**:7455–7459.

Littlefield, J., 1964, Selection of hybrids from matings of fibroblasts *in vitro* and their presumed recombinants, *Science* **10**:709.

Luben, R., and Mohler, M., 1980a, *In vitro* immunization as an adjunct to the production of hydridomas producing antibodies against the lymphokine osteoclast activating factor, *Mol. Immunol.* **17**:635–639.

Luben, R., and Mohler, M., 1980b, Use of *in vitro* immunization in production of monoclonal antibodies against osteoclast activating factor: A method with general applicability to lymphokines, in: *Biochemical Characterization of Lymphokines* (A. de Week, F. Kristensen, and M. Landy, eds.), Academic Press, New York.

Luben, R., Brazeau, P., Böhlen, P., and Guillemin, R., 1982, Monoclonal antibodies to hypothalamic growth hormone-releasing factor with picomoles of antigen, *Science* **218**:887–889.

Maizel, A., Sahasrabuddhe, C., Mehta, S., Morgan, J., and Ford, R., 1982, Characterization of B-cell growth factor, *Lymphokine Res.* **1**:9–14.

Maizels, R., and Dresser, D., 1977, Conditions for the development of IgM- and IgG-antibody-secreting cells from primed mouse splenocytes *in vitro, Immunology* **32**:793–801.

Marbrook, J., 1967, Primary immune response in cultures of spleen cells, *Lancet* **2**:1279.

Margulies, D., Keuhl, W., Scharll, M., 1976, Somatic cell hybridization of mouse myolom cells, *Cell* **8**:405.

Mather, J., Zhuang, L., Perez-Infante, V., and Phillips, D., 1982, Culture of testicular cells in hormone-supplemented serum-free medium, *Ann. N.Y. Acad. Sci.* **383**:44–68.

Meister, A., 1983, Selective modification of glutathione metabolism, *Science* **220**:472–478.

Melchers, F., Potter, M., and Warner, N., 1978, Lymphocyte hybridomas, *Curr. Topics Microbiol. Immunol.* **81**.

Milstein, C., Adetugbo, K., Cowan, N., Köhler, G., Secher, D., and Wilde, C., 1977, Somatic cell genetics of antibody-secreting cells: Studies of clonal diversification and analysis by cell fusion, *Cold Spring Harbor Symp. Quant. Biol.* **41**:793–803.

Mishell, R., and Dutton, R., 1966, Immunization of normal mouse spleen suspension *in vitro, Science* **153**:1004.

Mishell, R., and Dutton, R., 1967, Immunization of dissociated spleen cell cultures from normal mice, *J. Exp. Med.* **126**:423.

Mosier, D., 1981, Primary *in vitro* antibody responses by purified murine lymphocytes in serum-free defined medium, *J. Immunol.* **127**:1490–1493.

Murakami, H., Masui, H., Sato, G., and Raschke, W., 1981, Growth of mouse plasmacytoma cells in serum-free, hormone supplemented medium: Procedure for the determination of hormone and growth factor requirements for cell growth, *Anal. Biochem.* **114**:422–428.

Murakami, H., Masui, H., and Sato, G., 1982, Suspension culture of hybridoma cells in serum-free medium: Soybean phospholipids as essential components, in: *Growth of Cells in Hormonally Defined Media* (G. Sato, A. Pardee, and D. Sinbasku, eds.), Cold Spring Harbor Laboratories, Cold Spring Harbor, New York.

Nordin, A., 1978, The *in vitro* response to a T-independent antigen. I. The effect of macrophages and 2-mercaptoethanol, *Eur. J. Immunol.* **8**:776–781.

North, J., and Maizels, R., 1977, B-memory cells can be stimulated by antigen *in vitro* to become IgG antibody-secreting cells, *Immunology* **32**:771–776.

Ohmori, H., and Yahamoto, I., 1983a, Mechanism of augmentation of the antibody response *in vitro* by 2-mercaptoethanol in murine lymphocytes. II. A major role of the mixed disulfide between 2-mercaptoethanol and cysteine, *Cell. Immunol.* **79**:173–185.

Ohmori, H., and Yahamoto, I., 1983b, Mechanism of augmentation of the antibody response *in vitro* by 2-mercaptoethanol in murine lymphocytes. III. Serum-bound and oxidized 2-mercaptoethanol are available for the augmentation, *Cell. Immunol.* **79**:186–196.

Olosson, L., Kaplan, H., 1980, Human-human hybridoma producing monoclonal antibodies of predefined antigenic specificity, *Proc. Natl. Acad. Sci.* **77**:5429.

Pick, E. (ed.), 1981, *Lymphokines*, Volume 1–8: Academic Press, New York,

Pike, B., Vaux, D., Clark-Lewis, I., Schroder, J., and Nossal, G., 1982, Proliferation and differentiation of single-hapten-specific B lymphocytes is promoted by T-cell factor(s) distinct from T-cell growth factor, *Proc. Natl. Acad. Sci. U.S.A.* **79:**6350–6354.

Potter, M., 1972, Immunoglobulin-inducing tumors and myeloma proteins of mice, *Physiol. Rev.* **52:**631–719.

Reading, C., 1982, Theory and methods for immunization in culture and monoclonal antibody production, *J. Immunol. Methods* **53:**261–291.

Rumjanek, V., Hanson, J., and Morley, J., 1982, Lymphokines and monokines, in: *Immunopharmacology* (P. Sirois and M. Rola-Pleszczyhski, eds.), Elsevier Press, New York.

Schimpl, A., and Wecker, E., 1975, A third signal in B cell activation given by TRF, *Trans. Rev.* **23:**176.

Schreier, M., 1981, The antibody response *in vitro*: Dissection of a complex system, in: *Lymphokines*, Volume 2 (E. Pick, ed.), Academic Press, New York,

Schreier, M., and Lefkovits, I., 1978, Induction of suppression and help during *in vitro* immunization of mouse spleen T cells, *Immunology* **36:**743–753.

Schreier, M., and Nordin, A., 1977, An evaluation of the immune reponse *in vitro*, in: *B and T Cells in the Immune Response* (F. Loor and G. Rolants, eds.), John Wiley, New York.

Shiigi, S., Capwell, R., Grabstein, K., and Mischell, R., 1977, Sera and the *in vitro* induction of immune responses. III. Adjuvant obtained from gliding bacteria with properties distinct from enteric bacterial lipopolysaccharide, *J. Immunol.* **119:**679–684.

Shulman, M., Wilde, C., and Kohler, G., 1978, A better cell line for making hybridomas secreting specific antibodies, *Nature* (*London*) **276:**269–270.

Sidman, C., and Unanue, E., 1978, Control of proliferation and differentiation in B lymphocytes by anti-Ig antibodies and a serum-derived cofactor, *Proc. Natl. Acad. Sci. U.S.A.* **75:**2401–2405.

Soulillou, J., Carpenter, C., Lundin, A., and Strom, T., 1975, Augmentation of proliferation and *in vitro* production of cytotoxic cells by 2me in the rat, *J. Immunol.* **115:**1566–1571.

Spieker-Polet, H., and Polet, H., 1981, Requirement of a combination of a saturated and an unsaturated free fatty acid and a fatty acid carrier protein for *in vitro* growth of lymphocytes, *J. Immunol.* **126:**949–954.

Stähli, C., Staehelin, T., Miggiano, V., Schmidt, J., and Häring, P., 1980, High frequencies of antigen-specific hybridomas: Dependence on immunization parameters and prediction by spleen cell analysis, *J. Immunol. Methods* **32:**297–304.

Steinman, R., 1981, Dendritic cells, *Transplant.* **31:**151–155.

Tada, T., and Okumura, K., 1979, The role of antigen-specific T cell factors in the immune response, *Adv. Immunol.* **28:**1.

Taggart, R., Samloff, I., 1983, Stable antibody-producing murine hybromides, *Science* **219:**1128.

Trobridge, I., 1978, Inner species spleen-myeloma hybrid-producing monoclonal antibodies against mouse lymphocytes on T., *J. Exp. Med.* **148:**313.

Yelton, D., Diamond, B., Kwan, S., Scharff, M., 1978, Fusion of mouse myelin spleen cells, *Curr. Topics Micro Immun.* **81:**1.

Zmuda, J., and Friedenson, B., 1982, Changes in intracellular glutathione levels of stimulated and unstimulated lymphocytes in the presence of 2-mercaptoethanol or cysteine, *J. Immunol.* **130:**362–364.

CHAPTER 6

Kidney Cell Cultures in Hormonally Defined Serum-Free Medium

MARY TAUB

Introduction

Renal cell growth and function may be studied *in vitro* using both established kidney cell lines and primary kidney cultures. To a major extent, *in vitro* studies concerning renal functions have been primarily concerned with two established kidney epithelial cell lines, Madin Darby canine kidney (MDCK) and the pig kidney cell line LLC-PK_1. These two cell lines are well characterized with regard to their transport properties. However, *in vitro* studies concerning hormonal regulation of growth and transport have not been extensive. The hormonally defined, serum-free culture media for these cell lines should facilitate such studies. Furthermore, hormonally defined serum-free culture media should also prove to be invaluable in the study of primary kidney cultures.

Primary kidney epithelial cell cultures are important to study for several reasons. The primary cells, being taken directly from the animal, still bear a close resemblance to the kidney cells *in vivo*. Consequently, the primary cultures may be used to validate the physiologic significance of studies done with established kidney cell lines. Secondly, primary kidney cell cultures may also facilitate the study of medical disease states of the kidney.

MARY TAUB • Biochemistry Department, State University of New York at Buffalo, Buffalo, New York 14214.

This paper summarizes the *in vitro* studies concerning renal functions with established kidney cell lines. The advantages of using the hormonally defined, serum-free media for these cell lines are described. Furthermore, the importance of primary kidney cell cultures for such investigations is discussed. Hormonally defined, serum-free media are shown to be essential for *in vitro* physiologic studies concerning both established kidney cell lines and primary kidney cultures.

Kidney Cell Culture in Serum-Supplemented Medium: MDCK and LLC-PK1 as Model Systems

For a number of years the use of kidney epithelial cell cultures was limited to the study of virus–host relationships. Kidney epithelial cell cultures generally permit much higher levels of virus production than other cell types. The Madin Darby canine kidney (MDCK) cell line, which was established in 1958, was initially used for such viral studies. It is permissive for vaccinia, vesicular stomatitis virus (VSV), and influenza viruses (Gaush *et al.*, 1966).

The possibility that MDCK cells actually possessed the functional properties of kidney epithelial cells was first considered by Leighton *et al.* (1969). Leighton demonstrated that MDCK cells attached to tissue culture dishes in a polarized manner (i.e., the serosal surface of the cells faced toward the culture dish, and the mucosal surface faced the medium) (Leighton *et al.*, 1970). Adjacent cells formed tight junctions. At confluency the multicellular hemicysts or domes were observed. Rather than being focal lesions, the domes were shown to be groups of MDCK cells slightly elevated from the tissue culture dish, presumably due to the transport of salt and water across the cell layer (Fig. 1). Leighton's time-lapse microscopy studies indicated that dome formation was a dynamic process. Domes continuously formed and burst, such that the time-lapse pictures "resembled the bubbling surface of gently boiling oatmeal" (Leighton *et al.*, 1970). Dome formation was inhibited by ouabain (Abaza *et al.*, 1974) and stimulated by dibutyryl cAMP (Valentich *et al.*, 1979). Subsequently, dome formation was observed in epithelial cell lines originating from the kidney (Hull *et al.*, 1976), mammary gland (McGrath, 1975), and other tissues (Auersperg, 1969). Dome formation by baby mouse kidney cells in defined medium is illustrated in Fig. 2.

Electrophysiological studies indicated that dome formation by MDCK cells is indeed indicative of transepithelial solute transport.

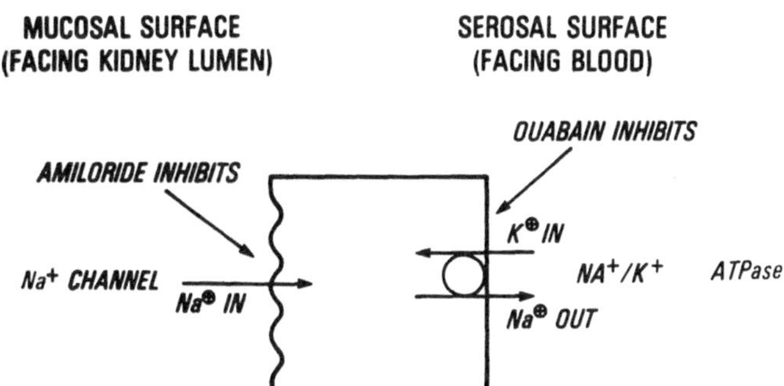

Figure 1. MDCK as a model of transporting kidney epithelial cells.

Misfeldt *et al.* (1976) grew MDCK cells to confluency on collagen-coated filter papers. Transport properties of the monolayers were then measured in a Ussing flux chamber. Madin Darby canine kidney cells in such a chamber exhibited an electrical resistance, and a potential across the cell layer, as well as transepithelial water flux (Misfeldt *et al.*, 1976; Cereijido *et al.*, 1978).

A number of biochemical studies suggested that the MDCK cells originate from the distal tubule or possibly the loop of Henle. The MDCK cells were shown to possess an arginine vasopressin-sensitive adenylate cyclase (Rindler *et al.*, 1979) typical of these nephron segments. Na^+ flux studies also indicated that Na^+ entered MDCK cells by means of an ATP-independent, saturable process, which was sensitive to inhibition by amiloride. Regulation of amiloride-sensitive Na^+ uptake by intracellular calcium was demonstrated in MDCK, although a hormone that initiates such regulation was not identified (Taub and Saier, 1979). The characteristics of MDCK are summarized in Table I.

The LLC-PK_1 cell line derived from pig kidney is another well-characterized kidney epithelial cell line. This cell line has been used by a number of investigators interested in proximal tubule functions. The LLC-PK_1 monolayers form multicellular domes (Hull, 1976; Mullin *et al.*, 1980) and exhibit sodium-dependent α-methylglucoside (αMG) uptake (Mullin *et al.*, 1980; Rabito and Ausiello, 1980). The αMG uptake occurs by an energy-dependent transport system, which is sensitive to inhibition by phlorizin. The phlorizin-sensitive sugar transport system in LLC-PK_1 has been localized to the brush border (Lever, 1982). A similar system has been observed in the rabbit kidney proximal tubule (Sacktor, 1977), as diagrammed in Fig. 3.

Misfeldt and Sanders (1981) have studied the electrophysiologic properties of LLC-PK_1 in a Ussing flux chamber. A potential difference

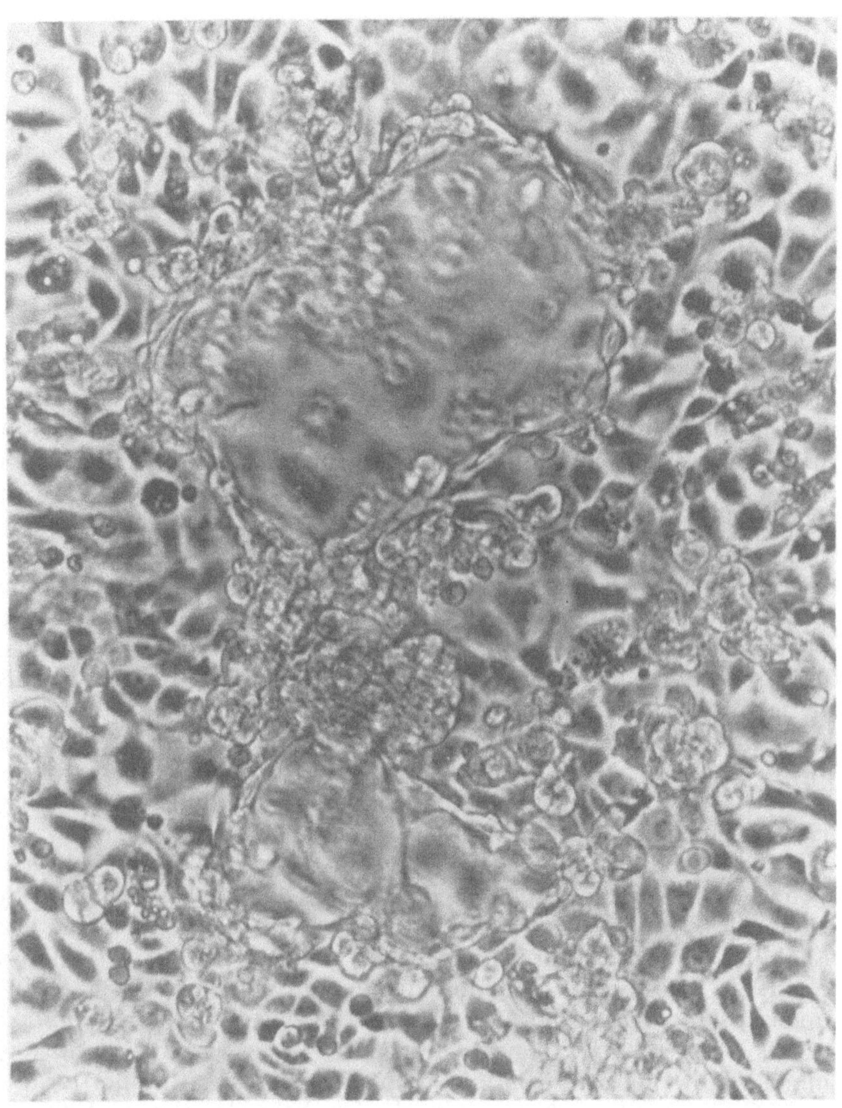

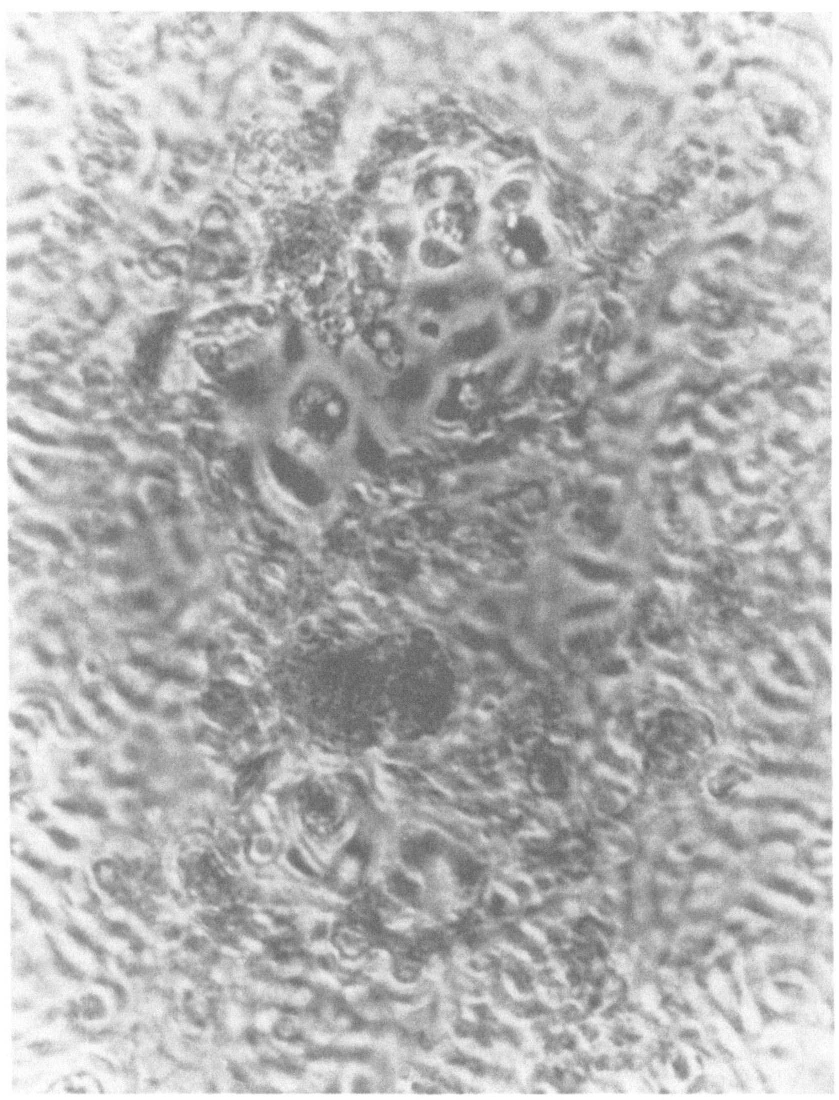

Figure 2. Dome formation by primary kidney cultures. Baby mouse kidney cells were prepared as described by Taub and Sato (1980). The cells were distributed at 5×10^4 cells per 35 mm dish into medium K-1. Domes were observed in confluent cultures. From Taub and Sato (1979).

Table I. MDCK Cells

Morphology (Leighton *et al.*, 1970)
Domes
Form polarized monolayers
Adjacent cells form tight junctions
Transport properties
Amiloride-sensitive Na^+ channel, $K_I = 5 \times 10^{-5}$ M (Rindler *et al.*, 1979b)
Furosemide-sensitive Na^+, K^+ cotransport (McRoberts *et al.*, 1982)
Transepithelial salt and water transport (Misfeldt *et al.*, 1976)
1.0 mV potential (Amiloride-sensitive, Cereijido *et al.*, 1978)
84 ohm $\times$ cm^2 resistance
Hormone responses
Arginine vasopressin-sensitive cAMP production (Rindler *et al.*, 1979a)
Norepinephrine sensitive (Taub *et al.*, 1979)
α and β receptors (Meier and Insel, 1982)
DOC receptors (Ludens *et al.*, 1978)
Growth properties
Form nodules in baby but not adult nude mice (Stiles *et al.*, 1976)
Grow in defined medium (Taub *et al.*, 1979)

(PD) of 2.8 mV and a transepithelial resistance of 211 ohm $\times$ cm^2 was measured, when D-glucose was present. Either phlorizin added to the mucosal bath or ouabain added to the serosal bath of the Ussing chamber caused a significant reduction in the PD. These observations suggest that the LLC-PK_1 cell line not only possesses a Na^+-dependent sugar transport system typical of the proximal tubule, but also exhibits transepithelial sugar transport as observed in the proximal segment of the nephron. The characteristics of LLC-PK_1 are summarized in Table II.

Advantages of Kidney Cell Culture

Since kidney cell cultures do express differentiated transport functions observed in the kidney tubule, a number of questions con-

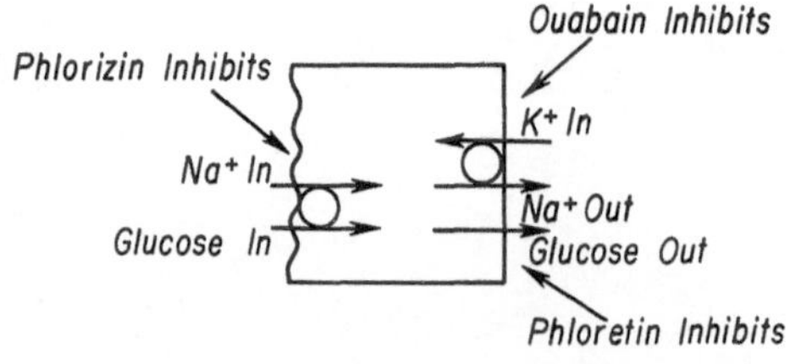

Figure 3. Model of transepithelial sugar transport by proximal tubule cells.

Table II. LLC-PK$_1$

Transport properties
Domes (Hull *et al.*, 1976)
Na^+-dependent sugar transport (Rabito and Ausiello, 1980; Mullin *et al.*, 1980)
Polarized amino acid transport (Rabito and Karish, 1982)
Transepithelial glucose and salt transport (Misfeldt and Sanders, 1981)
2.8 mV potential (phlorezin sensitive)
211 ohm × cm^2 resistance
Hormone responses (Goldring *et al.*, 1978)
Arginine vasopressin-sensitive cAMP production
Calcitonin-sensitive cAMP production
Growth properties
Do not form tumors in immunosuppressed animals (Hull *et al.*, 1976)
Grow in defined medium (Chuman *et al.*, 1982)

cerning the transport physiology of kidney cells may now be addressed *in vitro*. For example, the mechanisms underlying transepithelial salt transport may be examined, as well as the regulation of such transport processes by hormones. The role of such hormonal regulation of transport in growth regulation may be studied. Alterations in hormone responses that occur on malignant transformation of kidney cells may also be examined.

Kidney epithelial cell cultures have a number of advantages for such studies compared with *in vivo* studies or with studies that use kidney tissue taken directly from the animal. Unlike the *in vivo* situation, the cells *in vitro* have a defined environment that permits the effects of hormones on cells to be directly studied. Unlike tissue slices, large homogeneous populations of renal cells may be grown *in vitro* for biochemical studies. Somatic cell genetics may also be used to study the membrane transport of established cell lines *in vitro*. This approach was not previously possible in studies concerning renal function. Variants of MDCK that are resistant to killing by amiloride were the first example of the use of genetics to study epithelial transport *in vitro*. The amiloride-resistant variants were observed to have a decreased rate of Na^+ uptake via the Na^+ channel (Taub and Saier, 1981). Such variant cells may be used either to identify the Na^+ channel or to examine the mechanisms by which the Na^+ channel is regulated.

Kidney cell cultures also have the advantage over tissues taken directly from the animal in that the cells form a monolayer *in vitro*. Thus, the diffusion limitation problem of slices is avoided. The long-term effects of hormones on transport may be studied; the interrelationships between metabolism and transport may also be examined *in vitro*. For example, hormonal *events* mediated by cAMP may take hours to observe. Indeed, dome formation (and presumably ion transport) in MDCK cells is induced by cAMP. This induction process occurs over a 48-hour period. The cells in culture retain their viability during this extended period, and thus provide a convenient system to study the mechanism of induction of domes (and transepithelial transport) by hormones. Such *in vitro* studies, however, were previously limited by the presence of serum in the tissue culture medium.

Limitations on Transport Studies with Kidney Cell Culture Imposed by Serum

Serum has remained for many years the last undefined supplement in synthetic medium for cultured animal cells. The serum not only contains numerous hormones, but also has many other unidentified factors that may either interact synergistically or antagonistically with the hormones in the serum. Consequently, when adding hormones to the culture medium, their effects were often masked due to the serum. Furthermore, different lots of serum differ with respect to their hormone supplements. As a result, studies concerning hormonal effects with serum may be difficult to reproduce.

The removal of the serum and the addition of a number of defined supplements to the culture medium has permitted the effects of hormones on the functional properties of animal cells to be observed individually and reproducibly. The different supplements in our hormonally defined serum-free media can be manipulated, so that optimal hormone dosages can be quantitated; synergistic interactions between different supplements in the defined medium can also be observed.

The problems with serum and the importance of the use of hormonally defined serum-free medium with cultured kidney cells are illustrated with regard to the study of dome formation. A number of agents that act via cAMP increase the frequency of dome formation in MDCK cells. However, studies concerning the effects of these agents in serum-supplemented medium often are not highly reproducible. Confluent monolayers of MDCK cells may make domes at a very high frequency, or not at all, depending on the lot of serum used. Agents that induce dome formation may also have varying effects when different serum lots are used. These differences may be due to variations in the concentrations of undefined factors in serum that modulate dome formation. Such factors may mask the effects of hormones and prostaglandins on dome formation. For example, cAMP-dependent protein kinase activity was threefold higher in MDCK cells grown in hormonally defined serum-free medium (medium K-1), than in medium supplemented with serum (Taub *et al.*, 1982). Factors in serum may cause the activity of cAMP-dependent protein kinase to be reduced, ultimately preventing dome formation by such MDCK cells. Variation in the concentration of such factors in different serum lots may impede *in vitro* studies concerning the induction of dome formation.

These problems have been substantially alleviated by the use of hormonally defined medium. A serum-free medium supplemented with five factors (insulin, transferrin, T_3, hydrocortisone, and PGE_1) permits MDCK cells to grow over the long term and to form domes at confluency (Taub *et al.*, 1979). The growth rate of MDCK cells in this defined medium (medium K-1) is equivalent to the rate observed in serum-supplemented medium. All five factors in medium K-1 are required in order to obtain optimal growth, although a hormone deletion study indicated PGE_1 and transferrin affect growth of MDCK to a greater extent than the other three supplements. Prostaglandin E_1 at the physiologic dosage of 25 ng/ml, increased the growth rate and the frequency of dome formation by MDCK cells. Synergism was observed between PGE_1 and hydrocortisone in the induction of dome formation in serum-free medium. Such synergism had not been observed previously with cells in serum.

Limitations of Studies with Established Kidney Cell Lines: Importance of Primary Kidney Cell Culture

The serum has not only limited studies concerning hormonal regulation of renal functions, but has also limited the number of kidney cell culture systems that are available for study. Although a number of different segments are in the nephron, only two established kidney tubule cell lines have been studied in detail, MDCK and LLC-PK_1.

The MDCK has several properties of distal tubule cells including an arginine vasopressin-sensitive adenylate cyclase and an amiloride-sensitive Na^+ channel. However, high levels of amiloride are required to inhibit Na^+ uptake in MDCK (Ki = 1.7×10^{-5} M, Rindler *et al.*, 1979). In the distal tubule of the kidney, amiloride, at a concentration as low as 10^{-7} M, inhibits Na^+ reabsorption. Furthermore, the transepithelial resistance, which was measured (84 ohm $\times$ cm^2), was low. Thus, MDCK monolayers model "leaky epithelia," such as those present in the proximal tubule, by these criteria.

Similarly, the LLC-PK_1 cell line has an Na^+-dependent sugar transport system typical of the proximal tubule. However, observations made concerning the hormonal responses of LLC-PK_1 are at variance with the claim that this cell line is of proximal tubule origin. Arginine vasopressin and salmon calcitonin increased the cAMP content of LLC-PK_1, whereas parathyroid hormone (PTH) had no effect (Goldring *et al.*, 1978). This pattern of hormone response is typical of the medullary portion of the thick ascending limb of Henle. In contrast, PTH-sensitive cAMP production is typical of the proximal tubule. The hormone responses and transport properties of LLC-PK_1 are summarized in Table II.

Such observations do not necessarily indicate that LLC-PK_1 and MDCK differ from particular kidney tubule cell types actually present *in vivo*. A subpopulation of distal and proximal cells *in vivo* may be very similar to either MDCK or LLC-PK_1. On the other hand these cell lines may have originated from typical distal and proximal tubule cells *in vivo*, with the cell culture environment modifying the expression of their differentiated functions. The possibility that these cell lines have become genetically altered during their establishment and passage in culture also cannot be ignored.

Primary kidney cell cultures should prove to be an important means to evaluate the physiologic significance of observations made with MDCK and LLC-PK_1 cells. Primary kidney cells are taken directly from the animal, and thus have the potential of more closely resembling the cells *in vivo* than other available systems. Since the number of established kidney cell lines is small, primary kidney cell cultures may also be the only available means to study *in vitro* a number of the kidney cell types that are present in the nephron.

The serum in the cell culture medium has prevented the development of such primary cell culture systems. Serum contains many different components, which may permit the growth of many different cell types in the kidney. Particular serum lots may also contain components that are deleterious to the growth of the epithelial cells of interest. Those kidney epithelial cell lines that have become established in culture may have had to undergo a number of genetic changes, so as to survive indefinitely with serum *in vitro*. In any case, in primary kidney cultures, serum uniformly permits the extensive growth of fibroblasts so that the fibroblasts may eventually become the prevalent cell type. Primary kidney cell cultures have also exhibited extensive loss of differentiated functions (Burlington, 1959; Lieberman and Ove, 1958; Becker and Willis, 1979), which may result from the use of serum.

The possibility that fibroblast overgrowth accounts for the loss of differentiated renal functions in primary culture was examined by Gilbert and Migeon (1975). These investigators grew primary mouse kidney cultures in Dulbecco's modified Eagle's medium, in which D-valine substituted for L-valine. Epithelial cells in the kidney and other tissues contain the enzyme D-amino acid oxidase, which can convert D-amino acids to their essential L-enantimorph. However, fibroblasts lack D-amino acid oxidase and will not grow in this medium. Consequently, the D-valine medium may be used to maintain primary cultures of epithelial cells free from fibroblast overgrowth.

The growth properties and differentiated functions of the D-valine selected cells were examined. Only limited cell growth was obtained (Gilbert and Migeon, 1977), which could possibly be due to a low rate of conversion of D-valine to L-valine. The kidney primary cell cultures in D-valine medium retained alkaline phosphatase activity and the LDH isozyme profile typical of kidney cells. Carbonic anhydrase activity was also retained, although at lower levels than observed in newly isolated kidney cell suspensions. These observations suggest that decreased expression of renal functions in primary kidney cell cultures is not only the result of fibroblast overgrowth, but may also be due to the cell culture conditions.

Primary Kidney Culture and Hormonally Defined Medium

Previous problems encountered with primary kidney cell cultures can be alleviated to a significant extent by the use of hormonally defined serum-free medium. Primary cultures of kidney epithelial cells become overgrown with fibroblasts in serum-supplemented medium. However, the hormonally defined serum-free medium developed for the MDCK cell line (medium K-1) has been used to maintain primary cultures of kidney epithelial cells derived from a number of species without fibroblast overgrowth (Taub and Sato, 1980; Fig. 4).

The physiologic properties of primary cultures of baby mouse kidney epithelial cells were examined in detail. The baby mouse kidney primaries grew at 0.6 doublings per day both in medium K-1 and serum-supplemented medium. Epithelial colonies formed at low plating densities in K-1, whereas fibroblastic colonies were observed in serum-supplemented medium. A hormone deletion study indicated that each of the five supplements in medium K-1 was required for the formation of baby mouse kidney and MDCK colonies. Prostaglandin E_1 and

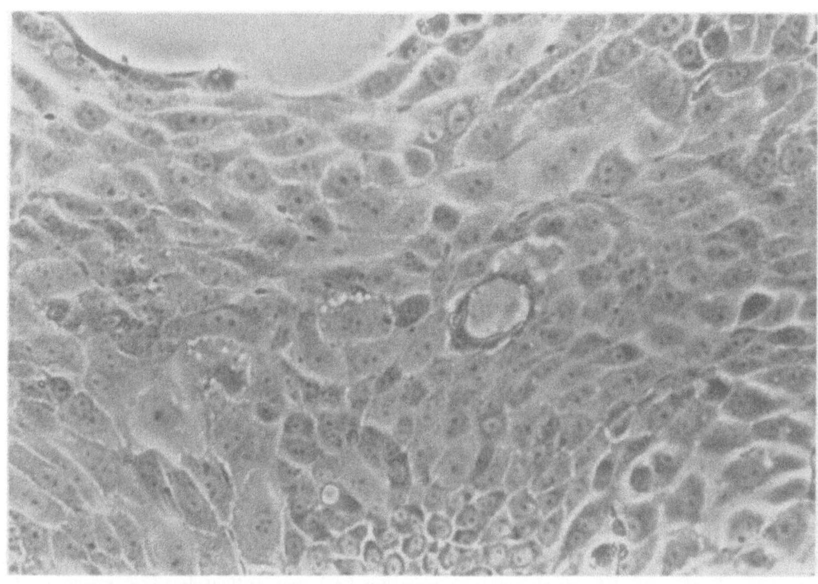

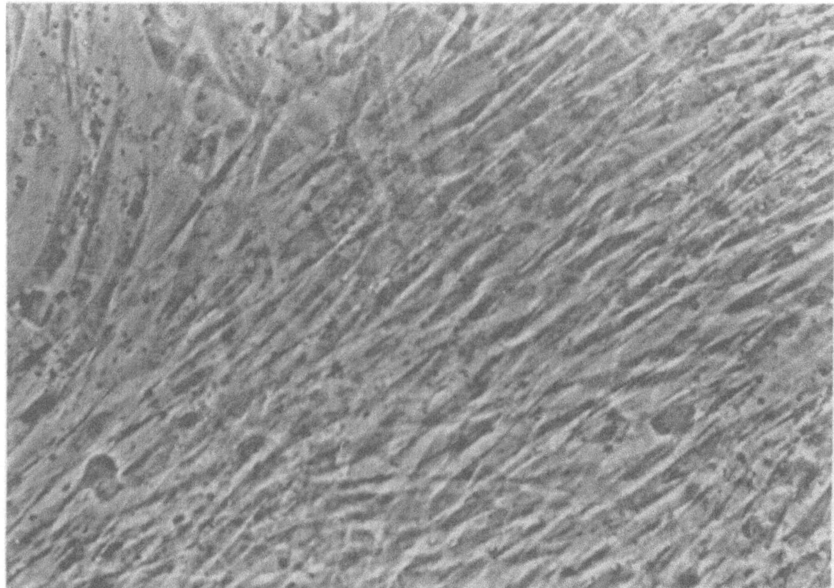

Figure 4. Primary baby mouse kidney cultures. Baby mouse kidney cells were prepared as described in Taub *et al.* (1979). The cells were distributed at 10^4 cells per 35 mm dish containing either (top) medium K-1 or (bottom) serum-free DME-F12 (50:50 mix) supplemented with 10% fetal calf serum. Eleven days later, representative microscope fields of the cultures were photographed.

transferrin had the most dramatic effect on growth (Taub and Sato, 1980). The baby mouse kidney primaries exhibited multicellular domes in medium K-1, which suggested that the monolayers contained transporting epithelial cells. Dome formation was not observed, however, in serum-supplemented medium. Amiloride-sensitive Na^+ uptake, a distal tubule marker, was observed in the baby mouse kidney primary cultures. However, Na^+-dependent α-methylglucoside uptake (a proximal tubule marker) was not detected. These studies illustrate that a hormonally defined serum-free medium developed for an established kidney epithelial cell line permits the growth of primary kidney cells with similar functional properties.

Similar observations have been made with other types of primary cell cultures in hormonally defined serum-free medium. The hormonally defined serum-free medium developed for the GH_3 line of rat pituitary carcinoma cells (Hayashi *et al.*, 1978) has been used successfully to maintain rat pituitary cells in primary culture (Bottenstein *et al.*, 1979). Similarly, the medium developed for the B104 rat neuroblastoma cell line (Bottenstein and Sato, 1978) and the medium developed for the RF-1 rat ovarian cell line (Bottenstein *et al.*, 1979) have been used to grow primary cultures of neural cells (Bottenstein *et al.*, 1980) and ovarian epithelial cells (Orly *et al.*, 1980), respectively, in the absence of fibroblast overgrowth. The survival of these cells in primary culture depended on the presence of each of the supplements in the hormonally defined medium. Furthermore, differentiated neural and ovarian functions were retained. Thus, hormonally defined media for established cell lines derived from a number of tissues may be used for appropriate primary cultures, with similar success to that obtained with primary kidney cultures.

Hormonally defined serum-free media have also been developed directly for the primary cells of interest. For example, Ambesi-Impiombato *et al.* (1980) have developed normal rat thyroid cell strains from primary cultures, using hormonally defined media. Functional testicular cell lines have also been developed from primary cultures in hormonally defined medium (Mather and Sato, 1979). Similarly, more distinct cell types from the kidney may be grown, and new functional studies may be done using hormonally defined serum-free media and primary kidney cell cultures.

Primary Culture of Specific Kidney Cell Types

Primary cell cultures have only been used to limited extent to study the functions of particular cell types in the kidney. The limited use of kidney primaries may be due to the heterogeneity of the epithelial cells in the cultures, as well as to the presence of fibroblasts. Due to the complex organization of the nephrons in the kidney, renal slices cannot be obtained that only contain a single kidney epithelial cell type. As a consequence, the cell cultures that are obtained from such slices also contain a number of different types of kidney epithelial cells.

Several approaches have been taken toward purifying and culturing the different cell types in the kidney. The different types of kidney cells may be separated from single cell suspensions based on differences in the densities of the cells. Large numbers of particular types of nephron segments may be purified based on their diameters. Alternatively, single nephron segments may be microdissected and subsequently cultured.

For example, the proximal tubule cells in a suspension of trypsinized rat kidney cells have been purified by means of Ficoll density gradient centrifugation (Kreisberg *et al.*, 1977a) or by free flow electrophoresis (Kreisberg *et al.*, 1977b). Such cells were identified as being proximal by their histochemically demonstrable alkaline phosphatase activity and their brush border. Such purified proximal tubule cells may then be cultured (Hemstreet *et al.*, 1981). However, the trypsin treatment of the tubular epithelial cells is deleterious to their subsequent attachment and growth *in vitro*. Consequently, the yield of cultured cells may be low.

The yield of cultured cells may be increased by plating purified renal tubules rather than single cells. The tubules and/or glomeruli have been purified using either sieves (Burlington and Cronkite, 1973; Quadracci and Striker, 1970) or Ficoll gradients (Torres *et al.*, 1978). Rabbit kidney tubules purified by such procedures have attached to the culture dish with high efficiency and yielded large numbers of cells.

However, the primary glomerular cultures obtained by such techniques have had several limitations. The percentage of cells in glomerular preparations that actually attach to the culture dish and grow *in vitro* is still low; furthermore the cells that finally grow from the glomerular explants may be any one of the three cell types in the glomerulus. Admittedly, cells from such primary populations may be cloned, and populations of purified cells can then be grown. However, the phenotypic stability of such cells has not been defined. The use of hormonally defined culture media should facilitate the development of primary glomerular cell culture systems (1) that are enriched for a particular glomerular cell type and (2) in which a high yield of cells is obtained from the original preparation.

The procedures described above were concerned with culturing large populations of kidney cells that are enriched for particular cell types. In order to insure that only the cell types of interest are being examined *in vitro*, the cells from microdissected nephron segments may also be cultured. Although limited material is obtained initially, investigators such as Handler and Burg have propagated large numbers of cells in culture from such microdissected nephron segments.

Established kidney cell lines that retain the differentiated functions of interest have not yet been obtained from such nephron segments. A number of such established kidney cell lines would be valuable to investigators interested in renal functions. Established cell lines derived from the three cell types in the glomerulus are not available, but would be important in studies concerning glomerular function. Kidney tubule cell lines that exhibit a high transepithelial resistance and potential are also needed for electrophysiologic studies. Established kidney cell lines developed from different nephron segments from the same animal kidney would be invaluable for comparative studies. In order to obtain such culture systems, defined methodologies for establishing particular types of kidney cell lines are needed. Investigators interested in particular renal cell types would then have the ability to develop appropriate cell lines from a kidney at will.

Use of Hormonally Defined Serum-Free Medium to Grow Specific Cell Types in the Kidney: The Concept of Selective Media

A strategy toward developing cell lines from the different nephron segments is to first optimize the primary culture conditions for the growth of particular kidney cell types. Selective media may be developed for the growth of proximal cells, distal cells, and cells from the loop of Henle. Such media could then be applied to unpurified kidney cell population to selectively grow the cells of interest.

A number of studies indicate that the development of selective media for particular cell types in the kidney is possible. The tubule epithelial cells in the different nephron segments have been shown to differ with regard to their hormonal responses, as well as their transport systems. For example, the effects of particular hormones such as parathyroid hormone and arginine vasopressin on cAMP production are only observed in restricted segments of the nephron (Morel, 1981).

Similarly, the cells in the different nephron segments have been shown to differ with regard to their growth responses to hormones and

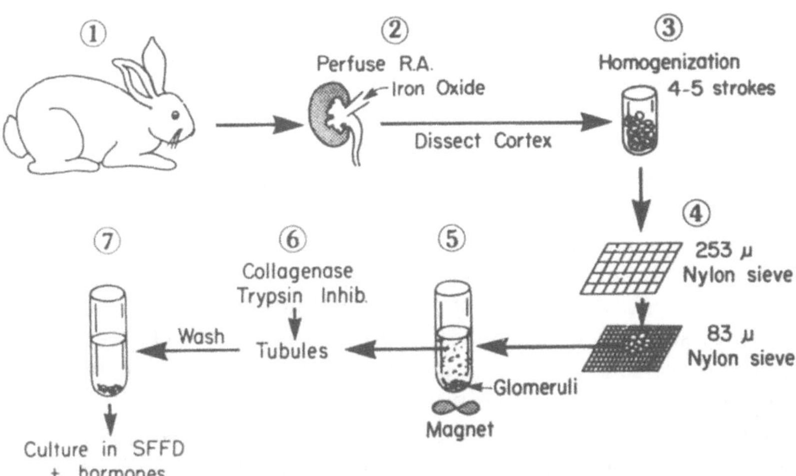

Figure 5. Preparation of primary cultures of rabbit kidney proximal tubules. Primary rabbit kidney cultures were prepared as diagrammed. Details are described in Chung *et al.* (1982).

growth factors. Primary cultures were initiated from proximal tubules purified from the rabbit kidney, as illustrated in Fig. 5 (Chung *et al.*, 1982). Rabbit kidney cultures were also initiated from unpurified nephron segments. The cells in both the proximal tubule cultures and the unpurified kidney cultures grew in response to insulin, transferrin, and hydrocortisone and assumed an epithelial morphology (Chung *et al.*, 1982). In addition, the cells in the unpurified kidney cultures also exhibited growth responsiveness to epidermal growth factor (EGF) and T_3; growth-stimulatory effects of EGF and T_3 were not observed in the proximal tubule cultures (Chung *et al.*, 1982). These studies suggested that at least two populations of epithelial cells exist in the kidney, which differ with regard to their growth responsiveness to hormones and growth factors.

In vitro growth studies with microdissected rabbit nephron segments are consistent with these observations. Horster (1979) observed that kidney cells originating from the proximal tubule, the cortical collecting duct, and the thick ascending limb of Henle, differ with regard to their *in vitro* growth requirements. However, cells from both the cortical collecting duct and the thick ascending limb of Henle exhibited DNA synthetic responses to epidermal growth factor and T_3 (Wilson and Horster, 1982). Thus, this study indicates that cells in the loop of Henle and collecting duct are among epithelial cells in the kidney that are growth responsive to epidermal growth factor, unlike proximal tubule cells.

These studies with primary rabbit kidney cells suggest furthermore that manipulation of the hormones and growth supplements in the media may result in the selective growth enhancement of particular cell types in the kidney. However, the importance of the basal media cannot be underrated in the development of selective media for the different cell types in the nephron. The metabolic capacities of the cells in the proximal tubule and the distal tubule differ as dramatically as their transport systems (Cohen and Kamm, 1976). Such differences may be expected to affect the nutritional requirements of the cells in these different nephron segments. Methodologies for the development of optimal basal media for the growth of particular types of differentiated cells have been described by Ham (1982) and by Ham and McKeehan (1979). Such methodologies may be applied for the growth of purified proximal and distal tubule cells in hormonally defined media. The choice of the appropriate substratum for cell attachment may also be important in facilitating the selective growth of the cells in the different nephron segments.

Use of Primary Kidney Cell Culture to Study Medically Related Problems of the Kidney

Primary kidney cell culture should prove to be invaluable for examining medically related problems of the kidney. In order to study such problems, kidney cell culture systems are needed that possess the normal growth and functional properties of cells originating from the different nephron segments. Such systems are needed, since particular segments of the nephron are often selectively affected in a number of medical disease states of the kidney. Nephrologists would like to define the reasons for these selective effects. Consequently, primary culture systems derived from cells in the different nephron segments are essential.

Primary cultures derived from the proximal tubule are of particular interest, since many disease states specifically affect the proximal tubule (Table III). The major form of kidney cancer, the clear cell carcinoma, is of proximal tubule origin (Oberling *et al.*, 1960). Similarly, during renal hypertrophy the proximal tubule cells grow and enlarge in size (Goss and Dittmer, 1969). In acute renal failure the first nephron segment affected is the proximal tubule (Flamenbaum, 1977). The proximal tubule has also been observed to be affected specifically in

Table III. Medical Disease Studies Affecting the Proximal Tubule

Renal hypertrophy (Lee and Attallah, 1977)
Clear cell carcinoma (Oberling *et al.*, 1960)
Acute renal failure (Flamenbaum, 1977)
Toxicity of drugs, heavy metals
Ischemia from heart failure
Immune complex diseases
Heymann nephritis (Kerjaschki and Farquhar, 1982)
Multiple myeloma (Martinez-Maldonado and Garayalde, 1981)
Inborn errors of metabolism
Cystinosis (Casky, 1981)

several immune complex diseases (Kerjaschki and Farquhar, 1982). The light chains of patients with multiple myeloma have been proposed to cause their deleterious affects on proximal tubule cells (Martinez-Maldonado and Garayalde, 1981).

Hormonally defined serum-free medium has been used as a means to grow primary cultures of rabbit proximal tubule cells (Chung *et al.*, 1982). The primary epithelial cells can be maintained for long periods without fibroblast overgrowth and they retain proximal tubule functions. Similar to the LLC-PK_1 cell line, the rabbit proximal tubule cultures possess a Na^+-dependent sugar transport system, typical of the proximal tubule. In addition, the primaries exhibit the hormone responses typical of proximal tubule cells. Cyclic AMP production was stimulated by PTH, whereas calcitonin and arginine vasopressin had no significant effect (Chung *et al.*, 1982). Consequently, the rabbit kidney proximal tubule cell cultures bear a closer resemblance to proximal cells *in vivo* than the LLC-PK_1 cell line. Thus, the primary cultures, rather than LLC-PK_1, can be used to study the effects of hormones on renal growth and transport functions.

Such characteristics are particularly important for studies concerning growth regulation in the kidney. In the normal physiologic state, hormones and other nutrient factors may regulate kidney epithelial cell growth. During renal hypertrophy, the cells in the proximal tubule of the kidney grow selectively. In order to understand this selective growth, an understanding of the *in vitro* growth requirements of the cells in the different nephron segments is important.

Several renal growth factors and hormones have been implicated as altering the growth and functional properties of kidney cells in such renal disorders. For example, renotrophin, a hypertrophy factor produced following nephrectomy, has been proposed to cause the growth stimulation in the proximal tubule (Lee and Attallah, 1977). Renotrophin has not yet been extensively purified, nor have detailed studies been done concerning the effects of this hypertrophy factor on kidney cells. The existence of another kidney-specific factor, natriuretic hormone, which regulates sodium reabsorption by the kidney, has also been suggested by a number of studies (Licht *et al.*, 1977). Cell culture systems derived from the cells in the different nephron segments are essential in the purification of such renal factors and in the determination of their effects on the tubule cells. The development of primary kidney cell culture systems using hormonally defined medium is, then, an important means of studying such medically related problems of the kidney.

References

Abaza, N. A., Leighton, J., and Schultz, S., 1974, Effects of ouabain on the function and structure of a cell line (MDCK) derived from canine kidney, *In Vitro* **10:**172–183.

Ambesi-Impiombato, F. S., Parks, L. A. M. and Coon, H. G., 1980, Culture of hormone-dependent functional epithelial cells from rat thyroids, *Proc. Natl. Acad. Sci. U.S.A.* **77:**3455–3459.

Auersperg, N., 1969, Histogenetic behavior of tumors. I. Morphologic variation *in vitro* and *in vivo* of two related human carcinoma cell lines, *J. Natl. Cancer Inst.* **43:**151–173.

Becker, J. H., and Willis, J. S., 1979, Properties of Na-K pump in primary cultures of kidney cells, *J. Cell. Physiol.* **99:**427–440.

Bottenstein, J. E., and Sato, G. H., 1978, Growth of a rat neuroblastoma cell line in serum-free supplemented medium, *Proc. Natl. Acad. Sci. U.S.A.* **76:**514–517.

Bottenstein, J. E., Hayashi, I., Hutchings, S., Masui, H., Mather, J., McClure, D. G., Ohasa, S., Rizzino, A., Sato, G., Serrero, G., Wolfe, R., and Wu, R., 1979, The growth of cells in serum-free hormone-supplemented media, in: *Methods in Enzymology*, Volume 45 (W. B. Jakoby, and I. H. Pastan, eds.), Academic Press, New York, pp. 94–109.

Bottenstein, J. E., Skaper, S. D., Varon, S. S., and Sato, G. H., 1980, Selective survival of neurons from chick embryo sensory ganglionic dissociates utilizing serum-free supplemented medium, *Exp. Cell Res.* **125:**183–190.

Burlington, H., 1959, Enzyme patterns in cultured kidney cells, *Am. J. Physiol.* **197:**68–70.

Burlington, H., and Cronkite, E. P., 1973, Characteristics of cell cultures derived from renal glomeruli, *Proc. Soc. Exp. Biol. Med.* **142:**143–149.

Casky, C. T., 1981, Inherited biochemical defects affecting the kidney, in: *The Kidney in Systemic Disease* (W. N. Suki and G. Eknoyan, eds.), John Wiley and Sons, New York, pp. 627–650.

Cereijido, M., Robbins, E. S., Dolan, W. J., Rotunno, C. A., and Sabatini, D. D., 1978, Polarized monolayers formed by epithelial cells on a permeable and translucent support, *J. Cell Biol.* **77:**853–880.

Chuman, L., Fine, L. G., Cohen, A. H., and Saier, M. H., 1982, Continuous growth of proximal tubular kidney epithelial cells in hormone-supplemented serum-free medium, *J. Cell Biol.* **94:**506–510.

Chung, S. D., Alavi, N., Livingston, D., Hiller, S., and Taub, M., 1982, Characterization of primary rabbit kidney cultures that express proximal tubule functions in a hormonally defined medium, *J. Cell Biol.* **95:**118–126.

Cohen, J. J., and Kamm, D. E., 1976, Renal metabolism: Relation to renal function, in: *The Kidney*, Volume I (B. M. Brenner and F. C. Rector, eds.), W. B. Saunders, Philadelphia, pp. 126–213.

Flamenbaum W., 1977, Pathophysiology of acute renal failure, in: *Pathophysiology of the Kidney* (N. A. Kurtzonan and M. Martinez-Maldonado, eds.), Charles C Thomas, Springfield, Illinois, pp. 795–841.

Gaush, C. R., Hard, W. L., and Smith, T. F., 1966, Characterization of canine kidney cells (MDCK), *Proc. Soc. Exp. Biol. Med.* **122:**931–935.

Gilbert, S. F., and Migeon, B. R., 1975, D-valine as a selective agent for normal human and rodent epithelial cells in culture, *Cell* **5:**11–17.

Gilbert, S. F., and Migeon, B. R., 1977, Renal enzymes in kidney cells selected by D-valine medium, *J. Cell. Physiol.* **92:**161–168.

Goldring, S. R., Dayer, J. M., Ausiello, D. A., and Krane, S. M., 1978, A cell strain cultured from porcine kidney increases cyclic AMP content upon exposure to calcitonin or vasopressin, *Biochem. Biophys. Res. Commun.* **83:**434–440.

Goss, R. S., and Dittmer, J. E., 1969, Compensatory renal hypertrophy: Problems and prospects, in: *Compensatory Renal Hypertrophy* (W. W. Nowinski and R. S. Goss, eds.), Academic Press, New York, pp. 299–307.

Ham, R. G., 1982, Importance of the basal nutrient medium in the design of hormonally defined medium, in: *Cold Spring Harbor Conferences on Cell Proliferation*, Volume 9, Book A, (G. H. Sato, A. B. Pardee, and D. A. Sirbasku, eds.), Cold Spring Harbor Laboratory, New York, pp. 39–60.

Ham, R. G., and McKeehan, W. L., 1979, Media and growth requirements, *Methods Enzymol.* **58:**44–93.

Handler, J. S., Perkins, F. M., and Johnson, J. P., 1980, Studies of renal cell functions using cell culture techniques, *Amer. J. Physiol.* **238:**F1–F9.

Hayashi, I., Larner, J., and Sato, G., 1978, Hormonal growth control of cells in culture, *In Vitro* **14:**23–30.

Hemstreet, G. P., Enoch, P. G., Fine, P. R., and Wheat, R., 1981, Lipid A induction of cytotoxic antibody to cultured syngeneic rat kidney tubular cells, *Kidney Int.* **19:**275–280.

Horster, M., 1979, Primary culture of mammalian nephron epithelia. Requirements for cell outgrowth and proliferation from defined explanted nephron segments, *Pflüg. Arch. Eur. J. Physiol.* **382:**209–215.

Hull, R. N., Cherry, W. R., and Weaver, G. W., 1976, The origin and characteristics of a pig kidney strain LLC-PK_1, *In Vitro* **12:**670–677.

Kerjaschki, D., and Farquhar, M. G., 1982, The pathogenic antigen of Heymann nephritis is a membrane glycoprotein of the renal proximal tubule brush border, *Proc. Natl. Acad. Sci. U.S.A.* **79:**5557–5561.

Kreisberg, J. I., Pitts, A. M., and Pretlow, T. G., 1977a, Separation of proximal tubule cells from suspensions of rat kidney cells in density gradients of Ficoll in tissue culture medium, *Am. J. Patho.* **86:**591–602.

Kreisberg, J. I., Sachs, G., Pretlow, T. G., and McGuire, R. A., 1977b, Separations of proximal tubule cells from suspension of rat kidney cells by free-flow electrophoresis, *J. Cell. Physiol.* **93:**169–172.

Lee, S. B., and Attallah, A. H., 1977, The renal prostaglandins, in: *Pathophysiology of the Kidney* (N. A. Kurtzmann and M. Martinez-Maldonado, eds.), Charles C Thomas, Springfield, Illinois, pp. 473–505.

Leighton, J., Brada, Z., Estes, L. W., and Justh, G., 1969, Secretory activity and oncogenicity of a cell line (MDCK) derived from canine kidney, *Science* **158:**472–473.

Leighton, J., Estes, L. W., Mansukhani, S., and Brada, Z., 1970, A cell line derived from normal dog kidney (MDCK) exhibiting qualities of papillary adenocarcinoma and of renal tubular epithelium, *Cancer* **26:**1022–1028.

Lever, J., 1982, Expression of a differentiated transport function in apical membrane vesicles isolated from an established kidney epithelial cell line, *J. Biol. Chem.* **257:**8680–8686.

Licht, A., Fine, L. G., and Bourgoigne, J. J., 1977, Natriuretic factor, a lasting enigma, in: *Contributions to Nephrology*, Volume 13: *Nonvasoactive Renal Hormones* (G. M. Eisenbach and J. Brod, eds.), S. Karger, Basel. pp. 3–11.

Lieberman, I., and Ove, P., 1958, Enzyme activity levels in mammalian cell culture, *J. Biol. Chem.* **233:**634–636.

Ludens, J. H., Vaughn, D. A., Mawe, R. C., and Fanestil, D. D., 1978, Specific binding of deoxycorticosterone by canine kidney cells in culture, *J. Steroid Biochem.* **9:**17–21.

Martinez-Maldonado, M., and Garayalde, G., 1981, Renal involvement in multiple myeloma, in: *The Kidney in Systemic Diseases* (W. N. Suki and G. Eknoyan, eds.), John Wiley, New York, pp. 197–209.

Mather, J. P., 1980, Establishment and characterization of two distinct mouse testicular epithelial cell lines, *Biol. Reprod.* **23:**243.

Mather, J. P., and Sato, G. H., 1979, The use of hormone-supplemented serum-free media in primary cultures, *Exp. Cell Res.* **124:**215.

McGrath, C. M., 1975, Cell organization and responsiveness to humans *in vitro*: Genesis of domes in mammary cell cultures, *Am. Zool.* **15:**231.

McRoberts, J. A., Erlinger, S., Rindler, M. J., and Saier, M. H., 1982, Furosemide-sensitive salt transport in the Madin-Darby Canine Kidney cell line, *J. Biol. Chem.* **254:**2260–2266.

Meier, K. E., and Insel, P. A., 1982, Clonal variation in the expression of catecholamine receptors in MDCK cells, *J. Cell Biol.* **95:**4169a.

Misfeldt, D. S., and Sanders, M. J., 1981, Transepithelial transport in cell culture: D-glucose transport by a pig kidney cell line ($LLC-PK_1$), *J. Mem. Biol.* **59:**13–18.

Misfeldt, D. S., Hamamoto, S. T., and Pitelka, D. R., 1976, Transepithelial transport in cell culture, *Proc. Natl. Acad. Sci. U.S.A.* **73:**1212–1216.

Morel, F., 1981, Sites of hormone action in the mammalian nephron, *Am. J. Physiol.* **240:**F159–F164.

Mullin, J. M., Weibel, J., Diamond, L., and Kleinzeller, A., 1980, Sugar transport in the $LLC-PK_1$ renal epithelial cell line: Similarity to mammalian kidney and influence of cell density, *J. Cell. Physiol.* **104:**375–389.

Oberling, C., Riviere, M., and Haguenau, F., 1960, Ultrastructure of the clear cells in renal carcinomas and its importance for the demonstration of their renal origin, *Nature* **186:**402–403.

Orly, J., Sato, G., and Erickson, G. F., 1980, Serum suppresses the expression of hormonally induced functions in cultured granulosa cells, *Cell* **20:**817–827.

Quadracci, L. J., and Striker, G. E., 1970, Growth and maintenance of glomerullar cells *in vitro*, Proc. Soc. Exp. Biol. Med. **135:**947–950.

Rabito, C. A., and Ausiello, D. A., 1980, Na^+ dependent sugar transport in a cultured epithelial cell line from pig kidney, *J. Mem. Biol.* **54:**31–38.

Rabito, C. A., and Karish, M. V., 1982, Polarized amino acid transport by an epithelial cell line of renal origin ($LLC\text{-}PK_1$), *J. Biol. Chem.* **257**:6802–6808.

Rindler, M. J., Chuman, L. M., Shaffer, L., and Saier, M. H., 1979a, Retention of differentiated properties in an established dog kidney epithelial cell line (MDCK), *J. Cell Biol.* **81**:635–648.

Rindler, M. J., Taub, M., and Saier, M. H., 1979b, Uptake of $^{22}Na^+$ by cultured dog kidney cells (MDCK), *J. Biol. Chem.* **254**:11431–11433.

Sacktor, B., 1977, The brush border of the renal proximal tubule and the intestinal mucosa, in: *Mammalian Cell Membranes*, Volume 4, *Membranes and Cellular functions* (G. A. Jamieson and D. M. Robinson, eds.), Butterworths, London, pp. 221–254.

Stiles, C. D., Desmond, W., Chuman, L. M., Sato, G., and Saier, M. H., 1976, Relationship of cell growth behavior *in vivo* to tumorigenicity in athymic nude mice, *Cancer Res.* **36**:3300–3305.

Taub, M., and Saier, M. H., 1979, Regulation of $^{22}Na^+$ uptake by calcium in an established kidney epithelial cell line, *J. Biol. Chem.* **254**:11440–11444.

Taub, M., and Saier, M. H., 1981, Amiloride-resistant Madin-Darby Canine Kidney (MDCK) cells exhibit decreased cation transport, *J. Cell Physiol.* **106**:191–199.

Taub, M., and Sato, G. H., 1979, Growth of kidney epithelial cells in hormone-supplemented serum-free medium, *J. Supranol. Struct. Cell. Biochem.* **11**:207–216.

Taub, M., and Sato, G., 1980, Growth of functional primary cultures of kidney epithelial cells in defined medium, *J. Cell. Physiol.* **105**:369–378.

Taub, M., Chuman, L., Saier, M. H., and Sato, G., 1979, Growth of Madin-Darby Canine Kidney epithelial cell (MDCK) line in hormone-supplemented serum-free medium, *Proc. Natl. Acad. Sci. U.S.A.* **76**:3338–3342.

Taub, M., Devis, P., and Hiller, S., 1984, Madin Darby Canine Kidney variant cells have altered cyclic AMP metabolism and altered responsiveness to PGE_1 in a hormonally defined medium, *J. Cell. Biochem.*, submitted.

Torres, V. E., Northrup, T. E., Edwards, R. M., Shah, S. V., and Dousa, T. P., 1978, Modulation of cyclic nucleotides in isolated rat glomeruli, *J. Clin. Invest.* **62**:1334–1343.

Valentich, J. D., Tchao, R., and Leighton, J., 1979, Hemicyst formation stimulated by cyclic AMP in dog kidney cell line MDCK, *J. Cell. Physiol.* **100**:291–304.

Wilson P. O., and Horster, M. F., 1982, Differential response to hormones of defined distal nephron epithelia in culture, *Am. J. Physiol.* **244**:C166–C174.

Chapter 7

Rat Hepatocytes in Culture

A Model for Studies of Growth Control during Experimental Chemical Hepatocarcinogenesis

IZUMI HAYASHI and BRIAN I. CARR

Introduction

The rat liver offers some unique characteristics for the study of cellular functions and tumorigenesis. These characteristics include the presence of a highly differentiated mass of cells capable of a multiplicity of chemical processes, as well as a capacity for synchronous regeneration in response to two-thirds partial hepatectomy. Liver tumors have been produced in response to a large number of chemical carcinogens of different structural types. Although much has been learned about the interactions of these carcinogens with cellular macromolecules, mechanisms by which carcinogens cause alterations in the control of growth are largely unknown. There are several regimens for the production of experimental hepatocellular carcinomas using defined chemical carcinogens. Some of these employ chronic exposure of the animal to a carcinogen (Preussmann, 1978; Wogan, 1976), and others employ a limited exposure of a carcinogen followed by a promoting influence (Pitot and Sirica, 1980; Forber, 1978, 1979). The regenerative capacity of the liver provided many investigators over the past thirty years with a tool to investigate the regulatory mechanisms involved in the control of cellular growth. There are other adult mammalian tissues that generate dividing cells, such as epidermis, the bone marrow, and the reproductive organs. In contrast to these tissues that renew their

IZUMI HAYASHI and BRIAN I. CARR • Division of Cytogenetics and Cytology and Department of Medical Oncology, City of Hope National Medical Center, Duarte, California 91010.

population throughout most of the adult life, hepatocytes are in a quiescent state and will divide only under special situations. An *in vivo* experimental model for such conditions is the partial hepatectomy.

Following a two-thirds partial hepatectomy of the adult rat liver, normally nonproliferating hepatocytes of the liver residue will undergo a burst of DNA synthesis and subsequent cell division. Within 20 days of the operation, the original tissue mass is restored (Higgins and Anderson, 1931). There are two general hypotheses concerning the mechanisms of hepatic growth after a partial hepatectomy: (1) removal of a suppressor agent (chalone) for hepatocyte proliferation (Saetren, 1956; Weiss and Kavanau, 1957), and (2) the activation of, or an increase in, hepatotrophic substance(s) (Bucher and Malt, 1971). Although persistent efforts have been made to identify hepatic chalones (Deschamps and Verly, 1975; Cook *et al.*, 1982), the nature of the hepatic suppressor agent is not clear, and the idea of negative regulation itself seems to be greeted with certain skepticism. This is probably due to the difficulties in establishing a good assay system for the detection of an inhibitory activity from crude starting materials.

There has been some progress in the characterization of the hepatotrophic substances. *In vivo* experiments in the last three decades have revealed insulin and glucagon as necessary stimulants for liver regeneration (Bucher and Swaffield, 1976). Using primary cultures of hepatocytes, epidermal growth factor (EGF) was found to be stimulatory for the induction of DNA synthesis (Richman *et al.*, 1976). The powerful stimulatory action of EGF on hepatocytes both *in vivo* (Bucher and Wands, 1977) and *in vitro* allows this growth factor to be used as a convenient tool for the study of hepatocyte growth control. Although a physiological role of EGF on hepatocytes has not been established *in vivo*, its activity at a nanomolar range suggests a possible physiological activity.

In the last decade, a collagenase perfusion technique for the dissociation of hepatocytes has allowed the highly reproducible establishment of primary monolayer cultures of rat hepatocytes with high viability that retain many normal functions (Seglen, 1976). Using such cultures, studies of hepatocyte growth regulation under the more defined conditions of the *in vitro* system have become possible. Unfortunately, the adult hepatocytes in primary cultures do not undergo cell division under culture conditions generally employed, and most of the *in vitro* work has focused on the controls of the induction of DNA synthesis (Strain *et al.*, 1982; Friedman *et al.*, 1981; Tomita *et al.*, 1981). Proliferation of hepatocytes in culture has been reported in the past (Leffert *et al.*, 1977; Takasoka *et al.*, 1975; Evans *et al.*, 1952). However, the specific markers for parenchymal hepatocyte were either not shown in these cultures, or when the biochemical analyses were done, they were not correlated with individual replicating cells. Since the cultures of primary hepatocytes contain nonparenchymal cells as contaminants, these experiments have not been universally accepted as proof for hepatocyte proliferation in cultures. Recently several groups have observed the proliferation of adult hepatocytes in low density cultures (Enat *et al.*, 1982; Michalopoulos *et al.*, 1982; Nakamura *et al.*, 1983) that may further the possibilities of the primary hepatocyte culture system in studying the regulation of liver growth *in vitro*. Some excellent reviews are available for past *in vivo* studies (Bresnick, 1971) and on the recent advances utilizing the *in vitro* culture system (Leffert *et al.*, 1978). In this chapter, we discuss some of our work on the control of DNA synthesis of normal hepatocytes in defined culture conditions, and the alterations in growth control during experimental chemical hepatocarcinogenesis.

Control of DNA Synthesis in Cultures of Adult Rat Hepatocytes

Factors Required for the Induction of DNA Synthesis

When single cell suspensions of rat hepatocytes obtained by the collagenase perfusion technique are inoculated into culture dishes, they attach within 3 h and form a monolayer. Since the addition of 10% calf serum, insulin (10 μg/ml) and hydrocortisone (10^{-6} M) enhance the attachment and subsequent survival of hepatocytes (McGowan *et al.*, 1981), they are added to the culture medium for the initial 3 h. At 3 h of culture the medium is changed to serum-free conditions. Unless otherwise stated, medium is changed daily thereafter during the course of an experiment. After a lag period of 24 h, DNA synthesis is induced in the presence of insulin, glucagon, and EGF (three factors) and reaches a peak at 72 to 96 h. In the absence of three factors, little DNA synthesis is observed. The timing of the induction and the appearance of the peak of DNA synthesis is not changed greatly with the culture conditions. However, the height of the peak of DNA synthesis may be increased by the addition of as little as 0.1% serum. At concentrations above 1%, the predominant action of human or rat serum on the DNA synthesis of normal rat hepatocytes is inhibitory (Fig. 1). In order to examine the nature of this additional activity in serum, we have started to survey the effects of some known hormones and factors on the DNA

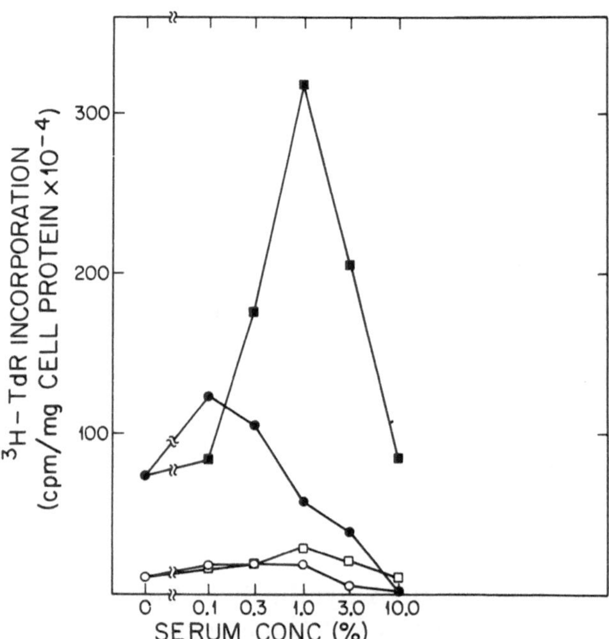

Figure 1. The effect of platelet depletion. Normal human serum (○, ●) and the serum prepared from platelet-depleted plasma (□, ■) were examined for their effect on the stimulation of normal adult rat hepatocyte DNA synthesis. Open symbols: DMESF; closed symbols: DMESF + three factors. Freshly obtained blood from normal volunteers was clotted to make normal serum. For platelet-poor serum, the blood was collected in ACD buffer, centrifuged immediately to remove cells and platelets, and incubated at 56 °C for 30 minutes to induce fibrin clotting.

synthesis under serum-free conditions. Table I lists the factors and their concentrations that we have thus far examined. Prolactin, 3,3′,5-triiodothyronine, glycyl-histidyl-lysine, and fibroblast growth factor produced some modest stimulation, but the extent of stimulation by any of these substances in the presence of three factors cannot account for the stimulation observed by the addition of serum. Most factors, however, had little effect (Table I).

DNA synthesis of hepatocytes in culture shows strong density dependence (Fig. 2, also Friedman *et al.*, 1981). The suppression of DNA synthesis at higher cell densities may be due to the production of chalonelike substance(s) by the cultured cells or may be due to the suboptimal condition of the culture. The gradual increase in DNA synthesis with the increase in inoculum size, on the other hand, suggests

Table I. The Effect of Hormones and Growth Factors on the DNA Synthesis of Normal Adult Rat Hepatocytes

Factor	Concentration	% Stimulation of DNA-S[a,b]
Prolactin	0.01–10 μg/ml	96/37
Growth hormone	0.1–10 μg/ml	NT[c]/none
Thyroid stimulating hormone	0.1–10 μgml	NT/none
Luteinizing hormone	0.1–10 μg/ml	NT/none
Somatostatin	0.1–100 ng/ml	48/19
Thyrotropin releasing hormone	1–100 ng/ml	NT/14
Luteotropin releasing hormone	1–100 ng/ml	NT/6
3,3′,5-triiodothyronine	10^{-12}–10^{-7} M	37/24
17β-estradiol	0^{-11}–10^{-7} M	NT/none
Prostaglandin E	1–50 ng/ml	NT/14
Gastrin	1–150 ng/ml	NT/none
7 S nerve growth factor	0.1–50 ng/ml	59/17
Multiplication stimulating activity	0.5–50 ng/ml	48/6
Fibroblast growth factor	1–100 ng/ml	67/83
Endothelial cell growth factor	1–100 μg/ml	NT/none
Platelet-derived growth factor	0.01–1 μg/ml	−23/−76
Interleukin II	1–100 U	NT/none
Transferrin	5–50 μg/ml	NT/none
Glycyl-histidyl-lysine	0.01–5 μg/ml	67/32

[a] % increase over DMESF.
[b] % increase over DMESF + three factors.
[c] NT = not tested.

that there may be a conditioning of culture medium by hepatocytes. The possibility that the cells are making their own "endogenous" stimulant is supported by the results of medium change versus no medium change experiments (Carr and Hayashi, 1983; Friedman *et al.*, 1981). In this type of experiment, DNA synthesis of hepatocytes in cultures in which the medium was not changed during the course of an experiment (four days) to allow the accumulation of the possible cell product was much higher than in the culture in which medium was changed daily (Fig. 2). The concept of an endogenous hepatocyte stimulant for DNA synthesis derives further support from another experiment in which the effect of the conditioned medium was examined. The cell-free conditioned medium, prepared after four days of undisturbed serum-free culture, stimulates the DNA synthesis of fresh hepatocyte cultures in a dose-dependent manner (Fig. 3). The fact that the addition of 20% conditioned medium alone can replace the effect of three factors suggests that one of the functions of three factors is to

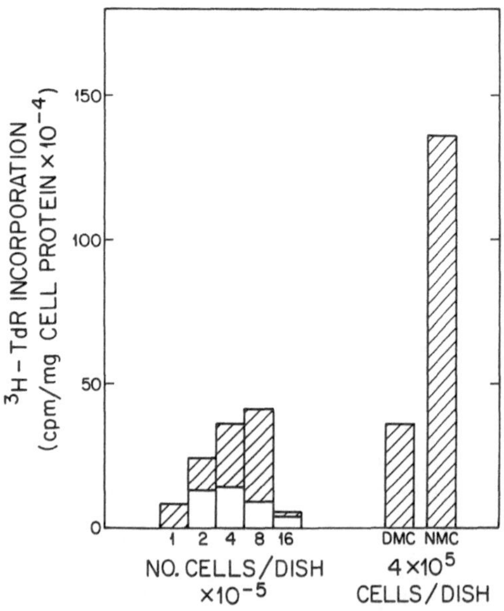

Figure 2. The effect of initial plating density and medium change on the DNA synthesis of normal adult rat hepatocytes.

The cells were plated at the densities indicated in the figure. Open bars: serum-free DME medium (DMESF); closed bars: DMESF + three factors (insulin, 0.2 μg/ml; glucagon 0.2 μg/ml; EGF, 10 ng/ml). DMC: the cultures were medium-changed daily; NMC: medium was changed to experimental condition at 3 h after the initiation of culture, and at the time of addition of [^{3}H]thymidine (72 h). No medium change was done between 3 and 72 h.

All of the experiments described in this chapter were performed using hepatocytes from adult male Fischer rats, weight 180 to 220 g. Single-cell suspensions of hepatocytes obtained by collagenase perfusion of the liver were collected by centrifugation, resuspended in the plating medium (DMESF supplemented with 10% calf serum, 10 μg/ml insulin, and 10^{-7} M hydrocortisone), and plated at 3×10^5 cells (unless otherwise indicated) into 35 mm Falcon tissue culture dishes. At 3 h, medium was changed to experimental conditions. Medium was changed daily. At 72 h, [^{3}H]thymidine (5 μCi/2 ml of medium/dish) was added. At 96 h, medium was removed, and the cells were freed from the dish with a rubber policeman, collected in glass centrifuge tubes, and washed twice with phosphate-buffered saline. Following incubation in 10% TCA for one hour, the cells were hydrolyzed in 1 ml of 1 N NaOH overnight at 37 °C. Protein concentration was determined using the Coomassie method (Sedmak and Grossberg, 1977), and the DNA synthesis was assayed by counting the [^{3}H]thymidine incorporated into acid-precipitable material. The results are expressed as cpm/mg of cell protein.

enhance the production of an endogenous stimulant. Whether the nature of this stimulant activity in the conditioned medium is the same as that found in serum and whether it could be replaced by the addition of a known, purified factor await further investigation.

Platelet-Derived Growth Factor as an Inhibitor of Normal Hepatocyte DNA Synthesis

In the course of examining the effects of hormones and factors for their effects on DNA synthesis in normal hepatocytes, it was observed

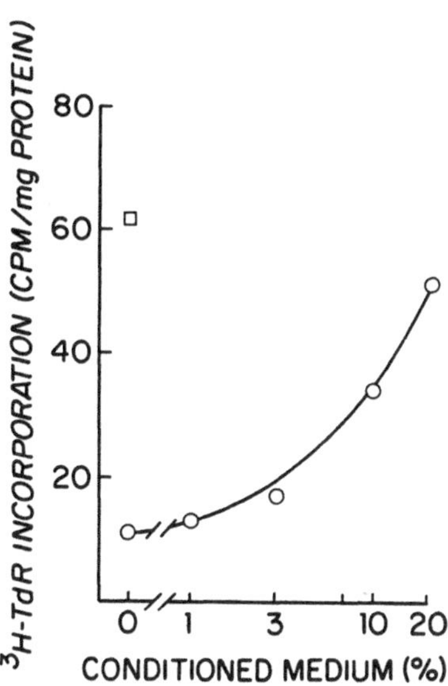

Figure 3. Effect of conditioned medium. The effect of conditioned medium on the stimulation of DNA synthesis in normal rat hepatocytes was examined. —O—: DMESF + conditioned medium; □: DMESF + three factors. The culture medium was changed daily. The conditioned medium was prepared as follows: Normal hepatocytes were cultured at optimal density in DMESF + three factors. On day 3, the culture medium was collected, centrifuged to remove floating cells, sterilized through a Millipore filter (0.2μm), and stored frozen until use. The addition to DMESF of the three factors equivalent to the amount in 20% conditioned medium does not stimulate DNA synthesis above the basal level.

that the addition of platelet-derived growth factor (PDGF) (Ross *et al.*, 1974; Antoniades *et al.*, 1979) caused a marked decrease in the DNA synthesis. Preparations of PDGF* inhibited the three-factor-induced DNA synthesis at concentrations that were 20-fold less than that required for the stimulation of DNA synthesis in 3T3 cells (C.D. Stiles, personal communication). Pure PDGF at 1 ng/ml inhibited the three-factor-induced normal hepatocyte DNA synthesis by 60%. This inhibitory effect of PDGF may well account for the strong inhibitory effect of human or rat serum. The stimulatory effect of serum on DNA synthesis is observed at concentrations as low as 0.1%, but this stimulation becomes overshadowed as the concentration of serum is increased; at 10%, the stimulation of DNA synthesis induced by the three factors is completely abolished. In order to investigate the nature of the inhibitory activity of serum, the effect of serum prepared from platelet-poor plasma (Pledger *et al.*, 1977) was examined. As shown in Fig. 1, the removal of platelets from human plasma prior to clot formation greatly reduced the inhibitory activity and unmasked the stimulatory effect of the serum. The residual inhibition at higher concentrations of platelet-poor serum may be due to an incomplete removal of PDGF or to other additional

* A preparation of PDGF and the pure PDGF were kind gifts of Dr. C. D. Stiles, Sidney Farber Cancer Institute, Boston, Massachusetts. A PDGF preparation from Collaborative Research was also used.

inhibitory substances in the serum. The result presented in Fig. 1 strongly indicates that the nature of the inhibitory activity in serum is largely due to PDGF.

The effect of PDGF on hepatocytes is the first known inhibitory effect of this growth factor in any system. Although an action of PDGF has not hitherto been observed except in cells of the connective tissue origin (Kohler and Lipton, 1974; Heldin *et al.*, 1977; Scher *et al.*, 1979), the effect of extremely low concentrations of PDGF in inhibiting normal hepatocyte DNA synthesis suggests a possible physiological role of PDGF on hepatocytes. How does PDGF inhibit DNA synthesis? There are several possible mechanisms. Platelet-derived growth factor may inhibit the actions of all or one of the three factors directly. Or PDGF may inhibit hepatocyte DNA synthesis through an independent pathway. In order to investigate these questions, some "window" experiments were performed to identify the requirements for the timing and the duration of the presence of three factors and PDGF. Figure 4 shows that the three factors need to be present from 0 to 72 h of culture for the induction of DNA synthesis. In contrast, PDGF needs to be present only from 48 to 96 h for the complete inhibition of DNA synthesis that was induced by the continuous presence of three factors. Although careful studies using pure factors are necessary, these results suggest that a direct interference of PDGF with the three factors at the cell surface level is unlikely.

Stimulatory Activity for Normal Hepatocyte DNA Synthesis in the Serum from Cancer Patients and Carcinogen-Fed Rats

Since the adult hepatocyte proliferation *in vivo* is observed both on partial hepatectomy and during experimental chemical hepatocarcinogenesis, it was of interest to examine the effect of sera from partially hepatectomized rats and from rats treated with chemical hepatocarcinogens. Since it has inhibitory actions, serum was examined for a DNA synthesis-stimulating activity at concentrations with minimum interference by the inhibitory activity (0.1% to 0.5%). As shown in Fig. 5, the sera from rats chronically treated with the hepatocarcinogen, 2-acetylaminofluorene (2-AAF), are increasingly stimulatory for normal hepatocyte DNA synthesis. The stimulatory activity of 2-AAF serum is already apparent at one week after the initiation of 2-AAF administration, and reaches a peak at three weeks. These results indicate that a stimulant activity appears in the blood at a very early time in hepatocarcinogenesis and long before the appearance of hyperplastic nodules.

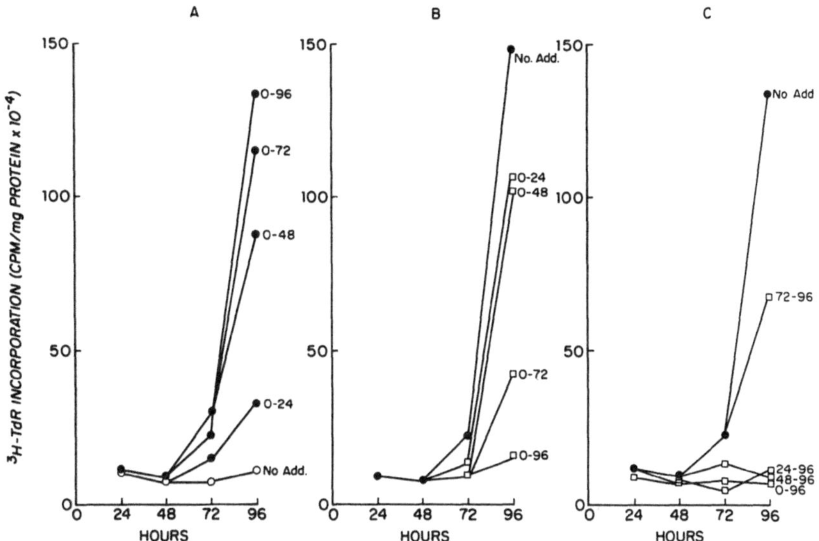

Figure 4. Timing of the effect of three factors and PDGF. To determine the minimum time that the three factors and PDGF are required to be present in order to exert their effects, three factors (A) or three factors and PDGF (B and C) were added to hepatocytes cultured in DMESF for the durations of time as indicated to the right of the graphs. DMESF: ○; DMESF + three factors: ●; DMESF + three factors + PDGF, 5 μl/ml: □. DNA synthesis was assayed daily by measuring the counts incorporated into acid precipitable material after 24-h labeling period with [^{3}H]thymidine.

It remains to be determined whether the stimulant activity is due to the direct action of 2-AAF in the diet or due to the production of a 2-AAF-induced growth factor. Whether sera from rats treated with other hepatocarcinogens have a similar effect is under study. Under identical assay conditions, we could find no stimulatory action by the serum (0.1% to 10% concentration) from rats taken between 1 and 48 h after a two-thirds partial hepatectomy.

The results of the stimulatory effect of sera from carcinogen-treated rats prompted us to examine the effect of sera from cancer patients. Sera were prepared from platelet-depleted plasma from cancer patients and the activity was compared with those of normal volunteers. Samples from some 200 cancer patients have been compared with 30 normal volunteer samples to date. Seventy percent of the samples from cancer patients showed 110% to 450% increase in the stimulation of normal rat hepatocyte DNA synthesis over the mean value for serum

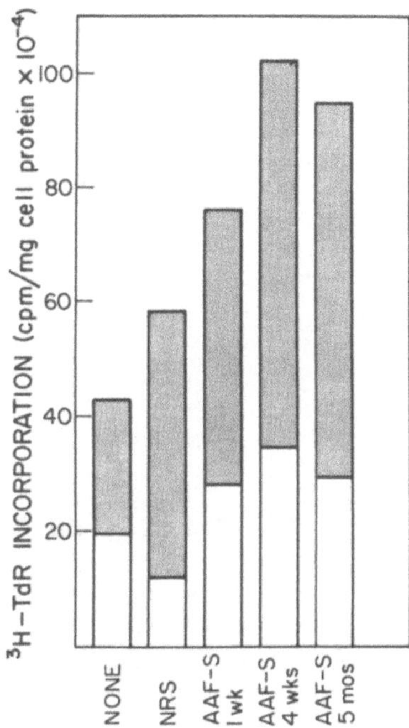

Figure 5. Effect of sera from 2-AAF-fed rats. Sera from rats fed the hepatocarcinogen, 2-AAF, were collected at one week, four weeks, and five months after the initiation of 2-AAF feeding. The sera (AAF-s) were examined for the stimulatory activity on the DNA synthesis in normal hepatocytes, and were compared with that of normal rat serum (NRS). All sera were tested at a concentration of 0.5%. Open bars: DMESF; closed bars: DMESF + three factors.

from normal volunteer sera. The stimulatory activity of sera from cancer patients does not show any tissue specificity, but seems to be correlated with the extent of the disease state (Nguyan *et al.*, 1983). Sera from patients suffering from inflammatory diseases of the chest and gut did not show any increase in the stimulatory activity over that shown by unaffected individuals. A representative result is shown in Fig. 6. Whether the activity found in the sera of cancer patients is similar to any of the tumor-associated growth factors recently reported (Todaro *et al.*, 1980; Nishikawa and Okitsu, 1979; Schumm and Webb, 1983), is being investigated.

Altered Response of Hepatocytes from Carcinogen-Treated Rats to PDGF

One of the central problems in the studies of growth control is the altered growth during carcinogenesis and cellular transformation. One

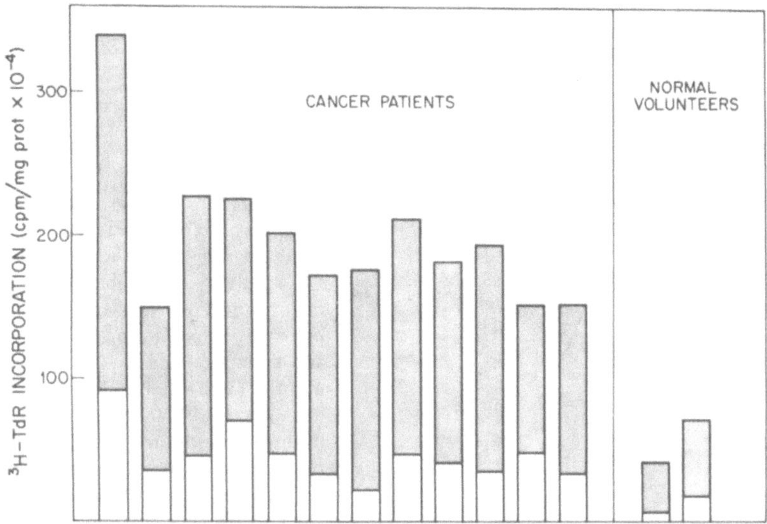

Figure 6. Effect of plasma-derived sera from tumor-bearing humans. Sera from cancer patients and normal volunteers were made from platelet-depleted plasma, as described in Fig. 1 legend. The DNA-synthesis-stimulating activity of sera on normal hepatocytes was examined at 3%, without (open bars) or with (closed bars) three factors.

possible explanation for this is the change in humoral environment. Hence the active search for transformation-associated growth factors by many laboratories.

The work described above is in accordance with this idea. However, equally probable but less studied is the alteration during carcinogenesis of the negative regulatory mechanisms of cells. We have observed such a change in the negative control in rat hepatocytes during chemical hepatocarcinogenesis. The response to the inhibitory effect of PDGF on DNA synthesis was examined in hepatocytes from rats treated with 2-AAF. After three months of 2-AAF treatment, the hepatocytes have completely lost their sensitivity to the inhibitory action of PDGF (Fig. 7). This loss of sensitivity to PDGF appears as early as one week after the commencement of feeding 2-AAF. In addition to their stimulatory activity at low concentrations, sera from carcinogen-fed rats are just as inhibitory as sera from normal rats at higher concentrations, indicating that the humoral environment of the negative regulator is unaltered during carcinogenesis. The results in Fig. 7 are further strong support that the change took place at the cellular level.

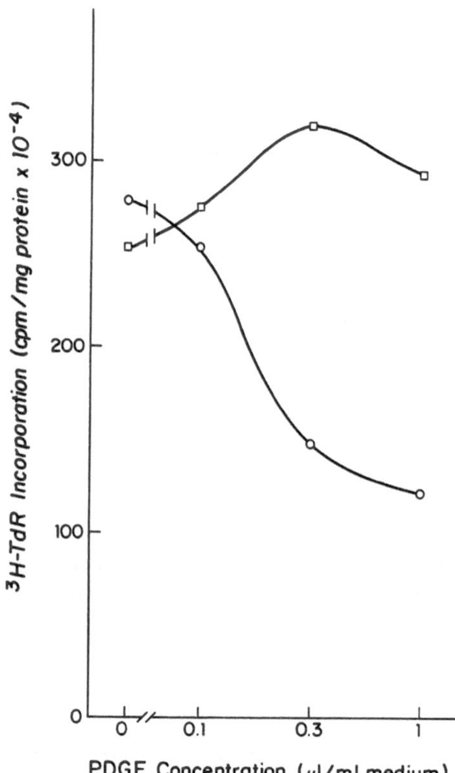

Figure 7. The effect of PDGF on hepatocytes from normal and 2-AAF-fed rats. PDGF was examined for its ability to inhibit the three-factor-induced DNA synthesis in hepatocytes from a normal (—O—) and from a rat fed 2-AAF for three months (—□—). PDGF was added to cultures with increasing concentrations as indicated. PDGF preparation used in this experiment was a kind gift of Dr. C. D. Stiles, Sidney Farber Cancer Institute, Boston, MA. The protein concentration of this preparation was 3.35 mg/ml, and was effective in inducing DNA synthesis in 3T3 cells at 25 μl/ml medium. In the absence of the three factors (DMESF alone), [^{3}H]thymidine incorporation for normal hepatocytes was 100 cpm/mg cell protein.

Discussion

The power of the cell culture system is in enabling the studies of the various aspects of cellular physiology in a defined environment, away from the complex and dynamic milieu of the *in vivo* system. The replacement of serum by known hormones and purified factors (Sato, 1975; Hayashi and Sato, 1976) is then a step toward the understanding of cellular physiology, and the process itself of replacing serum with defined components often reveals the hitherto unknown role of hormones and growth factors for the tissue under study (Barnes and Sato, 1980). The serum-free culture system, however, is not a goal in itself, but rather a tool to be used for answering pertinent questions.

In the case of the hepatocyte system, recent progress in the field has brought unique opportunities. The collagenase perfusion technique has allowed the production of mass primary cultures of parenchymal hepatocytes. Moreover, these hepatocytes, on attachment and spreading, can be maintained in a serum-free condition for the remainder of the experiment. The DNA synthesis of these hepatocytes in serum-free culture can be induced under a defined condition—by the addition of insulin, glucagon, and EGF. Although far from perfect, these aspects of hepatocyte culture fulfill one of the important advantages of the serum-free system: a defined basal condition.

Using such hepatocyte cultures, we have found that (1) serum has a stimulatory activity on normal hepatocyte DNA synthesis that cannot be replaced by increasing the amounts of insulin, glucagon or EGF, indicating that other as yet to be discovered stimulants occur under physiological conditions; (2) normal hepatocytes produce and release into the culture medium an endogenous stimulant for DNA synthesis; (3) PDGF inhibits the DNA synthesis of hepatocytes from normal, but not from carcinogen-fed rats; and (4) a stimulatory activity for normal hepatocyte DNA synthesis can be found in the sera of rats treated with chemical carcinogens and from tumor-bearing humans. All of these findings need to be pursued, particularly, the elucidation of the mechanisms of the actions of the factors thus far identified, and the purification and characterization of the stimulant activities found in the conditioned medium and sera from carcinogen-fed rats and tumor-bearing humans.

It is still widely accepted that there is an autonomous progression of cells through the cell cycle once the cells pass the G_1/S barrier. The case of hepatocytes in culture, which undergo DNA synthesis but not cell division, stands as contrary to this view. Although the work described in this chapter does not directly concern the control of proliferation, DNA synthesis is a necessary prelude to cell division. An examination of the changes in the regulation of hepatocyte DNA synthesis during carcinogenesis indicates that the control mechanisms for cell growth are complex. The controls concern not only the nature and/or the level of stimulants, but also of inhibitory factor(s) and critical alterations of response to these factors at the cellular level.

Our long-term objective is to use this unique *in vitro* system consisting of reproducible mass cultures of primary hepatocytes maintained in a defined, serum-free culture medium to study the alterations in growth control that occur during hepatocarcinogenesis induced by defined carcinogens *in vivo*.

References

Antoniades, H. N., Scher, C. D., and Stiles, C. D., 1979, Purification of human platelet-derived growth factor, *Proc. Natl. Acad. Sci. U.S.A.* **76:**1809–1813.

Barnes, D., and Sato, G. H., 1980, Serum-free cell culture: A unifying approach, *Cell* **22:**649–655.

Bresnick, E., 1971, Regenerating liver: An experimental model for the study of growth, in: *Methods in Cancer Research*, Volume 6 (H. Busch, ed.), Academic Press, New York, pp. 347–397.

Bucher, N. L. R., and Malt, R. A., 1971, *Regeneration of Liver and Kidney*, Little, Brown and Company, Boston, pp. 17–176.

Bucher, N. L. R., and Swaffield, M. N., 1976, Synergistic action of glucagon and insulin in regulation of hepatic regeneration, *Adv. Enz. Reg.* **13:**281–293.

Bucher, N. L. R., and Wands, J. R., 1977, Hormone cocktails to stimulate hepatocytes, *N. Engl. J. Med.* **296:**946.

Carr, B. I., and Hayashi, I., 1983, The controls of DNA synthesis in adult rat hepatocytes *in vitro*, in: *Isolation, Characterization and Use of Hepatocytes* Harris, R. A. and Cornell, N. W. (eds.), Elsevier, New York, pp. 183–188.

Cook, R. T., Roetman, K. J., and Steinmetz, G., 1982, Further purification of an inhibitory factor for DNA synthesis in regenerating rat liver, *Cell Biol. Int. Rep.* **6:**49–56.

Deschamps, Y., and Verly, W. G., 1975, The hepatic chalone; II. Chemical and biological properties of the rabbit liver chalone, *Biomedicine* **22:**195–208.

Enat, R., Ruiz-Opazo, N., Gatmaitan, Z., Leinwand, L., and Reid, L., 1982, Liver regeneration *in vitro*: Its dependence on synergies between hormones and extracellular matrix, *J. Cell Biol.* **95:**131a.

Evans, V. J., Earle, W. R., Wilson, E. P., Waltz, H. K., and Mackey, C. J., 1952, The growth *in vitro* of massive cultures of liver cells, *J. Natl. Cancer Inst.* **12:**1245–1265.

Forber, E., 1978, Experimental liver carcinogenesis: A perspective,in: *Primary Liver Tumors* (H. Remmes, H. M. Bolt, P. Bannasch, and H. Popper, eds.), University Park Press, London, pp. 357–375.

Forber, E., 1979, Response of the liver to carcinogens, in: *Toxic Injury of the Liver, Part A*, (E. Farber, and M. M. Fisher, eds.), Marcel Dekker, Inc., New York, pp. 445–467.

Friedman, D. L., Claus, T. H., Pilkis, S. J., and Pines, G. E., 1981, Hormonal regulation of DNA synthesis in primary cultures of adult rat hepatocytes, *Exp. Cell Res.* **135:**283–290.

Hayashi, I., and Sato, G. H., 1976, Replacement of serum by hormones permits growth of cells in a defined medium, *Nature* (*London*) **259:**132–134.

Heldin, C. H., Wasteson, A., and Westmarke, B., 1977, Partial purification and characterization of platelet factors stimulating the multiplication of normal human glial cells, *Exp. Cell Res.*. **109:**429–437.

Higgins, G. M., and Anderson, R. M., 1931, Experimental pathology of the liver, *Arch. Pathol.* **12:**186.

Kohler, N., and Lipton, A., 1974, Platelets as a source of fibroblast growth-promoting activity, *Exp. Cell Res.* **87:**297–301.

Leffert, H. L., Moran, T., Boorstein, R., and Koch, K. S., 1977, Procarcinogen activation and hormonal control of cell proliferation in differentiated primary adult rat liver cell cultures, *Nature* (*London*) **267:**58–61.

Leffert, H. L., Koch, K. S., Rubalcava, B., Sell, S., Moran, T., Boorstein, R., 1978, Hepatocyte growth control: *In vitro* approach to problems of liver regeneration and function, *J. Natl. Cancer Inst.* **48:**87–101.

McGowan, J. A., Strain, A. J., and Bucher, N. L. R., 1981, DNA synthesis in primary cultures of adult rat hepatocytes in a defined medium: Effect of epidermal growth factor, insulin and glucagon, and cyclic-AMP, *J. Cell. Physiol.* **108:**353–363.

Michalopoulos, G., Cianciulli, H. D., Novotny, A. R., Kligerman, A. D., Strom, S. C., and Jirtle, R. L., 1982, Liver regeneration studies with rat hepatocytes in primary culture, *Cancer Res.* **42:**4673–4682.

Nakamura, T., Tomita, Y., Noda, C., and Ichihara, A., 1983, Density dependent controls of growth and various cellular activities of mature hepatocytes in primary culture, in: *International Symposium on Isolation, Characterization and Use of Hepatocytes*, Harris, R. A. and Cornell, N. W. (eds.), Elsevier, New York, pp. 193–198.

Nguyan, A., Carr, B. I., Hayashi, I., and Terz, J., 1983, Plasma tumor-bearing patients stimulate DNA-synthesis: A possible new tumor marker system, *Proc. Aacr.* **24:**511.

Nishikawa, K., and Okitsu, C., 1979, Control of BALB/3T3 growth by factors present in tumor extract, in: *Cold Spring Harbor Conferences on Cell Proliferation*, Volume 6, *Hormones and Cell Culture* (G. H. Sato, and R. Ross, eds.), Cold Spring Harbor Laboratory, Cold Spring Harbor, New York, pp. 441–452.

Pitot, H. C., and Sirica, A. E., 1980, The stages of initiation and promotion in hepatocarcinogenesis, *Biochim. Biophys. Acta* **605:**191–215.

Pledger, W. J., Stiles, C. D., Antoniades, H. N., and Scher, C. D., 1977, Induction of DNA synthesis in BALB/c 3T3 cells by serum components: Reevaluation of the commitment process, *Proc. Natl. Acad. Sci. U.S.A.* **74:**4481–4485.

Preussmann, R., 1978, Hepatocarcinogens as potential risk for human liver cancer, in: *Primary Liver Tumors* (H. Remmes, H. M. Bolt, P. Bannasch, and H. Popper, eds.), University Park Press, London, pp. 11–29.

Richman, R. A., Claus, T. H., Pilkis, S. J., and Friedman, D. L., 1976, Hormonal stimulation of DNA synthesis in primary cultures of adult rat hepatocytes, *Proc. Natl. Acad. Sci. U.S.A.* **73:**3589–3593.

Ross, R., Glomset, J., Kariya, B., and Harker, L., 1974, A platelet-dependent serum factor that stimulates the proliferation of arterial smooth muscle cells *in vitro*, Proc. Natl. Acad. Sci. U.S.A. **71:**1207–1210.

Saetren, H., 1956, A principle of autoregulation of growth, *Exp. Cell Res.* **11:**229–232.

Sato, G. H., 1975, The role of serum in cell culture, in: *Biochemical Actions of Hormones*, Volume 3 (G. Litwack, ed.), Academic Press, New York, pp. 391–396.

Scher, C. D., Shepard, R. C., Antoniades, H. N., and Stiles, C. D., 1979, Platelet-derived growth factor and the regulation of the mammalian cell cycle, *Biochim. Biophys. Acta* **560:**217–241.

Schumm, D. E., and Webb, T. E., 1982, Putative transformation-dependent proteins in the blood plasma of tumor-bearing rats and cancer patients, *Cancer Res.* **42:**4964–4969.

Sedmak, J. J., and Grossberg, S. E., 1977, A rapid, sensitive, and versatile assay for protein using Coomassie Brilliant Blue G250, *Anal. Biochem.* **79:**544–552.

Seglen, P. O., 1976, Preparation of isolated rat liver cells, *Methods Cell Biol.* **13:**29–83.

Strain, A. J., McGowan, J. A., Bucher, N. L. R., 1982, Stimulation of DNA synthesis in primary cultures of adult rat hepatocytes by rat platelet-associated substance(s), *In Vitro* **18:**108–116.

Takasoka, T., Yasumoto, S., and Katsuta, H., 1975, A simple method for the cultivation of rat liver cells, *Jpn. J. Exp. Med.* **45:**317–326.

Todaro, G. T., Friling, C., and DeLarco, J. E., 1980, Transforming growth factors produced by certain human tumor cells: Polypeptides that interact with epidermal growth factor receptors, *Proc. Natl. Acad. Sci. U.S.A.* **77:**5258–5262.

Tomita, Y., Nakamura, G., and Ichihara, A., 1981, Control of DNA synthesis and ornithine decarboxylase activity by hormones and amino acids in primary cultures of adult hepatocytes, *Exp. Cell Res.* **135:**363–371.

Weiss, P. A., and Kavanau, J. L., 1957, A model of growth and growth control in mathematical terms, *J. Gen. Physiol.* **41:**1–47.

Wogan, G. N., 1976, The induction of liver cell cancer by chemicals, in: *Liver Cell Cancer* (H. M. Cameron, D. A. Linsell, and G. P. Warwick, eds.), Elsevier Scientific, Amersterdam, pp. 121–152.

CHAPTER 8

Intratesticular Regulation

Evidence for Autocrine and Paracrine Control of Testicular Function

JENNIE P. MATHER

Introduction

The testis was one of the first endocrine organs to be studied. Aristotle recognized the role of the testis in maintaining secondary sex characteristics, although he wrongly concluded that they were not necessary for fertility. In the 17th century, de Graaf concluded that semen originated in the testis. By the nineteenth century the spermatozoa were recognized as being essential for fertility and originating in the seminiferous epithelium (Setchell, 1978).

In the past century much of the work on the testis has been related to the role of testosterone in maintaining fertility and secondary sex characteristics. The recognition of extratesticular control of testicular function followed from experiments that showed that hypophysectomy resulted in testicular regression and infertility (Smith, 1927). Two different factors from the pituitary were shown to be required for the restoration of fertility in hypophysectomized animals. One, interstitial cell-stimulating hormone (ICSH), was later recognized to be identical to luteinizing hormone (LH). This hormone acts on the Leydig cells of the testis to stimulate testosterone secretion. The second pituitary hormone, follicle-stimulating hormone (FSH), has been shown to act on the seminiferous epithelium, in particular the Sertoli cells. The recognition that FSH was critical for maintaining testicular function

JENNIE P. MATHER • The Population Council and The Rockefeller University, New York, New York 10021.

and the identification of the Sertoli cell as its site of action led to an increased interest in the role of this cell in testicular function. Work in the last two decades has supported the idea that the Sertoli cell plays a critical role in maintaining spermatogenesis and may act as the mediator of many endocrine effects in the testis. The actions of FSH and LH in regulating testicular function have been reviewed (Means *et al.*, 1976, 1980; Fritz, 1978; Catt *et al.*, 1980; Dorrington and Fritz, 1974) and are beyond the scope of this chapter.

It has also been recognized that the testis produces substances that influence extratesticular function. As discussed above, the production of testosterone has long been known to affect secondary sex characteristics. These include influences on hair growth, skin, muscle, and male accessory sex organs. In addition, experiments indicate that the testis may produce a substance, inhibin, that regulates gonadotropin secretion by the pituitary (Setchell *et al.*, 1977; deJong, 1979).

Peritubular myoid cells surround the seminiferous epithelium as a single or multiple cell layer depending on the species. An extensive basal lamina exists between the Sertoli and myoid cells as well as outside the myoid cell layer, or between layers where multiple cell layers exist. Clermont (1958) demonstrated that myoid cells are responsible for the contractility of the testis. Until recently, there has been little further work on the role of these cells in regulating testicular function.

With the advent of methods to obtain preparations of testicular cells enriched for one cell type (Dufau *et al.*, 1978; Aldred and Cooke, 1980; Welsh and Wiebe, 1975), studies of the regulation of the physiology of individual cells in the testis *in vitro* became possible. This chapter will attempt to draw together data obtained from a number of different laboratories that have used many diverse types of culture systems. These data will be discussed in the following sections. Specific examples of each type of regulation will be presented from data obtained in this laboratory. A model of intratesticular regulation is presented that attempts to incorporate results from many different experimental approaches into an integrated whole.

Testicular Cell Culture

In vitro experiments with cultured testicular cells have been performed using a number of cell lines and primary culture systems. Several of these cell lines and their derivations and properties are listed in Table I. Data from these lines and from primary cultures enriched

Table I. Established Cell Lines Derived from Testicular Cells

Cell line	Cell type	Established from	Reference
R2C	Leydig, rat	Tumor	Shin, 1968
I10A	Leydig, mouse	Tumor	Shin, 1967
MA10	Leydig, mouse	Tumor	Ascoli, 1982
TM_3	Leydig, mouse	Primary culture 10- to 13-day-old animals	Mather, 1980
TM_4	Sertoli, mouse	10- to 13-day-old animals	Mather, 1980
TR-ST	Sertoli, rat	Tumor	Mather *et al.*, 1982a
TR-M	Peritubular myoid, rat	Tumor, (nontumorigenic)	Mather and Phillips, 1983b
TR-1	Endothelial, rat	Primary culture 30-day-old animals)	Mather *et al.*, 1982a

for Sertoli, Leydig, and peritubular myoid cells will be discussed below by individual cell type.

Sertoli Cells

In a sense, Sertoli cells were one of the first cell types to be cultured in serum-free medium. The morphological characteristics of Sertoli cells are distinctive among the cell types of the testis. Their anatomical location as the epithelial cell lining the inside of the tubule and the tight junctions between Sertoli cells made it possible to use mechanical dissociation and differential enzymatic digestion to obtain primary cultures highly enriched for Sertoli cells (Aldred and Cooke, 1980; Tung and Fritz, 1977) or organ cultures from the tubules (Steinberger *et al.*, 1964; Vernon *et al.*, 1971). These cultures were grown in a defined medium without serum. Sertoli cells will survive in such serum-free medium for prolonged periods of time in culture. The addition of vitamins (Steinberger, 1975) and the hormones known to affect Sertoli cell function *in vivo*, FSH and testosterone (Tung and Fritz, 1977; Fritz, 1978; Griswold *et al.*, 1977), led to improved cell function *in vitro*.

The recognition that serum may contain factors inhibitory for cell growth and function (Hayashi and Sato, 1976; Mather and Sato, 1979; Hyashi *et al.*, 1977; Bottenstein *et al.*, 1979) led to attempts to establish nontransformed, permanent cell lines derived from primary cultures of testes from normal immature mice in the absence of serum. A clonal cell line (TM_4) was established that maintained many of the properties of Sertoli cells (Mather, 1980). The definition of the requirements for the growth of this cell line in serum-free medium at low cell densities increased our understanding of the factors regulating Sertoli cell function. These factors have since been shown to improve Sertoli cell survival and function in primary cultures of Sertoli cells (Mather and Sato, 1979; Karl and Griswold, 1980; Rich *et al.*, 1983).

The early use of serum-free culture made it possible to identify and purify a number of Sertoli cell-secreted proteins. Androgen-binding protein (ABP) was one of the first such products to be identified (Ritzen *et al.*, 1971; Hansson and Djoseland, 1972; Vernon *et al.*, 1974). This protein has been purified and studied in great detail (Musto *et al.*, 1980; Larrea *et al.*, 1981; Bardin *et al.*, 1981). The development of a radioimmunoassay for ABP (Gunsalus *et al.*, 1978) and *in vitro* culture of Sertoli cells in defined medium have made detailed studies of the hormonal regulation of ABP secretion possible (Mather and Phillips, 1983b; Rich *et al.*, 1983).

More recently, it has been recognized that Sertoli cells secrete a number of proteins commonly found in plasma (Wright *et al.*, 1981), including transferrin (Skinner and Griswold, 1980) and ceruloplasmin (Wright *et al.*, 1981). The use of two-dimensional gel electrophoresis to separate Sertoli cell-secreted proteins has led to the recognition that these cells secrete a large number of proteins (Kissinger *et al.*, 1982; Wright *et al.*, 1981; DePhilip and Kierszenbaum, 1982) and that these proteins are highly regulated during the development of the germ cells (Parvinen *et al.*, 1983; Wright *et al.*, 1983). Identification of these proteins and elucidation of their role in testicular function is a major challenge for future research in this area.

In addition to protein secretion, Sertoli cells also secrete large amounts of lactate and pyruvate (Robinson and Fritz, 1982), myoinositol (Robinson and Fritz, 1979), and sulfoproteins including heparin and chondroitin sulfate-like material (Elkington and Fritz, 1980).

The unique ability of these cells among testicular cell types to survive and function in a serum-free environment has aided the rapid progress in the area of defining the products of Sertoli cell secretory activity. Conversely, the extensive secretion of transport proteins, peptides, and other materials by these cells may be, in part, responsible for their ability to survive in a simple defined medium. This point will be further discussed below in the section on Sertoli cells.

Leydig Cell Culture

Most of our understanding of Leydig cell response *in vitro* has come from the use of freshly isolated cell suspensions of interstitial cells or purified Leydig cells (Aldred and Cooke, 1980; Dufau *et al.*, 1978). Many studies using such isolated cell suspensions have contributed to our understanding of gonadotropin binding, the regulation of gonadotropin receptors, and the regulation of steroidogenesis in Leydig cells. However, studies requiring longer time periods (i.e., days rather than hours) are impossible to perform in such cell suspensions as the cells will not survive for long periods of time.

Attempts to establish clonal cell lines from Leydig cell tumors have been successful (Shin, 1967; Shin *et al.*, 1968). However, these cells lost much of their gonadotropin response and androgen secretory capabilities in culture and were not widely used in studies of Leydig cell function. More recently, other attempts to establish clonal cell lines from a mouse Leydig tumor (Ascoli, 1981) and normal primary cultures of interstitial cells from immature mice, TM_3 (Mather, 1980), have led to the establishment of cell lines that maintain part, if not all, of these functions. The TM_3 cell line was used to define the hormone, vitamin, and growth factor requirements for the growth of these cells in serum-free defined medium (Mather, 1980; Mather *et al.*, 1982a).

The requirements of the TM_3 line were then used to support the growth/survival of Leydig cells in primary culture (Mather and Sato, 1979). Such defined media with various modifications have been successfully used to prolong the survival and function of mouse (Mather *et al.*, 1981; Murphy and Moger, 1982), rat (Mather *et al.*, 1981; Hseuh, 1980; Browning *et al.*, 1983), and porcine (Mather *et al.*, 1981; Mather *et al.*, 1982b) Leydig cells in culture. There seems to be considerable species variability in the maintenance of Leydig cell function in cultures derived from different species. More work is required to define the conditions for the maintenance of function in mouse or rat cultures much beyond one week of culture. Porcine Leydig cell cultures, however, can be maintained for several weeks in hormone- and vitamin-supplemented cultures (Mather and Phillips, 1983a). Such long-term cultures have been used to study the regulation of gonadotropin receptors (Mather *et al.*, 1982b) and androgen (Haour *et al.*, 1983) and prostaglandin E_2 and $F_{2\alpha}$ secretion (Haour *et al.*, 1981) by hormones, vitamins, and transport proteins (Mather *et al.*, 1982c; Benahmed *et al.*, 1981).

Peritubular Myoid Cell Culture

Peritubular myoid cells were present *in vitro* in some of the earliest attempts at culturing testicular cells (deKrester *et al.*, 1971; Bressler and Ross, 1973). Attempts have been made to obtain cultures enriched for peritubular myoid cells by passaging tubular cultures from 20-day-old rats in serum-containing medium (Tung and Fritz, 1977). Under these conditions, the Sertoli cells will not replicate, whereas the peritubular cells will. However, there are presently no criteria other than morphology for identifying a peritubular myoid cell in culture. It is thus impossible to precisely define the percentage of myoid cells present in such cultures. Possible contaminating cell types in these cultures could originate from interstitial fibroblasts, vascular or lymphatic endothelial cells, or vascular smooth muscle. All of these cell types should be capable of surviving and proliferating, to some degree, in serum-supplemented medium. We have altered the cell purification method used for the testis by first dispersing the tubules in 1 M glycine, followed by a brief collagenase/dispase digestion. The glycine removes and lyses the interstitial tissue and lymphatic endothelium surrounding the tubule, reducing possible contaminating cell types (Mather and Phillips, 1983b). We feel this increases the purity of the myoid-enriched cell population obtained, although the lack of markers still prevents assessment of contaminating cell types.

We have recently established a clonal nontumorigenic cell line in this laboratory, TR-M, that we feel has maintained many of the properties reported for myoid cells *in vivo* and in primary myoid-enriched cultures (Mather *et al.*, 1982a; Mather *et al.*, 1983b). It is hoped that this line will prove useful in elucidating the role of peritubular myoid cells. The identification of cell-specific functions and antigens for this cell line can then be compared with the properties of myoid cells *in vivo* and in primary culture, and should aid both in definitively identifying the origin of the cell line and in providing markers for identifying and quantifying myoid cells in primary culture.

Culture of Germinal Cells

Germ cell survival and development has been reported in organ cultures (Steinberger, 1975; Vernon *et al.*, 1971) and in cultures of segments of seminiferous tubules from defined stages of the seminiferous cycle (Parvinen *et al.*, 1983). Some germ cells will adhere to Sertoli cell monolayers and survive for some time when maintained in these cocultures. However, the fact that not all stages of spermatocytes adhere to Sertoli cells and the complexity of these cultures, which contain spermatocytes in many stages of development, have made quantification of survival and development in such cultures difficult. The role of the Sertoli cell in maintaining spermatocyte survival in such cultures will be discussed in the section on Sertoli–germ cell interaction.

Highly purified populations of spermatocytes at different stages of development can be obtained (Meistrich *et al.*, 1981). However, no progress has been made in keeping these isolated cells viable in culture for more than 24 h. Defining the conditions required for germ cell survival and development *in vitro* remains one of the challenges of testicular cell culture.

Autocrine Regulation of Testicular Cells

The recent discovery of transforming growth factors has drawn attention to autocrine regulation of cells *in vitro* (Marquardt and Todaro, 1982). However, the phenomenon of autocrine regulation is not limited to transforming growth factors or to transformed cells. Studies on the cells of the intestinal tract indicate that these cells may produce and respond to autocrine and/or paracrine factors (Soll, 1981). In this section, we will review evidence that suggests that some testicular cell types regulate their own function. We will discuss data for true autocrine regulation: that is, the production of factors that may bind to specific receptors and regulate cell function. We will also review data for other types of autoregulation such as the production of transport proteins that bind a nutrient and transport it back into the producing cell via specific receptor-mediated mechanisms (e.g., transferrin and iron) and feedback inhibition of metabolic pathways by secreted products. These cell products may not act exclusively on the producing cell, but may also act on other cell types in the testis. The intercellular actions of these factors will be discussed in the section on cell–cell interaction in the testis.

Sertoli Cells

The culture of Sertoli cells, both primary cultures and established cell lines, has provided a body of experimental data on the products produced by Sertoli cells and the factors required for the growth of established Sertoli-derived cell lines in low-density cultures (Mather, 1980, 1982; Mather *et al.*, 1982a; Zhuang and Mather, 1983; Perez-Infante and Mather, 1982).

The cell lines respond to hormones produced in the pituitary (FSH) and in other extratesticular sources [hydrocortisone, insulin, epidermal growth factor (EGF)] as well as to vitamins (vitamins A and E) provided in the diet. In addition, these cells respond to products produced by

Table II. Comparison of Response of Primary Cultures of Sertoli Cells and Sertoli-Derived Cell Lines to Humoral Factors and Their Production of These Factors

	Response			Production	
Factor	TM_4	TR-ST	Primary	TM_4/TR-ST	Primary
Transferrin	G	G	R	+	+
Ceruloplasmin	G	G	?	?	+
Somatomedin	G	G	R	?	+
Growth factor	G	G	R	+	+
Testosterone	G	G	R	−	−
Hydrocortisone	G	G	R	−	−
Progesterone	G	G	R	?	?
Estradiol	G	G	R	−	+
Vitamin A	G	G	R	−	−
Vitamin E	G	G	R/S	−	−
Prostaglandins	G	G	R	?	?
HDL	G	G	?	?	?

The data from which this table is derived are discussed in the text. Indication of responses are G: growth; R: response of secreted protein such as ABP of TF secretion and/or cAMP production; S: affects growth and/or survival; ?: not tested; +: secreted; −: not secreted.

the Leydig cells, such as testosterone and prostaglandins. However, the TM_4 and TR-ST cell lines also show responses to factors produced by Sertoli cells themselves *in vitro* such as transferrin, ceruloplasmin and somatomedins (Table II). These factors are among the most critical for the growth of these cells in serum-free medium. Figure 1 shows the effects of individually omitting each of eight factors that have growth stimulatory effects on the TM_4 Sertoli cell line. It is immediately apparent that three of these factors, insulin (Ins), transferin (TF), and EGF are major requirements for the growth of these cells in low-density cultures.

The strong dependence of the established cell lines on extraneous factors, in contrast to the ability of primary Sertoli cell cultures to survive without additional factors in serum free medium, was of interest. Was the dependence of the established lines due to a fundamental change in the environmental conditions pertinent to the culture of these cells or to different culture conditions? A comparison of the properties of primary and established cultures is shown in Table III. The ability of established cell lines to divide would tend to rapidly dilute out any cellular products that are taken up from the cellular environment (i.e., serum *in vitro* or products of the circulation or local environment *in vivo*). In addition, the cell lines are generally plated at low densities ($\sim 7 \times 10^2$ cells/cm^2) compared with primary cultures (~ 4

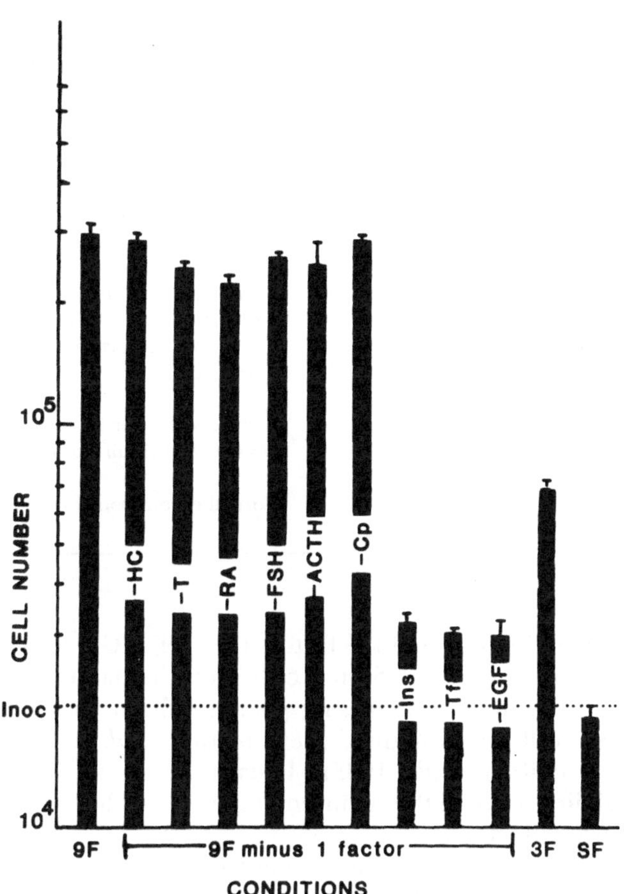

Figure 1. The growth of the TM_4 cell line in serum-free (SF) medium supplemented with insulin (Ins), hydrocortisone (HC), testosterone (T), retinoic acid (RA), FSH, adrenocortical stimulating hormone (ACTH), ceruloplasmin (Cp), TF and EGF, and growth when each of the nine factors is omitted individually. 3F is (Ins, TF, EGF). Cells were plated at the indicated inoculum and counted on day 4. The omission of six of the nine factors has only a minor effect even though all of these factors give significant growth stimulation in the presence of 3F alone. However, if Ins, TF, or EGF are omitted, there is less than one cell doubling even in the presence of the other eight factors.

$\times$ 10^4 cells/cm^2). When the established cell lines were plated at increasing cell densities, cell survival and cell growth (at higher densities) were seen in the absence of any added factors (Table IV). Conversely, primary Sertoli cell cultures plated at very low cell densities in serum-free medium survived for only a few days and then died (unpublished

Table III. Comparison of Primary Cultures and Established Cell Lines *In Vitro*

Clonal cell line	Primary culture
Contains only one cell type	Contains more than one cell type—composition may vary from preparation to preparation
Cells in logarithmic growth	Cells divide slowly or not at all
Cells plated at low densities	Cells plated at high densities
Cells removed with trypsin (short, @ 1 min, exposure)	Cells prepared by mechanical or enzymatic digestion—[time of exposure to enzymes may be long (20′ to 1 h)]
"Carryover" of substances from culture medium controllable; rapidly diluted by cell division	"Carryover" of factors from *in vivo* may be extensive
Permanent cell line	Culture viable for limited time period
Cells removed from *in vivo* environment for prolonged periods of time (years)	Cells recently in *in vivo* environment
May have chromosomal alterations (i.e., polyploidy)	Normal chromosomal number

results). The addition of the three factors (Ins, TF, EGF) also improves cell survival and growth (for Sertoli cells from immature animals) in primary culture, as well as increasing plating efficiency and growth in the established cell lines (Mather, 1980; Mather *et al.*, 1982a; Mather and Sato, 1979; Rich *et al.*, 1983). Thus, it seems that there is no fundamental difference in the requirement of the cell lines and primary cultures for these factors.

Carryover of nutrients and hormones from the *in vivo* environment also seems to play a role in the maintenance of cells in primary cultures. The effect of adding Ins, TF, and EGF (3F) to Sertoli cell cultures from immature (5 to 10 days old) animals could be seen within the first four days of culture. The addition of steroids (progesterone, testosterone, hydrocortisone), vitamin E, and FSH also improved cell viability over that seen in 3F alone, but the effects were seen only after the second week of culture (Rich *et al.*, 1983).

Transferrin has been shown to be produced by Sertoli cells *in vitro* (Skinner and Griswold, 1980) and is, in fact, produced by the cell lines TM_4 and TR-ST (Perez-Infante and Mather, unpublished results). A somatomedinlike, insulinlike growth factor substance has also been reported to be produced by Sertoli cultures (Johnsonbaugh *et al.*, 1982). Somatomedin C can substitute for Ins as a growth stimulator in the TM_4 cell lines (Mather and Furlanetto, 1983), although it is not yet known whether the TM_4 cells produce this substance. To date, no data

Table IV. TM_4 Cell Growth and Survival in Serum-Free Medium As a Function of Inoculum Density

Inoculum cell number/ 60 mm dish	Cell number on day 4 % of inoculum
	TM_4
10^3	0%
5×10^3	0%
10^4	0%
2.5×10^4	136%
5×10^4	153%
10^5	185%
2.5×10^5	180%

Cells were plated in 60 mm dishes (28 cm^2) in serum-free F12/DME (1:1) [Mather (1980)] medium with no added hormones. After 4 days in culture, cell numbers were obtained using a Coulter counter. Results are expressed as percent of inoculum: 100% would thus represent survival but no growth.

exist on EGF production by Sertoli cells. However, the requirement of the TM_4 cells for this growth factor at low densities and their ability to grow at high densities without it would suggest that these cells are capable of producing EGF or a related growth factor that can substitute for EGF as a growth stimulator.

Since Sertoli cells produce a large number of proteins, it seems likely that other factors in addition to transferrin, ceruloplasmin, somatomedins, and growth factors may both be produced by, and act on, Sertoli cells as well as other cell types in the testis. It has also been suggested that plasminogen activator (PA) that is produced by Sertoli cells at a specific stage of the seminiferous cycle (Lacroix *et al.*, 1977, 1981) may act to disrupt Sertoli–Sertoli cell tight junctions to allow the movement of developing germ cells from the adluminal to the luminal compartment of the seminiferous tubule. Thus, although not strictly autocrine regulation, this protein would be another substance that is both produced by and acts on the Sertoli cell.

The Sertoli cell function *in vivo* is cyclical (Leblond and Clermont, 1952; Parvinen and Vanha-Perttula, 1972). The sensitivity of the cell to external stimulation by hormones and the rate of secretion of Sertoli cell products such as PA (Lacroix *et al.*, 1981), transferrin, androgen-binding protein (Mather *et al.*, 1983b; Ritzen *et al.*, 1982), and CP2 (Wright *et al.*, 1983) all vary as a function of the stage of developing germ cells (Parvinen, 1982). The high degree of autocrine regulation in Sertoli cells may play an important role in the cycling of Sertoli cell function. These factors may also play a role during development. It is possible that some factors may have an autocrine role at one stage of development and a paracrine role at another, or both. More work is needed to define the control of the production of these factors and to define their site of action before a complete understanding of this complex regulatory process will be possible.

Leydig Cells

Recent studies on Leydig cell function *in vitro* indicate that Leydig cell products may also affect Leydig cell function. Leydig cells secrete prostaglandins (E_2, $F_{2\alpha}$). This secretion is increased by hCG stimulation both *in vivo* and *in vitro* (Haour *et al.*, 1979, 1981; Mather *et al.*, 1982c). These cells contain receptors for prostaglandins (Haour *et al.*, 1979), and exogeneously added prostaglandins can influence Leydig cell survival and hCG receptor levels *in vitro* (Mather and Haour, 1981; Grotjan *et al.*, 1978). Although the exact mode of action of prostaglandins in controlling Leydig cell physiology is not known, it seems likely from the above observations that these Leydig cell products may play a role in regulating the function of these cells.

Recent work has also suggested that steroidogenic products secreted by Leydig cells may act via a feedback mechanism to inhibit the steroidogenic pathway in the cells. The exact steroid(s) involved in this inhibition are not known at this time. Estrogens administered *in vivo* have been shown to inhibit Leydig cell function (Hseuh *et al.*, 1978; Brinkman *et al.*, 1980; Moger, 1980). However, whereas unpurified porcine Leydig cell function was inhibited by estrogen, highly purified Leydig cells were not (Benahmed *et al.*, 1982). This suggests estrogens may act on Leydig cells via a primary action on another testicular cell type. In another study, gonadotropin-stimulated testosterone secretion was decreased by synthetic androgens and increased by antiandrogens *in vitro*, suggesting that the androgen receptor plays a role in this autoregulation (Adashi and Hseuh, 1981).

The TM_3 Leydig-derived cell line (Mather, 1980) can be grown in a serum-free defined medium with hormone supplements. Under these conditions, cells were labeled with [^{35}S]methionine and the secreted proteins separated by two-dimensional gel electrophoresis. More than 20 proteins can be separated in such a fashion (Mather and Wright, unpublished data). Very little is known concerning the identity or site of action of these proteins.

After binding of gonadotropins by Leydig cells (hCG or LH), the gonadotropin is internalized. *In vivo* (Haour and Saez, 1977), in cell suspensions (Cigorraga *et al.*, 1978) and in a Leydig tumor-derived cell line (Ascoli, 1982), a down-regulation of receptors is seen after exposure to high levels of gonadotropins. In addition, in adult rats, desensitization of steroidogenic response to hCG stimulation occurs in a transient fashion (Cigorraga *et al.*, 1978). This desensitization does not occur in Leydig cell cultures from immature pigs (Mather *et al.*, 1982b) or fetal rats (Warren *et al.*, 1982). In an established cell line derived from a Leydig cell tumor, this desensitization was blocked by the addition of low-density lipoprotein (LDL) as an extracellular source of steroid precursors (Freeman and Ascoli, 1982). Thus, the Leydig cell is capable of regulating its own response to gonadotropin at the receptor and postreceptor level and this regulation changes during development. The factors responsible for this acquisition of sensitivity to down-regulation by gonadotropins during maturation have not yet been identified, although it has been suggested that the development of estrogen-mediated inhibition of gonadotropin activity may be involved (Huhtaniemi *et al.*, 1982).

Cell–Cell Interaction in the Testis

The close approximation of different cell types in the testis and the anatomical relationships of these different cell types, as well as the site of action of some known secretory products of testicular cells, strongly suggest that paracine regulation plays an important role in testicular function. Such interactions of closely apposed cell types is extremely difficult to study *in vivo*. The development of *in vitro* culture systems, and most especially serum-free systems, has allowed investigators in several laboratories to design experimental systems in which to study these complex interrelationships. The evidence for cell–cell interactions will be discussed for individual pairs of testicular cells. However, it is evident that the product(s) of one cell type may influence several cell types and vice versa, allowing for an extremely complex network of paracrine regulation in the testis.

Sertoli–Leydig Cell Interaction

The best studied of the Sertoli-Leydig cell interactions is the effect of testosterone on Sertoli cell function. The importance of testosterone in regulating Sertoli cell function was first recognized from *in vivo* experiments with hypophysectomized rats (Smith, 1944). If testosterone was given immediately after hypophysectomy, it was adequate to partially maintain fertility. The discovery of androgen receptors in Sertoli cells led to the hypothesis that testosterone is acting directly on this cell type (Fritz, 1978). Testosterone is capable of stimulating ABP secretion by Sertoli cells *in vitro* and of stimulating the growth of the Sertoli-derived cell lines (Mather *et al.*, 1982a). However, other hormones and growth factors are required for optimal Sertoli cell function. The effects of androgens on Sertoli cells have been reviewed elsewhere (Fritz, 1978).

Rat Sertoli cells are also capable of producing estrogen from testosterone (Dorrington *et al.*, 1978). As discussed in the section on Leydig cells, estrogens modulate Leydig cell function. It has thus been suggested that estrogen production by Sertoli cells might act as a local regulatory agent.

As mentioned above, prostaglandins are produced by Leydig cells in response to hCG. Prostaglandins can affect ABP secretion in primary cultures of rat Sertoli cells (Rich and Mather, unpublished results) and inhibit the growth of the TM_4 and TR-ST Sertoli cell lines (Mather *et al.*, 1982a). These results suggest that Sertoli cells may be a target of local prostaglandin production in the testis.

The testes have also been shown to contain pro-opiomelanocortin derivatives (Sharpe *et al.*, 1980). Immunohistochemical staining for β-endorphin and ACTH localizes these substances in the Leydig cell (Tsong *et al.*, 1982a,b). The amount of staining does not decrease after hypophysectomy, suggesting that these hormones may be produced locally in the testis. The TM_4 cell line has a growth response to ACTH (Mather, 1980). Both the TM_4 and TR-ST cell lines, as well as primary cultures of Sertoli cells, show an increased conversion of [^{3}H]adenine to [^{3}H]cAMP in the presence of synthetic ACTH and MSH (Salomon, Mather, and Bardin unpublished results). These data suggest that the Sertoli cells may be the site of action of the pro-opiomelanocortin derivatives, and that these substances may play a role in the paracrine regulation of testicular function.

Sertoli cells may also produce factors that regulate Leydig cell function. The testis contains and LH-RH-like factor. The LH-RH-like factor is thought to be produced by Sertoli cells (Sharpe *et al.*, 1981). Leydig cells specifically bind LHRH, and LHRH can alter gonadotropin receptor levels and steroidogenesis of Leydig cells *in vitro* (Sharpe and Cooper, 1982a,b). The LH-RH-like activity may then represent a Sertoli cell-produced substance that regulates Leydig cell function.

Local damage to the seminiferous tubule caused by implanting steroid capsules causes a change in adjacent Leydig cells (Aoki and Fawcett, 1978). In rats with seminiferous tubule damage induced by hydroxyurea, vitamin A deficiency or fetal irradiation changes were seen in Leydig cell function (Rich *et al.*, 1979). Unilateral cryptorchidism that damages the seminiferous tubules also causes changes in Leydig cells (Risburger *et al.*, 1981). More recent work has shown that Leydig cells adjacent to seminiferous tubules in stage 7 to 8 of the spermatogenic cycle were morphologically distinct from those adjacent to stage 9 to 14 or perivascular Leydig cells (Bergh, 1983). All of these data together strongly suggest that the tubules locally influence Leydig cell activity *in vivo*. It is not known whether these changes are mediated by the above-mentioned hormones or by other, as yet unidentified, regulatory factors produced by the Sertoli cells. As previously mentioned, Sertoli cells produce a number of proteins whose role is as yet unknown. These cell–cell interactions are obviously difficult to study *in vivo*. The regulating substances need act only over a very small distance and may thus not be released into the general circulation. The use of defined cell culture systems should prove valuable in furthering our understanding of Leydig–Sertoli cell interactions and identifying the factors involved in these interactions.

Hseuh (1980) has cultured whole testes from hypophysectomized adult rats. In these cultures, testosterone production initially falls and then reappears after eight days of culture. Presumably some factor, or factors, produced by another testicular cell type is responsible for the reappearance of Leydig cell function in these cultures. This system presents intriguing possibilities for discovering other testicular factors that are involved in Leydig–Sertoli cell interactions.

Sertoli–Myoid Cell Interaction

These two cell types are closely associated in the testis and, with the basal lamina between, form the seminiferous tubule. This basement membrane is a complex structure that contains collagen elastic fibers and fibrous components, as well as amorphous material at the base of the Sertoli cells (Hermo *et al.*, 1977; Denduchis *et al.*, 1975; Neaves, 1977). *In vivo* labeling data suggest that the collagen fibers are produced by the peritubular myoid cells (Pfeiffer *et al.*, 1981). The TR-M cell line established in this laboratory is thought to be of peritubular myoid cell origin (Mather and Phillips, 1983b). Monolayer cultures of this cell line produce an extensive extracellular matrix beneath the cells. This matrix contains material recognized by antifibronectin and antilaminin antisera (Wolpe, Hightower, and Mather, unpublished results), collagen, a material that morphologically resembles elastin, and a fibrous noncollagenous material (Mather and Phillips, 1983b). These components are all found in basement membranes (Hay, 1981). The TR-ST and TM_4 cell lines are also capable of producing extracellular matrix material (Mather and Phillips, unpublished results). Sertoli cells have been shown to produce heparinlike and chondroitin sulphatelike material (Elkington and Fritz, 1980) that are common components of basement membrane. It thus seems likely that both Sertoli and myoid cells contribute portions of the basal lamina of the seminiferous epithelium.

Extracellular matrix can regulate the function and hormone responsiveness of a number of cell types *in vitro* (Gospodarowicz and Ill, 1980; Reid, 1982). The extracellular matrix produced by the TR-M cell line has been used as a substrate for the culture of primary Sertoli cells. Cells grown on this matrix produce more ABP and are more responsive to FSH (stimulation of ABP) than those grown on plastic or plastic coated with collagen, laminin, or fibronectin (Mather and Phillips, 1983b). The removal of peritubular myoid cells and portions of the basal lamina also decreases ABP and transferrin secretion by dissected fragments of the seminiferous tubule (Mather *et al.*, 1983b). Reports from two laboratories indicate that coculture (primary culture) of Sertoli and peritubular myoid cells increases ABP secretion by Sertoli cells (Tung and Fritz, 1980; Hutson and Stocco, 1981). Coculture of primary Sertoli cells with the TR-M cell line also increases ABP secretion, although not to as great an extent as that seen in cells cultured on TR-M-produced matrix (Mather and Phillips, 1983b).

In addition to insoluble matrix components, the TR-M cell line also secretes a large number (> 30) of soluble proteins (Mather *et al.*, 1983b). Myoid cells in primary culture also secrete a large number of proteins with similar migration pattern in two-dimensional gel electrophoresis (Kissinger *et al.*, 1982; Wright *et al.*, 1983). The lack of markers for myoid cells *in vitro* and the inability to obtain primary cultures consisting only of myoid cells make it difficult to quantitate protein secretion by myoid cells in these cultures. Metabolic cooperation between cultured primary myoid and Sertoli cells has been reported (Hutson, 1983). The above results make it seem very likely, however, that peritubular myoid cells influence Sertoli cell function directly via secreted products and/or indirectly via insoluble components of the basement membrane.

Experiments with the TR-ST and TR-M cell lines suggest that Sertoli cells might also produce products that influence myoid cell

Table V. Growth of TR-M Cells in Serum and Serum-Free Conditions

Supplement	Cell number
None	6.0×10^4
Ins, TF, EGF	7.1×10^4
Ins, TF, EGF, testosterone	6.9×10^4
Ins, TF, EGF, prostaglandin $F_{2\alpha}$	7.1×10^4
Ins, TF, EGF, hydrocortisone	6.9×10^4
Ins, TF, EGF, progesterone	7.3×10^4
Ins, TF, EGF + 50% TR-ST conditioned medium	20.4×10^4
Serum-supplemented medium	36.1×10^4

Cells were inoculated at 5×10^4 cells per dish and counted on day 4. Conditioned medium was medium with Ins, TF and EGF conditioned for 24 h on confluent monolayers of TR-ST cells and filtered through a 0.22 μ filter.

function. The TR-M cell line will survive, but not grow in serum-free medium supplemented with insulin, transferrin, and EGF (3F). However the addition of conditioned medium from TR-ST cells grown in 3F-supplemented serum-free medium stimulates the growth of these cells to a growth rate near that seen in serum-supplemented medium (Table V). These data suggest that the TR-ST (Sertoli) cells are producing products that stimulate the growth of TR-M cells *in vitro*. The cell lines, grown in defined culture, could thus prove an ideal system in which to identify factors produced by Sertoli cells that influence peritubular myoid cell function.

Sertoli–Germ Cell Interaction

The spermatogonia are present in the outer layer of the seminiferous tubule in direct contact with the basal lamina and the basal portion of adjacent Sertoli cells. These stem cells are outside the Sertoli–Sertoli cell tight junction and thus exposed to extratubular factors, perhaps originating from peritubular myoid and Leydig cells, as well as those from the general circulation. As the germ cells develop, they move through the Sertoli cell junctions into the lumen of the seminiferous tubule where they complete meiosis and spermiogenesis, still in close association with the surrounding Sertoli cells (Fawcett, 1975). The Sertoli–Sertoli cell tight junctions form a barrier to diffusion of even relatively small molecules into the tubule (Dym and Fawcett, 1970; Neaves, 1977). The tubular environment must therefore consist exclusively of molecules that can pass through the Sertoli cell or are products of the Sertoli cell itself. This cell type must then play a major role in regulating the environment of the developing germ cell. As mentioned above, Sertoli cells secrete a large number of proteins. In addition, Sertoli cells secrete myoinositol, lactate, and pyruvate (Robinson and Fritz, 1982). Lactate rather than glucose has been shown to be essential for the support of RNA synthesis and survival of isolated germ cells (Jutte *et al.*, 1982).

The secretion of a number of Sertoli cell proteins occurs in a cyclical fashion in the adult. The seminiferous epithelium can be divided into a number of stages on the basis of the stages of development of the germ cells associated with the Sertoli cells. In the rat these stages occur in a cyclical progression along the length of the tubule with any one segment progressing through all stages every 40 days (Leblond and Clermont, 1952; Parvinen and Vanha-Perttula, 1972). The pattern of protein secretion (Wright *et al.*, 1983) as well as the secretion of specific proteins such as ABP (Ritzen *et al.*, 1982; Mather *et al.*, 1983b), transferrin (Mather *et al.*, 1983b), and plasminogen activator (Lacroix *et al.*, 1981) vary as a function of the stage of the seminiferous epithelium (for a review see Parvinen, 1982). Although autocrine regulation of Sertoli cells may play a role in controlling this cycling as discussed in the section on Sertoli cells, it also seems likely that the germ cells are capable of signaling the Sertoli cells as to their stage of development in order to stimulate or inhibit the secretion of specific products.

This hypothesis is supported by work on the control of plasminogen activator (PA) secretion *in vitro*. Plasminogen activator secretion by primary cultures of Sertoli cells from 20-day-old rats can be stimulated by FSH or by phagocytosis of small particles such as killed bacteria or residual bodies of developing germ cells (normally phagocytosed by adjacent Sertoli cells *in vivo*). Follicle-stimulating hormone and residual bodies together have a synergistic effect on PA secretion (Lacroix *et al.*, 1982). It was thus suggested that ingestion of residual bodies from developing germ cells *in vivo* might amplify the effect of FSH, present continuously in the testis, at a specific stage of the seminiferous cycle when the PA activity would be required.

Other Sertoli cell products that have been proposed as regulators of germ cell development include meiosis-inducing substance (MIS), meiosis-preventing substance (MPS) (Parvinen *et al.*, 1982), and seminiferous growth factor (SGF) (Feig *et al.*, 1980).

Leydig–Myoid Cell Interaction

Myoid cells concentrate androgen *in vivo* (Sar *et al.*, 1975) and the TR-M cell line contains androgen receptors (Nakla, Bardin, Mather,

Janne, in preparation). Androgen is required for morphological maturation of this cell type at puberty (Hovatta, 1972). It thus seems that myoid cells are one of the targets of the Leydig cell-produced androgen (Fritz, 1978). The TR-M cell line shows increased conversion of [^{3}H]adenosine to [^{3}H]-cAMP in the presence of PGE_1 (Mather and Salomon, unpublished data). This suggests that myoid cells may be a target of Leydig cell-produced prostaglandins in the testis.

Leydig cells are found in close proximity to peritubular myoid cells. In some species, cells morphologically similar to Leydig cells are actually found within the myoid cell layer of the seminiferous tubule. This close association would provide an opportunity for an interaction of these two cell types more extensive than that mediated solely by testosterone. However, there is no experimental evidence for such interactions at this time. Identification and purification of the many proteins produced by these two cell types *in vitro* (and discussed in the sections on Leydig cell culture and peritubular myoid cell culture) might provide much needed information on Leydig–myoid cell interactions.

Summary and Model of Intratesticular Regulation

The data presented above are summarized in Fig. 2. The information presented in this figure is drawn from diverse morphological, biochemical, physiological, and cell biology experiments using animals, cell suspensions, primary cultures, and established cell lines as experimental systems. Some of the information used to construct the model comes from indirect or circumstantial evidence or from several very different experimental systems, as discussed above. However, this model of integrated testicular function may provide a starting point for fruitful discussion and further experimentation. It seems likely that the model, complex as it seems, may be only a simplified version of the true *in vivo* regulation. However, it does suggest the great advances in our understanding of intratesticular regulation that have occurred in the last decade. It is obvious that the regulation of cell function within the testis is extremely complicated involving autocrine and paracrine, as well as endocrine regulatory systems.

The availability of defined culture has contributed a great deal to our understanding of the processes depicted in Fig. 2. These systems should continue to be of prime importance in future work in this field that will be required to define the mechanism of interaction suggested by earlier studies discussed above and provide further evidence for the

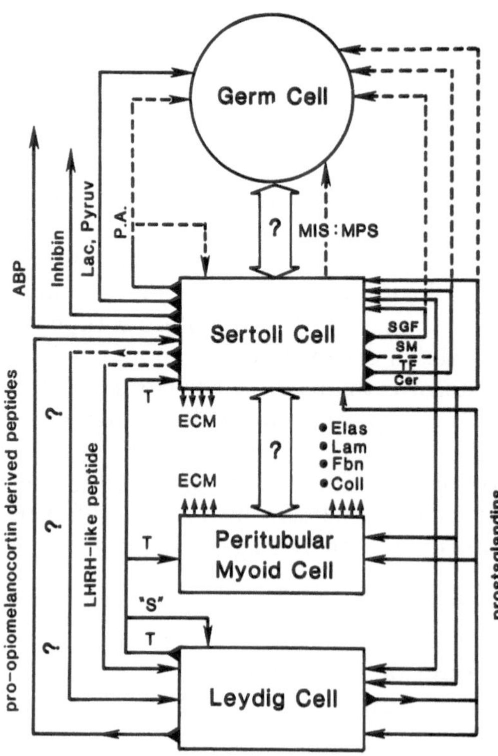

Figure 2. Model of intratesticular regulation. Solid lines indicate evidence is available for pathway, dotted line indicates indirect evidence for production or action of a factor. Question marks indicate evidence that cell–cell interactions exist but the factor(s) mediating such actions have not yet been identified. Abbreviations: MIS, meiosis-inducing substance; MPS, meiosis-preventing substance; PA, plasminogen activator; Lac, lactate; pyruv, pyruvate; SGF, seminiferous growth factor; SM, somatomedin; Cer, ceruloplasmin; T, testosterone; "S", steroids; ECM, extracellular matrix material; Elas, elastin; Lam, laminen; Fn, fibronectin; Coll, collagen. The evidence for such interactions is discussed in sections 3 and 4 of the text.

existence of those interactions that have been suggested by indirect evidence.

A single cell in culture is still an extremely complex living system. The development of functional clonal cell lines and the definition of hormone-supplemented conditions for the growth of these cells has done much to broaden our understanding of how cell function is regulated. Although a clonal cell line may be an artificial system compared with the *in vivo* situation, it is also a rational one. As we learn how to recreate, piece by piece, the proper environment required for a cell to function *in vitro*, we gain a profound insight into the multiple factors controlling cell function. Experience with cell lines derived from a number of different tissues including thyroid (Ambesi-Impiombato *et al.*, 1980), kidney (Chapter 6, this volume), and liver (Chapter 11, this volume) indicate that these cells are capable of expressing a broad range of differentiated functions in the proper *in vitro* conditions. The prevailing attitude of a decade ago that established cell lines represented artificial and highly undifferentiated systems is seen to be largely the result of culturing these cells in an artificial and inappropriate environment (i.e., serum-containing medium) on plastic culture dishes.

ACKNOWLEDGMENTS. This work was supported by National Institute of Health Grants #HO016149-01 and Program Project Grant #HD13541-01. I would like to thank Alica Byer and Florence Kaczorowski for technical assistance and Linda McKeiver for her help in the preparation of the manuscript.

References

Adashi, E. Y., and Hseuh, A. J., 1981, Autoregulation of androgen production in a primary culture of rat testicular cells, *Nature* (*London*) **293:**737–738.

Aldred, L. F., and Cooke, B. A., 1980, The deleterious effect of mechanical dissociation of rat testis on the functional activity and purification of Leydig cells using Percoll gradients, *Int. J. Andrology* **5:**191–195.

Ambesi-Impiombato, F. S., Parks, L. A. M., and Coon, H. G., 1980, Culture of hormone dependent functional endothelial cells from rat thyroids, *Proc. Natl. Acad. Sci. U.S.A.* **77:**3455–3459.

Aoki, A., and Fawcett, D. W., 1978, Is there a local feedback from the seminiferous tubules affecting activity of the Leydig cells, *Biol. Reprod.* **19:**144–158.

Ascoli, M., 1981, Regulation of gonadotropin receptors and gonadotropin responses in a clonal strain of Leydig tumor cells by epidermal growth factor, *J. Biol. Chem.* **256:**179–183.

Ascoli, M., 1982, Internalization and degradation of receptor-bound human choriogonadotropin in Leydig tumor cells, *J. Biol. Chem.* **257:**13306–13311.

Bardin, C. W., Musto, N. A., Gunsalus, G. L., Kotite, N., Cheng, S.-L., Larrea, F., and Becker, R., 1981, Extracellular androgen binding proteins, *Annu. Rev. Physiol.* **43:**189–198.

Benahmed, M., Dellamonica, C., Haour, F., and Saez, J. M., 1981, Specific low density lipoprotein receptors in pig Leydig cells. Role of this lipoprotein in cultured Leydig cells steroidogenesis, *Biophys. Biochem. Res. Commun.* **99:**1123–1130.

Benahmed, M., Bernier, M., Ducharme, S. R., and Saez, J. M., 1982, Steroidogenesis of cultured purified pig Leydig cells: Secretion and effects of estrogens, *Mol. Cell. Endocrinol.* **28:**705–716.

Bergh, A., 1983, Paracrine regulation of Leydig cells by the seminiferous tubules, *Int. J. Andrology.* **6:**57–65.

Bottenstein, J., Hayashi, I., Hutchings, S. H., Masui, H., Mather, J., McClure, D. B., Okasa, S., Rizzino, A., Sato, G., Serroro, G., Wolfe, R., and Wu, R., 1979, The growth of cells in serum-free medium, in: *Methods in Enzymology*, (W. B. Jakoby and I. H. Pastan, eds.) Volume LVIII, Academic Press, New York, pp. 94–109.

Bressler, R. S., and Ross, M. H., 1973, On the character of the monolayer outgrowth and the fate of the peritubular myoid cells in cultured mouse testis, *Exp. Cell. Res.* **78:**295–302.

Brinkman, A. O., Leemborg, F. G., Roodnat, E. M., deJong, F. H., and VanderMolen, H. J., 1980, A specific action of estradiol on enzymes involved in testicular steroidogenesis, *Biol. Reprod.* **23:**801–809.

Browning, J. V., Heindel, J. J., and Grotjan, H. E., 1983, Primary culture of purified Leydig cells isolated from adult rat testis, *Endocrinology* **112**:543–549.

Catt, K. J., Harwood, J. P., Clayton, R. N., Davies, T. F., Chan, V., Katikineni, M., Nozu, K., and Dufau, M. L., 1980, Regulation of peptide hormone receptors and gonadal steroidogenesis, *Recent Prog. Horm. Res.* **36**:577–720.

Cigorraga, S. B., Dufau, M. L., and Catt, K. J., 1978, Regulation of luteinizing hormone receptors and steroidogenesis in gonadotropin-desensitized Leydig cells, *J. Biol. Chem.* **253**:4297–4304.

Clermont, Y., 1958, Contractile elements in the limiting membrane of the seminiferous tubules of the rat, *Exp. Cell Res.* **15**:438–440.

deJong, F. H., 1979, Inhibin—Fact or artifact?, *Mol. Cell. Endocrinol.* **13**:1–10.

deKrester, D. M., Catt, K. J., Dufau, M. L., and Hudson, B., 1971, Studies on rat testicular cells in tissue culture, *J. Reprod. Fertil.* **24**:311–318.

Denduchis, B., Lustig, L., Gonzalez, N. N., and Mancini, R. E., 1975, Studies on the nature of extracellular components of rat seminiferous tubular wall. I. Isolation and chemical characterization of basement membrane, *Biol. Reprod.* **13**:274–281.

DePhilip, R. M., and Kierszenbaum, A. L., 1982, Hormonal regulation of protein synthesis, secretion, and phosphorylation in cultured rat Sertoli cells, *Proc. Natl. Acad. Sci. U.S.A.* **79**:6551–6555.

Dorrington, J. H., and Fritz, I. B., 1974, Cell types influenced by FSH in the rat testis, in: *Gonadotropins and Gonadal Functions* (N. R. Moudgal, ed.), Academic Press, New York, pp. 500–512.

Dorrington, J. H., Fritz, I. B., and Armstrong, D. T., 1978, Control of testicular estrogen synthesis, *Biol. Reprod.* **18**:55–64.

Dufau, M. L., Horner, K. A., Hayashi, K., Tsuruhara, T., Conn, P. M., and Catt, K. J., 1978, Actions of choleragen and gonadotropin in isolated Leydig cells, *J. Biol. Chem.* **253**:3721–3729.

Dym, M., and Fawcett, D. W., 1970, The blood-testis barrier in the rat and the physiological compartmentation of the seminiferous epithelium, *Biol. Reprod.* **3**:308–326.

Elkington, J. S. H., and Fritz, I. B., 1980, Regulation of sulphoprotein synthesis by rat Sertoli cells culture, *Endocrinology* **107**:970–976.

Fawcett, D. W., 1975, Ultrastructure and function of the Sertoli cell, in: *Handbook of Physiology: Endocrinology,* Volume V (D. W. Hamilton and R. O. Greep, eds.), American Physiological Society, Washington, D.C., pp. 21–56.

Feig, L. A., Bellve, A. R., Erickson, N. H., and Klagsburn, M., 1980, Sertoli cells contain a mitogenic polypeptide, *Proc. Natl. Acad. Sci. U.S.A.* **77**:4774–4778.

Freeman, D. A., and Ascoli, M., 1982, Desensitization of steroidogenesis in cultured Leydig tumor cells: Role of cholesterol, *Proc. Natl. Acad. Sci. U.S.A.* **79**:7796–7800.

Fritz, I. B., 1978, Sites of action of androgens and follicle stimulating hormone on cells of the seminiferous tubule, in: *Biochemical Action of Hormones,* Volume V (G. Litwack, ed.), Academic Press, Inc., New York, pp. 249–281.

Gospodarowicz, D., and Ill, C., 1980, Extracellular matrix and control of proliferation of vascular endothelial cells, *J. Clin. Invest.* **65**:1351–1364.

Griswold, M. D., Solari, A., Tung, P. S., and Fritz, I. B., 1977, Stimulation by follicle-stimulating hormone of DNA synthesis and of mitosis in cultured Sertoli cells prepared from testes of immature rats, *Mol. Cell. Endocrinol.* **7**:151–165.

Grotjan, H. E., Heindel, J. J., and Steinberger, E., 1978, Prostaglandin inhibition of testosterone production induced by luteinizing hormone, dibutyryl cyclic AMP or 3-isobutyl-1-methyl-xanthine in dispersed rat testicular interstitial cells, *Steroids* **32**:307–322.

Gunsalus, G. L., Musto, N. A., and Bardin, C. W., 1978, Immunoassay of androgen binding protein in blood: A new approach for the study of the seminiferous tubule, *Science* **200:**65–66.

Hansson, V., and Djoseland, O., 1972, Preliminary characterization of the 5α-dihydrotestosterone binding protein in the epididymal cytosol fraction. *In vivo* studies, *Acta Endocrinol.* **71:**614–624.

Haour, F., and Saez, J. M., 1977, Regulation by hCG of gonadotropins receptors in testicular Leydig cells. Evidence for a down regulation, *Mol. Cell Endocrinol.* **7:**17–24.

Haour, F., Kouznetzoua, B., Dray, F., and Saez, J. M., 1979, hCG-Induced prostaglandin E_2 and $F_{2\alpha}$ release in adult rat testis: Role in Leydig cell desensitization to hCG, *Life Sci.* **24:**2151–2259.

Haour, F., Mather, J., Kouznetzova, B., and Dray, F., 1981, Role of prostaglandins in Leydig cell stimulation by hCG and Leydig cell function, in: *Reproductive Processes and Contraception* (K. W. McKerns, ed.), Plenum Press, New York, pp. 691–713.

Haour, F., Bommelaer, M. C., Bernier, M., Sanchez, P., Saez, J., and Mather, J. P., 1983, Androgen production in primary culture of immature porcine Leydig cells, *Mol. Cell Endocrinol.* **30:**73–84.

Hay, E. D., 1981, Collagen and embryonic development, in: *Cell Biology of Extracellular Matrix* (E. D. Hay, ed.), Plenum Press, New York, pp. 379–410.

Hayashi, I., and Sato, G., 1976, Replacement of serum by hormones permits growth of cells in a defined medium, *Nature (London)* **259:**132–134.

Hayashi, I., Hutchings, S., Mather, J., and Sato, G., 1977, Replacement of serum by hormones and its implication for primary cell culture, in: *The International Congress Series*, Volume No. 408 (E. Von Wasielewski and W. H. Chick, eds.),Excerpta Medica, Amsterdam, Holland pp. 23–27.

Hermo, L., Lalli, M., and Clermont, Y., 1977, Arrangement of connective tissue components in the walls of seminiferous tubules of man and monkey, *Am. J. Anat.* **148:**433–446.

Hovatta, O., 1972, Effect of androgens and antiandrogens on the development of the myoid cells of the rat seminiferous tubule organ culture, *Z. Zellforsh* **131:**299–308.

Hseuh, A. J. W., 1980, Gonadotropin stimulation of testosterone production in primary culture of adult rat testis cells, *Biochem. Biophys. Res. Commun.* **97:**506–512.

Hseuh, A. J. W., Dufau, M. L., and Catt, K. J., 1978, Direct inhibitory effect of estrogen on Leydig cell function of hypophysectomized rats. *Endocrinology* **103:**1096–1102.

Huhtaniemi, I. T., Nozu, K., Warren, D. W., Dufau, M. L., and Catt, K. J., 1982, Acquisition of regulatory mechanisms for gonadotropin receptors and steroidogenesis in the maturing rat testis, *Endocrinology* **111:**1711–1720.

Hutson, J. C., 1983, Metabolic cooperation between Sertoli cells and peritubular cells in culture, *Endocrinology* **112:**1375–1381.

Hutson, J. C., and Stocco, D. M., 1981, Peritubular cell influence on the efficiency of androgen binding protein by Sertoli cells in culture, *Endocrinology*. **108:**1362–1368.

Johnsonbaugh, R. E., Ritzen, E. M., Hall, K., Parvinen, M., and Wright, W. W., 1982, Secretion of somatomedin-like compound from Sertoli cells *in vitro,* in: *Proceedings of the Second European Workshop on Molecular and Cellular Endocrinology of the Testis*, Rotterdam, The Netherlands.

Jutte, N. H. P. M., Grootegoed, J. A., Rommerts, F. F. G., Clausen, O. P. F., and Molen, H. J. van der, 1982, Regulation of survival of rat pachytene spermatocytes by lactate supply from Sertoli cells, *J. Reprod. Fertil.* **65:**431–438.

Karl, A. F., and Griswold, M. D., 1980, Actions of insulin and vitamin A on Sertoli cells, *Biochem. J.* **186:**1001–1003.

Kissinger, C., Skinner, M. K., and Griswold, M. D., 1982, Analysis of Sertoli cell-secreted proteins by two-dimensional gel electrophoresis, *Biol. Reprod.* **27**:233–246.

Lacroix, M., Smith, F., and Fritz, I. B., 1977, Secretion of plasminogen activator by Sertoli cell-enriched cultures, *Mol. Cell. Endocrinol.* **9**:227–236.

Lacroix, M., Parvinen, M., and Fritz, I. B., 1981, Localization of testicular plasminogen activator in discrete portions (stages VII and VIII) of the seminiferous tubule, *Biol. Reprod.* **25**:143–146.

Lacroix, M., Smith, F. E., and Fritz, I. B., 1982, Changes in levels of plasminogen activator activity in normal and germ-cell depleted testes during development, *Mol. Cell. Endocrinol.* **26**:259–267.

Larrea, F., Musto, N., Gunsalus, G., Mather, J., and Bardin, C. W., 1981, Origin of the subunit composition of androgen binding protein of the rat testis, *J. Biol. Chem.* **256**:12566–12573.

Leblond, C. P., and Clermont, Y., 1952, Definition of the stages of the cycle of the seminiferous epithelium in the rat, *Ann. N.Y. Acad. Sci.* **55**:548–573.

Marquardt, H., and Todaro, G. J., 1982, Human transforming growth factor: Production by a melanoma cell line, purification, and initial characterization, *J. Biol. Chem.* **257**:2220–2225.

Mather, J. P., 1980, The establishment and characterization of two distinct mouse testicular epithelial cell lines, *Biol. Reprod.* **23**:243–250.

Mather, J. P., 1982, Ceruloplasmin, a copper-transport protein, can act as a growth promoter for some cell lines in serum-free medium, *In Vitro* **18**:990–996.

Mather, J. P., and Furlanetto, R., 1983, Insulin and somatomedin as growth promoters of cells in serum-free medium, in: *Insulin-like Growth Factors/Somatomedins: Basic Chemistry, Biology and Clinical Importance* (E. M. Spencer, ed.), Walter de Gruyter & Co., New York, (in press).

Mather, J. P., and Haour, F., 1981, Hormone response of testicular cells in culture: Established cell lines and primary cultures, in: *Functionally Differentiated Cell Lines* (G. Sato, ed.), Alan R. Liss, Inc., New York, pp. 93–108.

Mather, J. P., and Phillips, D. M., 1983a, Culture of testicular somatic cells, in: *Methods in Molecular and Cell Biology* (D. Barnes, D. Sirbasque, and G. Sato, eds.), Alan R. Liss, Inc. New York, (in press).

Mather, J. P., and Phillips, D. M., 1983b, Establishment of a peritubular myoid-like cell line and interactions between established cell lines in culture, (*submitted*).

Mather, J. P., and Sato, G., 1979, The use of hormone-supplemented serum-free media in primary culture, *Cell Res.* **124**:215–224.

Mather, J. P., Haour, F., and Saez, J. M., 1981, Maintenance of gonadotropin receptors and steroidogenic responsiveness in primary cultures of interstitial cells from the mouse, rat and pig, *Steroids* **38**:35–44.

Mather, J. P., Perez-Infante, L-Z., Zhuang, V., and Phillips, D. M., 1982a, Culture of testicular cells in hormone-supplemented serum-free medium, *Ann. N.Y. Acad. Sci.* **383**:44–68.

Mather, J. P., Saez, J. M., and Haour, F., 1982b, The regulation of gonadotropin receptors and steroidogenesis in cultured porcine Leydig cells, *Endocrinology* **110**:933–940.

Mather, J. P., Saez, J. M., Dray, F., and Haour, F., 1982c, Hormone–hormone and hormone–vitamin interactions in the control of growth and function of Leydig cells *in vitro*, in: *Cold Spring Harbor Conferences on Cell Proliferation,* Volume 9, Cold Spring Harbor, New York, pp. 1117–1128.

Mather, J. P. Saez, J. M., Dray, F., and Haour, F., 1983a, Vitamin E prolongs survival and function of porcine Leydig cells in culture, *Acta Endocrinol.* **102**:470–475.

Mather, J. P., Gunsalus, G. L., Musto, N. A., Cheng, C. Y., Parvinen, M., Wright, W., Margioris, A., Liotta, T., Perez-Infante, V., Becker, D. T., and Bardin, C. W., 1983b,

The hormonal and cellular control of Sertoli cell secretion, *J. Steroid. Biochem.* **19:**41–51.

Means, A. R., Fakunding, J. L., Huckins, C., Tindall, D. J., and Vitale, R., 1976, Follicle-stimulating hormone, the Sertoli cell, and spermatogenesis, *Recent. Prog. Horm. Res.* **32:**477–528.

Means, A. R., Deman, J. R., Tash, J. S., Tindall, D. J., Sickle, M. Van, and Welsh, M. J., 1980, Regulation of the testis Sertoli cell by follicle stimulating hormone, *Annu. Rev. Physiol.* **42:**59–70.

Meistrich, M. L., Longtin, J., Brock, W. A., Grimes, S. R. Jr., and Mace, M. L., 1981, Purification of rat spermatogenic cells and preliminary biochemical analysis of these cells, *Biol. Reprod.* **25:**1065–1077.

Moger, W. H., 1980, Direct effects of estrogens on the endocrine function of the mammalian testis, *Can. J. Physiol. Pharmacol.* **58:**1011–1022.

Murphy, P. R., and Moger, W. H., 1982, Short-term primary culture of mouse interstitial cells: Effects of culture conditions on androgen production, *Biol. Reprod.* **27:**38–47.

Musto, N. A., Gunsalus, G. L., and Bardin, C. W., 1980, Purification and characterization of androgen binding protein from the rat epididymis, *Biochemistry* **19:**2853–2860.

Neaves, W. B., 1977, The blood–testis barrier, in: *The Testis,* Volume IV (A. D. Johnson and W. R. Gomes, eds.),Academic Press, New York, pp. 126–163.

Parvinen, M., 1982, Regulation of the seminiferous epithelium, *Endocrinol. Rev.* **3:**404–417.

Parvinen, M., and Vanha-Perttula, T., 1972, Identification and enzyme quantitation of the stages of the seminiferous epithelial wave in the rat, *Anat. Rec.* **174:**435–450.

Parvinen, M., Byskov, A. G., Andersen, C. Y., and Grimsted, J., 1982, Is the spermatogenic cycle regulated by MIS and MPS?, *Ann. N.Y. Acad. Sci.* **383:**483–484.

Parvinen, M., Wright, W. W., Phillips, D. M., Mather, J. P., Musto, N. A., and Bardin, C. W., 1983, Rat spermatogenesis *in vitro*: Completion of meiosis and early spermiogenesis, *Endrocrinology.* **112:**1150–1152.

Perez-Infante, V., and Mather, J. P., 1982, The role of transferrin in the growth of testicular cell lines in serum-free medium, *Exp. Cell. Res.* **142:**325–332.

Pfeiffer, J. C., Alger, E. A., and Boccabella, A. V., 1981, Origin of collagen in the peritubular boundary tissue of the rat testis, *Anat. Rec.* **199**(3):200A.

Reid, L., 1982, Regulation of growth and differentiation of mammalian cells by hormones and extracellular matrix, in: *From Gene to Protein: Translation into Biotechnology* (F. Ahmad, J. Schultz, E. Smith, and W. Whelan, eds.), Academic Press, New York, pp. 53–73.

Rich, K. A., Kerr, J. B., and DeKrester, D. M., 1979, Evidence for Leydig cell dysfunction in rats with seminiferous tubule damage, *Mol. Cell Endocrinol.* **13:**123–135.

Rich, K. A., Bardin, C. W., Gunsalus, G. L., and Mather, J. P., 1983, Age-dependent secretion of androgen binding protein by cultured Sertoli cells, *Endocrinology.* **113**(6):2284–2293.

Risburger, G. P., Kerr, J. B., and DeKrester, D. M., 1981, Evaluation of Leydig cell function and gonadotropin binding in unilateral and bilateral cryptorchidism: Evidence for local control of Leydig cell function by the seminiferous tubule, *Biol. Reprod.* **24:**534–540.

Ritzen, E. M., Nayfeh, S. N., French, F. S., and Dobbins, M. C., 1971, Demonstration of androgen binding components in rat epididymis cytosol and comparison with binding components in prostate and other tissues, *Endocrinology.* **89:**143–151.

Ritzen, E. M., Biotani, L., Parvinen, M., French, F. S., and Feldman, M., 1982, Stage-dependent secretion of ABP by rat seminiferous tubules, *Mol. Cell. Endocrinol.* **25:**25–33.

Robinson, R., and Fritz, I. B., 1979, Myoinositol biosynthesis by Sertoli cells, and levels of myoinositol biosynthetic enzymes in testis and epididymis, *Can. J. Biochem.* **51:**962–967.

Robinson, R., and Fritz, I. B., 1982, Metabolism of glucose by Sertoli cells in culture, *Biol. Reprod.* **24:**1032–1045.

Sar, M., Stumpf, W. E., McLean, W. S., Smith, A. A., Hansson, V., Nayfeh, S. N., and French, F. S., 1975, Localization of androgen target cells in the rat testis, in: *Current Topics in Molecular Endocrinology,* Volume 2 (F. S. French, V. Hansson, E. M. Ritzen, and S. N. Nayfeh, eds.),pp. 311–322.

Setchell, B. P., Davies, R. V., and Main, S. J., 1977, Inhibin, in: *The Testis,* Volume IV (A. D. Johnson, W. R. Gomes, N. L. Vandemark, eds.),Academic Press, New York, pp. 190–238.

Setchell, B. P., 1978, *The Mammalian Testis,* Cornell University Press, Ithaca, New York.

Sharpe, B., Pekary, A. E., Meyer, N. V., and Hersham, J. M., 1980, β-endorphin in male rat reproductive organs, *Biochem. Biophys. Res. Commun.* **95:**618–623.

Sharpe, R. M., and Cooper, I., 1982a, The mode of action of LHRH agonists on the rat Leydig cell, *Mol. Cell Endocrinol.* **27:**199–211.

Sharpe, R. M., and Cooper, I., 1982b, Stimulating effects of LH-RH and its agonists on Leydig cell steroidogensis *in vitro, Mol. Cell. Endocrinol.* **26:**141–150.

Sharpe, R. M., Fraser, H. M., Cooper, I., and Rommerts, F. F. G., 1981, Sertoli-Leydig cell communication via an LH-RH-like factor, *Nature (London)* **290:**785–787.

Shin, S., 1967, Studies on interstitial cells in tissue culture: Steroid biosynthesis in monolayers of mouse testicular interstitial cells, *Endocrinology.* **81:**440–448.

Shin, S., Yoshihiro, Y., and Sato, G. H., 1968, Studies on interstitial cells in tissue culture. II. Steroid biosynthesis by a clonal line of rat testicular cells, *Endocrinology.* **82:**614–616.

Skinner, M. K., and Griswold, M. D., 1980, Sertoli cells synthesize and secrete transferrin-like protein, *J. Biol. Chem.* **255:**9523–9525.

Smith, P. E., 1927, The disabilities caused by hypophysectomy and their repair, *J. Am. Med. Assoc.* **88:**158–161.

Smith, P. E., 1944, Maintenance and restoration of spermatogenesis in hypophysectomized rhesus monkeys by androgen administration, *Yale J. Biol. Med.* **17:**281–287.

Soll, A. H., 1981, Physiology of isolated canine parietal cells: Receptors and effectors regulating function, in: *Physiology of the Gastrointestinal Tract* (L. R. Johnson, ed.), Raven Press, New York, pp. 673.

Steinberger, A., 1975, Studies on spermatogenesis and steroidogenesis in culture, *Am. Zool.* **15:**273–278.

Steinberger, A., Steinberger, E., and Perloff, W. H., 1964, Mammalian testis in organ culture, *Exp. Cell. Res.* **36:**19–27.

Tsong, S. D., Phillips, D. M., Halmi, N., Liotta, A. S., Margioris, A., Bardin, C. W., and Krieger, D. T., 1982a, ACTH and β-endorphin-related peptides are present in multiple sites in the reproductive tract of the male rat, *Endocrinology.* **110:**2204–2206.

Tsong, S. D., Phillips, D. M., Halmi, N., Krieger, D. T., and Bardin, C. W., 1982b, β-Endorphin is present in the male reproductive tract of five species, *Biol. Reprod.* **27:**764–775.

Tung, P. S., and Fritz, I. B., 1977, Isolation and culture of testicular cells: A morphological characterization, in: *Techniques of Human Andrology* (E. S. E. Hafez, ed.), Elsevier, North Holland, Amsterdam, pp. 125–146.

Tung, P. S., and Fritz, I. B., 1980, Interactions of Sertoli cells with myoid cells *in vitro, Biol. Reprod.* **23:**207–217.

Vernon, R. G., Go, V. L. W., and Fritz, I. B., 1971, Studies on spermatogenesis in rats, II. Eivdence that carnitine acetyltransferase in a marker enzyme for the investigation of germ cell differentiation, *Can. J. Biochem.* **49:**761–767.

Vernon, R. G., Kopec, B., and Fritz, I. B., 1974, Observations on the binding of androgens by rat testis seminiferous tubules and testis extracts, *Mol. Cell. Endocrinol.* **1:**167–187.

Warren, D. W., Dufau, M. L., and Catt, K. J., 1982, Hormonal regulation of gonadotropin receptors and steroidogenesis in cultured fetal rat testis, *Science* **218:**375–377.

Welsh, M. J., and Wiebe, J. P., 1975, Rat Sertoli cells: A rapid method for obtaining viable cells, *Endocrinology.* **96:**618–624.

Wright, W. W., Musto, N. A., Mather, J. P., and Bardin, C. W., 1981, Sertoli cells secrete both testis-specific and serum proteins, *Proc. Natl. Acad. Sci. U.S.A.* **78:**7565–7569.

Wright, W. W., Parvinen, M., Musto, N. A., Gunsalus, G. L., Phillips, D. M., Mather, J. P., and Bardin, C. W., 1983, Identification of stage-specific proteins synthesized by rat seminiferous tubules, *Biol. Reprod.* **29:**257–270.

Zhuang, L.-Z., and Mather, J. P., 1983, The establishment and growth in serum-free culture of a rat Sertoli cell line, (submitted).

CHAPTER 9

Attachment Factors in Cell Culture

DAVID BARNES

Introduction

Cells *in vivo* are subject to several classes of environmental factors that affect proliferative and differentiative potential. Among these are nutrients, hormones, binding proteins that modulate the actions of nutrients and hormones, extracellular enzymes, and extracellular matrix, stroma or basement membranes (Barnes and Sato, 1980b,c). Although the importance of the extracellular substratum *in vivo* in the regulation of cellular growth and differentiation has been recognized for some time (Hay, 1981), only recently has it become possible to explore *in vitro* the complicated and often subtle relationships between cells and substrata and to examine the effects of cell–substratum interactions. Several important insights led to improved experimental design of investigations into the role of the cell substratum in culture. First, it was recognized that components of the serum supplement to conventional culture media provided a substratum for cells in culture, and that the nature and effects of these and other factors affecting cell differentiation and proliferation could best be examined in serum-free cell culture (Fisher *et al.*, 1958; Grinnell, 1978; Barnes and Sato, 1980b,c). Secondly, cells in culture were discovered to synthesize and deposit extracellular matrix components that sometimes are organized in elaborate and complex simulations *in vitro* of basement membrane or interstitial stroma *in vivo* (Hay, 1981; Kleinman *et al.*, 1981). Finally,

DAVID BARNES • Department of Biological Sciences, University of Pittsburgh, Pittsburgh, Pennsylvania 15260.

it was realized that at least some of the interactions between cells and substrata were primarily the result of interactions of cells with discrete molecular entities present within the complex mixture of molecules making up the substratum, and that these molecules could be purified free of the extracellular matrix and demonstrated to retain biological activity and their biochemical properties studied in serum-free model substratum systems *in vitro* (Yamada and Olden, 1978; Timpl *et al.*, 1979; Hewitt *et al.*, 1980; Hay, 1981; Kleinman *et al.*, 1981).

The extracellular matrix is a complex and organized mixture of a number of components: collagen, glycosaminoglycan, proteoglycan, and glycoprotein (Hay, 1981). Several new approaches, including the use of serum-free cell culture techniques, have allowed the identification of some of the components of extracellular matrix that are active in cell–substratum interactions, and also have led to studies of the mechanisms of action of these factors. In turn, the identification and availability of purified, biologically active components of the extracellular matrix and other factors that promote attachment and spreading of cells on culture vessels have allowed the development of well-defined, serum-free culture media for cell types that probably cannot be grown in the absence of members of this class of medium supplements (Barnes and Sato, 1980b,c). The attachment factors useful in cell culture that will be reviewed in this chapter are collagen (Kleinman *et al.*, 1981), fibronectin (Yamada and Olden, 1978), laminin (Timpl *et al.*, 1979), chondronectin (Hewitt *et al.*, 1980), serum-spreading factor (Barnes *et al.*, 1980, 1983), epibolin (Stenn, 1981), and fetuin (Fisher *et al.*, 1958). In each case, in addition to a summary of the effects and use of the factor in culture, a brief review of the procedures for isolation, biochemical properties, and biological activities of the material *in vivo* is also included. An appreciation of the *in vivo* physiology of each attachment factor, as far as it is understood, is important because when the opportunity arises to use two or more different attachment factors in cell culture, singly or in combination, it is useful to consider in making choices where and how each attachment factor exerts its activity in the whole animal.

Also included in this chapter is a brief summary of observations made in experiments in which the extracellular matrix deposited by some cells in culture or isolated from tissues is used as an attachment and spreading-promoting substratum that approximates to some extent the natural extracellular matrix to which cells are exposed *in vivo* (Gospodarowicz *et al.*, 1978; Rojkind *et al.*, 1980). Experiments of this type possess the obvious advantage of presenting to the cells to be studied a more physiologically relevant substratum than one composed of single, purified components of the extracellular matrix, combinations of these components, or other attachment factors. However, experiments of this type also possess the obvious disadvantage of introducing into the culture system undefined biologically active components deposited in the matrix of the cells, or bound by the matrix from the medium. Finally, as a summary and conclusion, some of the most recent and interesting experiments dealing with the complex relationships of cell–substratum interactions are discussed as possible indications of future directions in the field.

Collagen

Biochemical and Biological Properties of Collagen

Collagen is the most abundant protein in stroma or basement membrane, and an obvious candidate as a molecule capable of acting in culture to replace the extracellular matrix with which cells interact in their normal environments *in vivo* (Hay, 1981; Kleinman *et al.*, 1981). Collagens may be classified into five types based on biochemical and other criteria. All collagen types are composed of protein subunits of distinctive amino acid composition compared with most other proteins found in nature; about one third of the amino acids are glycine and about one fifth are proline or hydroxyproline. Hydroxylysine is also present and is involved in crosslinking of collagen molecules, as well as acting as a site of covalent attachment of galactose and glucosylgalactose, the only sugars found on the molecule. In the course of collagen synthesis in the cell, three of the subunit chains (designated alpha 1, alpha 2, etc., based on biochemical differences among the chains) combine to form a triple-helical procollagen molecule that is processed and transported to extracellular sites of deposition and converted by specific proteolytic cleavage to insoluble tropocollagen. This molecule is crosslinked and arranged in an orderly way into large, discrete collagen fibrils or less distinct, fibrous superstructures. Interstitial collagens, types I, II, and III, are biochemically similar to each other and somewhat different from basement membrane collagens, types IV and V, in relative amounts of some of the characteristic amino acids and sugar. Type IV collagen, for instance, contains considerably more 3-hydroxyproline, hydroxylsine, and sugar residues and less alanine and proline than the interstitial types (Hay, 1981; Kleinman *et al.*, 1981). The collagen types also are individually distinguishable by immunological methods (Rennard *et al.*, 1980).

The Use of Collagen in Cell Culture

It has been recognized for some time that the extracellular matrix, and collagen in particular, is involved in the determination of cell shape, adhesiveness, and proliferative and differentiative potential *in vivo* and that synthesis and deposition of the different collagen types is developmentally regulated in the intact, maturing organism (Ketley *et al.*, 1976; Stenman *et al.*, 1979; Leivo *et al.*, 1980; Wicha *et al.*, 1980; Hay, 1981; Kleinman *et al.*, 1981). However, many of the most definitive experiments regarding the role of collagen in regulation of cellular processes have come from experiments *in vitro* using cultured cells. The use of collagen in culture vessels is somewhat difficult, mainly due to the insoluble nature of the polymerized fibrous forms, and interpretation of results from experiments in which collagen gels are used sometimes must be qualified by the possibility that the observed effects may be the results of activity of other molecules, such as fibronectin, laminin, chondronectin, or proteoglycans associated with the preparations under the cell culture conditions used (Klebe, 1974; Engvall *et al.*, 1978; Klebe *et al.*, 1978; Gold and Pearlstein, 1980; Ruoslahti and Engvall, 1980; Hayman *et al.*, 1982a; Leivo *et al.*, 1982).

Collagen preparations generally are handled as solutions in acidic buffers, and gels are formed on culture vessels by exposure to increased ionic strength, pH, and temperature or to ammonia vapors (Hauschka and Konigsberg, 1966; Kleinman *et al.*, 1979a; Murakami and Masui, 1980; Yang *et al.*, 1980a,b; Barnes *et al.*, 1981; Sun-Yang *et al.*, 1981). Gels are sometimes air dried after polymerization. Depending on the procedures used for preparation, the collagen substratum may represent molecules in a range of states from native, hydrated collagen fibrils that are quite similar to that formed *in vivo* to highly denatured material with little or no biological activity in the absence of other factors. An extreme example of such a substratum is gelatin, a form of denatured collagen that has been found to be of some limited use *in vitro* as a replacement for the native substratum. In addition to the growth of cells on collagen applied to the vessel as a coat or gel on the bottom of the culture vessel, systems also have been developed in which cells are maintained on floating collagen gels, embedded in collagen gels, or grown on copolymers of collagen and hydroxyethylmethacrylate, collagen-coated cellulose sponges or collagen-coated nylon meshes (Russo *et al.*, 1976; Emerman and Pitelka, 1977; Sirica *et al.*, 1979; Civerchia-Perez *et al.*, 1980; Yang *et al.*, 1980a,b). Until recently, virtually all of the collagen preparations used in cell culture were of the interstitial type, isolated from skin, tendon, or muscle. The availability of a murine tumor that produces a large amount of basement membrane (type IV) collagen has allowed the large-scale isolation and use of this collagen type in culture experiments in the last several years (Timpl *et al.*, 1978). Manipulations of basement membrane collagen for use in culture are in general similar to those used for the preparation of gels of the other collagen types, although the biological effects of basement membrane collagen in some cases is quite different from that of interstitial collagen.

Effects of Collagen on Adhesion, Proliferation, and Differentiation of Cells in Culture

Collagen mediates cell attachment in culture by both direct and indirect mechanisms. Direct mechanisms involve noncovalent attachment of the cell membrane to the extracellular collagen matrix; indirect mechanisms involve attachment of cells through interaction with attachment proteins like fibronectin, laminin, and chondronectin that are noncovalently bound to the collagen matrix (Grinnell and Minter, 1978; Kleinman *et al.*, 1978; Postlethwaite *et al.*, 1978; Murray *et al.*, 1979; Timpl *et al.*, 1979; Wicha *et al.*, 1979; Grotendorst *et al.*, 1981; Rubin *et al.*, 1981; Hewitt *et al.*, 1980, 1982; Ruoslahti *et al.*, 1982; Engvall and Ruoslahti, 1977). The binding of attachment factors to collagen occurs through sites that are specific for this interaction on both the attachment factors and collagen molecules, and the interaction between the cells and collagen-bound attachment factors also is likely to occur through specific sites on the cell membrane and the attachment factors (Kleinman *et al.*, 1979b; Grinnell, 1980; Pierschbacher *et al.*, 1981, 1983; Ruoslahti *et al.*, 1981; Yamada *et al.*, 1981). Direct interactions in culture between some cell types, such as hepatocytes, and collagen may represent a less specific binding of the cell membrane to low-affinity sites present in large numbers on the collagen molecules (Rubin *et al.*, 1981). Other cell types such as mammary epithelial, epidermal, and smooth muscle cells exhibit a preferential binding to some collagen types, or bind to some collagen types directly and to others through collagen-bound attachment factors (Murray *et al.*, 1979; Wicha *et al.*, 1979; Grotendorst *et al.*, 1981). Fibroblasts seem to bind to all collagen types, but interact with denatured or dried collagen or gelatin through the collagen-bound attachment factor, fibronectin (Klebe, 1974; Grinnell and Minter, 1978; Postlethwaite *et al.*, 1978; Timpl *et al.*, 1979; Gold and Pearlstein, 1980). In addition to acting directly and indirectly as an attachment-promoting substratum for fibroblasts, collagen and collagen fragments also are chemotactic for this cell type (Postlethwaite *et al.*, 1978).

A collagen substratum is important for the maintenance and growth of some cell types in culture. Because many cell types synthesize collagen and attachment factors *in vitro* and because some attachment factors are present in serum used in conventional culture media, it is difficult to evaluate the general importance of collagen to the health and differentiative and proliferative capacity of cells cultured with serum. Certainly it may be demonstrated in serum-free medium, a situation in which at least the contribution of exogenous factors from serum can be eliminated, that collagen is capable of stimulating cell growth for some epithelial cell types (Murakami and Masui, 1980; Barnes and Sato, 1980b; Barnes *et al.*, 1981; Van der Bosch, 1984). Perhaps the most dramatic effects of a collagen substratum in culture may be observed on mammary epithelia (Wicha *et al.*, 1979; Salomon *et al.*, 1981; Kidwell *et al.*, 1983). Under appropriate conditions in culture, it can be shown that cells of this type taken from several species are dependent for extended proliferation on the presence of collagen in the cultures, either provided as an exogenous substratum or synthesized *in vitro* by the cells themselves, and that some of the factors identified as growth promoting in culture for this cell type are acting to affect the amount of collagen synthesized by the cells or the amount of collagen degraded in the culture system.

In addition to affecting the proliferation of mammary and other cell types in culture, collagen substrata may also elicit marked effects on the differentiative state of the cells. Mammary epithelia exhibit considerable morphological differentiation when maintained in culture on collagen gels attached to plastic, collagen-coated cellulose sponges or collagen membranes floating free in the medium, whereas little differentiation is seen if the same cell type is maintained attached to the plastic tissue culture vessel in the absence of collagen (Russo *et al.*, 1976; Emerman and Pitelka, 1977; Yang *et al.*, 1980a,b; Sun-Yang *et al.*, 1981). Similarly, hepatocytes maintained on collagen-coated nylon meshes exhibit morphological and biochemical properties characteristic of differentiated liver *in vivo* for prolonged periods of time, with some changes to characteristics associated with fetal liver observed coordinate with increased DNA synthesis in the cultures (Sirica *et al.*, 1979). Floating collagen gels have also been used as an effective means of maintaining differentiated liver cells *in vitro*, although it is not clear why floating gels are more effective in this respect than gels attached to the dishes in the culture vessels (Michalopoulos and Pitot, 1975; Strom and Michalopoulos, 1982). Development of muscle cells in culture is characterized by a striking requirement for a collagen substratum, although several different collagen types will function in this way, and fibronectin as a collagen-associated attachment factor is involved in at least part of the differentiative process (Hauschka and Konigsberg, 1966; Ketley *et al.*, 1976; Furcht *et al.*, 1978; Podleski *et al.*, 1979; John and Lawson, 1980). Development of differentiated epidermal cells and chondrocytes *in vitro* is also enhanced by a collagen matrix; type I or type II collagen is preferred by chondrocytes and type IV by epidermal cells (Murray *et al.*, 1979; Pennypacker *et al.*, 1979; Weiss and Reddi, 1980; Dessau *et al.*, 1981).

Synthesis and Deposition of Collagen by Cells in Culture

In addition to the situations in which collagen can be shown to affect directly the proliferative or differentiative state of the culture, many situations probably exist in which collagen synthesized by cells in the culture exerts effects in the absence of any added collagen from exogenous sources. The dependence of some culture systems on feeder layers of fibroblastic or other cell types probably reflects to some degree a requirement for collagen or other components of the substratum synthesized *in vitro* by feeder cells. Furthermore, growth-stimulatory effects of some medium supplements in culture appear to be due in some cases to the ability of these factors to influence production of collagen by cells in the culture. Effects of growth factors on mammary cells and other cell types in culture, as well as growth-stimulatory effects of nutritional supplements such as ascorbic acid, may be in many cases related to effects of the supplements on collagen production by the cells of the culture (Murakami and Masui, 1980; Barnes and Sato, 1980b; Salomon *et al.*, 1981; Kidwell *et al.*, 1983).

Collagens or procollagens of all types, as well as most of the other components of extracellular matrix *in vivo* identified thus far, have been found to be synthesized by one or another cell type under various culture conditions (Hay, 1981; Kleinman *et al.*, 1981; Grotendorst *et al.*, 1982). Collagens found associated with normal fibroblastic or endothelial cells or sarcomas in culture are primarily types I and III, although some type IV collagen can also be detected (Vaheri *et al.*, 1978; Hedman *et al.*, 1979; Alitalo *et al.*, 1981; Gospodarowicz *et al.*, 1981; Tseng *et al.*, 1981). Chondrocytes have been shown to secrete and deposit type II collagen in culture, and epidermal cells have been found associated with several types including type V (Stenn *et al.*, 1979; Dessau *et al.* 1981; Alitalo *et al.*, 1982). This type of collagen is also seen in cultures of some tumor cells (Alitalo *et al.*, 1981). Type IV collagen is generally found in cultures of normal epithelial or carcinoma cells (Furcht *et al.*,

1978; Wicha *et al.*, 1979; Leivo *et al.*, 1982; Kidwell *et al.*, 1983). Collagens or procollagens that are of unusual composition or incompletely processed and not generally observed *in vivo* are also present in cultures of some cell types (Crouch *et al.*, 1978a; Alitalo *et al.*, 1982).

Studies of embryos, cultured embryonic cells, or embryonal carcinoma cells in culture have established that synthesis of collagen types I, III, and IV appear early in development and in temporal sequence and location that identifies initiation of synthesis of collagen and other substratum components as important markers of early differentiative events, and suggests a role for these molecules in subsequent differentiative direction (Adamson *et al.*, 1979; Leivo *et al.*, 1980, 1982; Strickland *et al.*, 1980; Hogan and Tilly, 1981). Induction of differentiation in undifferentiated embryonal carcinoma cells in culture and coordinate induction of synthesis of specific collagen types have been achieved primarily through the use of inducers such as retinoic acid and cyclic AMP analogues (Strickland *et al.*, 1980). These events *in vivo* are presumably under the influence of other, unidentified local inducers or diffusable factors. Effects of some diffusable hormones or growth factors on the ability of some cell types in culture to synthesize collagen have been observed. Epidermal growth factor (EGF) or EGF-like peptides stimulate collagen production of mammary epithelia and hepatocytes, and other factors are also effective on mammary epithelia: insulinlike peptides, alpha-2 macroglobulin (possibly with associated hormones or hormonelike factors bound to the molecule), dibutryl cyclic AMP, ascorbic acid, and prostaglandin E_1 (Salomon *et al.*, 1981; Kumegawa *et al.*, 1982; Kidwell *et al.*, 1983). Fibroblast growth factor (FGF) affects both the type and organization of collagens synthesized in culture by vascular endothelial cells, and glucocorticoids stimulate the accumulation of extracellular collagenous matrix in cultured mammary and fibroblastic cells (Furcht *et al.*, 1979; Salomon *et al.*, 1981; Tseng *et al.*, 1982; Kidwell *et al.*, 1983).

The Role of Collagen in Neoplastic Transformation

Considerable evidence suggests that neoplastic transformation in cultured cells is associated with decreased or altered ability to recognize, synthesize, or deposit extracellular matrix components, including collagen (Hay, 1981; Kleinman *et al.*, 1981). Virus-transformed chick fibroblasts synthesize collagen at a reduced rate, and such cells transformed with a virus that is temperature-sensitive for transformation deposit collagen on extracellular matrix at the nonpermissive temperature for transformation but not at the permissive temperature (Green *et al.*, 1966; Vaheri *et al.*, 1978). Decreased levels and changes in types of collagens synthesized are observed when virally or chemically transformed rat or mouse fibroblasts are compared with their nontransformed counterparts in culture, and some evidence exists that the patterns observed *in vitro* are maintained by the tumors arising from these cells *in vivo* (Sandmeyer *et al.*, 1981). Human fibroblasts transformed with SV40 are reported to synthesize collagen and maintain reasonable internal cellular levels, but are unable to establish an external collagenous matrix characteristic of untransformed human fibroblasts in culture (Vaheri *et al.*, 1978; Furcht *et al.*, 1979). Changes in the activities of several of the enzymes required for proper collagen processing have been reported to occur in association with transformation, and may account by several different mechanisms for an inadequate or abnormal deposition of matrix by transformed fibroblasts (Hay, 1981; Kleinman *et al.*, 1981).

Similar changes in collagen synthesis or deposition associated with transformed epitheloid cells in culture have been reported, but clear examples also exist in which production of collagen characteristic of the untransformed epitheloid cell is maintained in the corresponding carcinoma cell *in vitro* (Wicha *et al.*, 1979; Kidwell *et al.*, 1980, 1982, 1983; Keski-Oja *et al.*, 1982). For instance, transformed mouse mammary epithelial cells synthesize and secrete type IV collagen, as do normal mammary epithelial cells. The mammary tumor cells, however, have lost the ability to recognize selectively type IV collagen for attachment and, unlike the normal cells that prefer type IV, will attach equally well to other collagen types (Wicha *et al.*, 1979; Kidwell *et al.*, 1980). Also unlike normal mammary cells, the mammary carcinoma cells produce protein factors that stimulate collagen production in the normal cells, and presumably also act in this manner on the transformed cells (Kidwell *et al.*, 1982, 1983). In this way the transformed cells may gain some autonomy from exogenously produced factors, such as EGF, that may be required for stimulation of collagen production in the normal cells (Salomon *et al.*, 1981). Mouse mammary tumor cells also produce collagenase activity that is reduced by treatment of the cells with glucocorticoids, a phenomenon that results in a glucocorticoid-induced increase in extracellular collagenous matrix in these cells in culture (Salomon *et al.*, 1981). Similar effects are also observed on glucocorticoid treatment of transformed human fibroblasts, and may occur by a similar mechanism (Furcht *et al.*, 1979). Secretion of such collagneases, particularly those that degrade type IV collagen, and other enzymes capable of degrading components of the extracellular matrix may be important determinants of the metastatic ability of tumor cells *in vivo* (Liotta *et al.*, 1979, 1980).

Fibronectin

Biochemical and Biological Properties of Fibronectin

Like collagen, fibronectin is a protein that one might expect to be involved in cellular adhesion *in vivo*, and a likely candidate as a physiologically significant attachment protein *in vitro*. Fibronectin is found *in vivo* in serum or plasma, and is also a component of extracellular matrix or basement membrane and a component of cellular plasma membranes (Mosesson and Umfleet, 1970; Hynes, 1976; Grinnell, 1978; Yamada and Olden, 1978; Kleinman *et al.*, 1981; Ruoslahti *et al.*, 1982). The simplest structure of the native form of the molecule is a dimer of about 440,000 mol. wt. composed of two disulfide-linked subunits, each of about 220,000 mol. wt. (Mosesson and Umfleet, 1970; Yamada and Olden, 1978; Crouch *et al.*, 1978b; Engel *et al.*, 1981). Fibronectin is a glycoprotein, containing about 5% to 10% carbohydrate. The actual form of the molecule that may be active in culture or *in vivo* may vary because not only can fibronectin self-associate to form larger aggregates, but the individual subunits and fragments of these subunits retain some of the biological specificities of the intact molecule (Vuento *et al.*, 1980; Yamada *et al.*, 1980; Pierschbacher *et al.*, 1981, 1983; Ruoslahti *et al.*, 1981). The interactions in which fibronectin is probably involved both *in vivo* and in culture are likely to be quite complicated because this attachment factor has been shown to possess specific binding sites for a large number of molecules that are biologically relevant to cell–substratum interactions. In addition to its cell-binding capability, fibronectin can also bind collagen, heparin, hyaluronic acid, fibrin and fibrinogen, actin, and staphylococci (Klebe, 1974; Engvall and Ruoslahti, 1977; Stathakis and Mosesson, 1977; Engvall *et al.*, 1978; Klebe *et al.*, 1978; Gold and Pearlstein, 1980; Johansson and Höök, 1980; Ruoslahti and Engvall, 1980; Vuento *et al.*, 1980; Yamada *et al.*, 1980; Ruoslahti *et al.*, 1981).

Fibronectin was identified as a major component of plasma some years before its role as a mediator of cell–substratum interactions was recognized. The plasma form of fibronectin, cold-insoluble globulin (CIg), was shown to be a protein, distinct from fibrinogen or fibrin, that was present along with fibrin in blood clots (Mosesson and Umfleet, 1970; Stathakis and Mosesson, 1977). Later, fibronectin was also isolated from serum independently by other laboratories on the basis of the ability of the molecule to promote cell attachment and cell spreading on glass, plastic, or collagen-coated cell culture vessels, although it was not recognized immediately that the attachment and spreading-promoting activity isolated from serum was the same molecule as CIg (Klebe, 1974; Grinnell, 1976, 1978; Grinnell and Hays, 1978; Klebe *et al.*, 1978). Fibronectin is also found inside and, under the appropriate conditions, on the surface of platelets, and may be one of the components involved in the adhesion of platelets to collagen (Bensusan *et al.*, 1978). Some time after its characterization in the literature as the plasma protein CIg, fibronectin was identified by several laboratories as a component of the cell surface that was present in decreased amounts in transformed cells when compared with nontransformed controls (Hynes, 1973, 1976; Yamada and Olden, 1978). It was also observed that this cell membrane component was antigenically related to a protein present in serum (Ruoslahti and Vaheri, 1974). Subsequently, fibronectin was identified as a major glycoprotein in some extracellular matrix and basement membranes *in vivo*, as well as a component of the analogous matrix formed by many cell types in culture (Mosher *et al.*, 1977; Birdwell *et al.*, 1978; Chen *et al.*, 1978b; Grinnell, 1978; Yamada and Olden, 1978; Hedman *et al.*, 1979; Stenman *et al.*, 1979; Courtoy *et al.*, 1980; Hay, 1981; Kleinman *et al.*, 1981). It is unclear if the plasma, cell surface, and cell-derived matrix forms of fibronectin are all products of the same gene; minor biochemical differences between plasma and cell-derived forms have been reported, as well as differences in biological activities of the molecules (Hayashi and Yamada, 1981; Yamada *et al.*, 1982). A form of fibronectin that is more extensively glycosylated than the cell surface and plasma forms also has been isolated from amniotic fluid (Crouch *et al.*, 1978b).

In the intact molecule, the two subunits of fibronectin are joined at one end by disulfide linkages, producing a molecule that appears primarily as two flexible, rodlike subunit arms with little globular character, each fairly independently available for the multiple-binding capabilities of the protein (Yamada and Olden, 1978; Engel *et al.*, 1981; Kleinman *et al.*, 1981; Ruoslahti *et al.*, 1982; Yamada *et al.*, 1982). Studies using fragments of the fibronectin subunits that retain one or more capability for binding cells, fibrin, actin, collagen, hyaluronic acid, or gelatin have allowed the orientation of some of the distinct binding sites on the protein relative to each other, to the amino and carboxy terminals, and to the area of disulfide linkages (McDonald and Kelley, 1980; Vuento *et al.*, 1980; Yamada *et al.*, 1980; Ruoslahti *et al.*, 1981; Pierschbacher *et al.*, 1981, 1983). Some of the active fragments used in these studies were isolated from plasma preparations or were generated by proteases that one might expect fibronectin to encounter *in vivo*, such as thrombin, plasmin, and elastase, and it is possible that complexes containing fragments generated from the native molecule by proteolytic activity at the site of a blood clot or in the areas of immunological response or tissue remodeling may act as a temporary or initial substratum affecting cell adhesion and migration (Yamada and Olden, 1978; McDonald and Kelley, 1980; Kleinman *et al.*, 1981). The physiological significance of the presence of biologically active fragments of fibronectin *in vivo* has not yet been examined in detail.

Fibronectin forms an extensive extracellular matrix in cultures of many types of cells (Mosher *et al.*, 1977; Birdwell *et al.*, 1978; Chen *et al.*, 1978b; Crouch *et al.*, 1978b; Jaffe and Mosher, 1978; Vaheri *et al.*, 1978; Hedman *et al.*, 1979; Alitalo *et al.*, 1981; Kleinman *et al.*, 1981). The source of the fibronectin in the matrix may be from both endogenous material synthesized by the cells and exogenous fibronectin from the culture medium, in cases in which serum-supplemented medium is used (Hayman and Ruoslahti, 1979). The areas in which fibronectin is found *in vivo* strongly suggest that the molecule acts as a mediator of cell attachment in the intact animal. Although fibronectin binds preferentially to type III (interstitial) collagen, it is capable of binding to all collagen types and to denatured collagen or gelatin (Klebe, 1974; Engvall and Ruoslahti, 1977; Engvall *et al.*, 1978; Grinnell and Minter, 1978; Gold and Pearlstein, 1980; Johansson and Höök, 1980), and is found associated with basement membrane in some areas, as well as in areas of deposition of interstitial collagen (Yamada and Olden, 1978; Stenman *et al.*, 1979; Courtoy *et al.*, 1980; Weiss and Reddi, 1980; Hay, 1981; Kleinman *et al.*, 1981). The major source of the fibronectin of extracellular matrix *in vivo* is probably the fibroblastic cells in the immediate area (Yamada and Olden, 1978; Hay, 1981; Kleinman *et al.*, 1981). The source of plasma fibronectin is less clear; fibronectin is synthesized by vascular endothelial cells (Birdwell *et al.*, 1978; Jaffe and Mosher, 1978), but other sources may also be involved. Changes in amounts or in the areas of deposition of extracellular fibronectin are likely to be invovled in cell differentiation and migration in the developing organism, and aberrations in amounts or functional capability of the molecule may be involved in pathological states in the adult (Furcht *et al.*, 1978; Yamada and Olden, 1978; Podleski *et al.*, 1979; Penneypacker *et al.*, 1979; Stenman *et al.*, 1979; West *et al.*, 1979; Weiss and Reddi, 1980; Hay, 1981; Kleinman *et al.*, 1981; Sieber-Blum *et al.*, 1981). Both the free and the platelet-associated forms of fibronectin in plasma may play a role in clot formation and wound healing; attachment and spreading of vascular smooth muscle and endothelial cells, as well as the adhesive capability of platelets and fibroblasts, are enhanced by fibronectin (Bensusan *et al.*, 1978; Grinnell, 1978; Kleinman *et al.*, 1981; Weinstein *et al.*, 1982).

The Use of Fibronectin in Cell Culture

Although fibronectin may be isolated from cultured cells or culture medium, from extracellular matrix deposited *in vivo* or *in vitro*, or from amniotic fluid, plasma is the principal source for large-scale preparation for use as a cell culture substratum (Mosesson and Umfleet, 1970; Crouch *et al.*, 1978b; Yamada and Olden, 1978; Ruoslahti *et al.*, 1982). Fibronectin for cell culture use can be prepared in an easy, one-step column procedure that takes advantage of the affinity of the molecule for gelatin (Engvall and Ruoslahti, 1977; Klebe *et al.*, 1980; Ruoslahti *et al.*, 1982). Plasma is passed through a gelatin-sepharose column, and the column washed extensively. Plasma fibronectin can then be eluted with buffered urea-containing solutions or solutions of other denaturants or chaotrophic agents (Engvall and Ruoslahti, 1977; Klebe *et al.*, 1980; Ruoslahti *et al.*, 1982). Problems encountered with contaminating sepharose-binding plasma proteins that also elute with urea can be avoided by a prerun of the plasma through underivatized sepharose, and problems encountered with isolation of fibronectin fragments that may be present in the plasma and possess gelatin-binding activity but not other activities can be minimized by passing material isolated on the gelatin-sepharose column through a heparin-sepharose column (Ruoslahti *et al.*, 1982). Fibronectin can be eluted from the heparin-sepharose column with 0.5 M NaCl. Selection for molecules that have both gelatin-binding and heparin-binding activities results in a preparation that is likely to be exclusively intact fibronectin (Ruoslahti *et al.*, 1982). Because some fibronectin is lost in the clot when plasma is converted to serum and activation of proteases in the clotting cascade may result in degradation of the molecule, it is best to use plasma rather than serum as an initial source of the protein (Ruoslahti *et al.*, 1982). Similar procedures may be applied to preparations of cell or tissue extracts or culture media for isolation of fibronectin from these sources, but yields are likely to be considerably less than that obtained from plasma. The isolated fibronectin may be stored in urea-containing solutions, or high pH, calcium-containing solutions. The latter is particularly useful for fibronectin isolated from cell surfaces; this form of the molecule undergoes extensive aggregation and is somewhat less soluble at neutral pH than is the plasma form (Yamada and Olden, 1978).

It is important to be aware in working with fibronectin for cell culture purposes that the protein has a high affinity for glass and some types of plastic (e.g., polystyrene), and may be lost from solution to the walls of the container in significant amounts if stored in dilute form (Barnes and Sato, 1980b; Ruoslahti *et al.*, 1982). Filtration for purposes of sterilization can also result in considerable loss of protein to the filter if small amounts or dilute solutions are filtered. Although fibronectin has a tendency to aggregate at higher concentrations, it is possible to work with the material fairly easily at concentrations in the range of 0.5 to 1.0 mg/ml (Barnes and Sato, 1980b,c). Adherence to containers or filters and aggregation of the molecule can be reduced by inclusion of 1 M urea in the concentrated stock solutions. Solutions should be stored frozen in aliquots to minimize freezing and thawing. Because fibronectin will adhere well to the polystyrene surface of the culture vessel, cells may be exposed to the fibronectin substratum by pretreating the culture vessel with a fibronectin solution in the range of 10 μg/ml or less made by diluting a small volume of a concentrated fibronectin stock into culture medium or a buffered saline solution (Orly and Sato, 1979; Barnes and Sato, 1980b,c; Murray *et al.*, 1980; Rizzino and Crowley, 1980; Grinnell and Feld, 1982). Incubations as short as 1 h at 37 °C are effective at producing an attachment-promoting fibronectin substratum on the culture cell vessel. The incubation solution may be removed and the plate washed before cells are added without diminishing the activity of the substratum; by this procedure exposure of cells to potentially toxic substances in the stock solutions of the fibronectin, such as urea, is avoided.

Effects of Fibronectin on Adhesion, Proliferation, and Differentiation of Cells in Culture

Although some cell types apparently are capable of using collagen directly as a substratum, fibronectin, binding to both collagen and cells, acts as a substratum for many cell types plated *in vitro* on collagen matrices (Hay, 1981; Kleinman *et al.*, 1981). The fibronectin may come from serum added to the culture medium (CIg) or may be material synthesized by the cells themselves (Birdwell *et al.*, 1978; Crouch *et al.*, 1978b; Grinnell and Feld, 1979; Hayman and Ruoslahti, 1979; Murray *et al.*, 1980). The molecule was, in fact, identified as a collagen-binding and cell attachment-promoting protein in serum some time before it was recognized that the protein responsible for the attachment-promoting activity was similar or identical to CIg and the cell membrane and extracellular matrix forms of fibronectin (Klebe, 1974; Grinnell and Minter, 1978; Klebe *et al.*, 1978; Gold and Pearlstein, 1980). Although fibronectin binds to all collagen types (Engvall *et al.*, 1978; Ruoslahti and Engvall, 1980) and mediates the attachment of many cell types to collagen or to plastic cell culture vessels (Grinnell, 1978; Yamada and Olden, 1978; Kleinman *et al.*, 1981), it is not effective as an attachment-promoting protein for all types of cells in culture (Timpl *et al.*, 1979; Hewitt *et al.*, 1980, 1982; Barnes and Sato, 1980b,c; Vlodavsky and Gospodarowicz, 1981).

The site on the molecule that promotes cell attachment has been localized to a limited area within the molecule between the heparin-binding and collagen-binding sites (Pierschbacher *et al.*, 1981, 1983; Ruoslahti *et al.*, 1981; McDonald and Kelley, 1980; Vuento *et al.*, 1980; Yamada *et al.*, 1980). The structure on the cell surface that is capable of recognizing fibronectin is not yet clear, although it has been suggested that a ganglioside may be involved (Kleinman *et al.*, 1979b; Yamada *et al.*, 1981). Cell lines variant from the parent strain in the ability to recognize and respond to fibronectin have been isolated; these cells cannot attach to a fibronectin substratum, but are capable of attachment and spreading on some other substrata (Harper and Juliano, 1980, 1981). It may be that these cells no longer express a fibronectin receptor on their surface.

A substratum composed of a coat of fibronectin or other appropriate molecule has been shown to be necessary for proliferation in serum-free media of a number of cell types of both fibroblastic and epithelial origin (Orly and Sato, 1979; Barnes and Sato, 1980b,c; Orly *et al.*, 1980; Rizzino and Crowley, 1980; Barnes, 1982a,b). Although the reason that fibronectin is required presumably relates to the requirement of these cell types for anchorage to a surface in order to progress through the cell cycle, it is not clear why anchorage is necessary, or exactly what role fibronectin may play in the process of cell growth and division. Some insight into the possible role of fibronectin in the proliferative cycle of the cell comes from the work of Orly and Sato (1979). Using the RF1 diploid, nontumorigenic epithelial cell line established from rat ovary, these investigators found that in a serum-free medium composed of nutrients, hormones, and transferrin, fibronectin was required in order for the cells to undergo cytokinesis. In the absence of fibronectin and presence of the other components of the serum-free medium, these cells were capable of nuclear division but not cellular division. This phenomenon resulted in cultures that contained a high percentage of multinucleate cells. If cells were plated in the complete serum-free medium, including fibronectin, cell division occurred.

Orly and Sato (1979) suggest that fibronectin as a substratum may be required for the proper formation and function of the contractile ring microfilaments. It is possible that the actin-binding property of fibronectin may be involved in this phenomenon, since actomyosin complexes and actin filaments have been implicated as playing a role in cell motility, including cytokinesis. In the capacity described here, fibronectin as a growth-promoting molecule should be thought of as a "permissive" factor, mediating effects of other growth-promoting factors, such as hormones, rather than as a protein that promotes growth independently (Orly and Sato, 1979). Fibronectin added to a basal nutrient medium, in the absence of growth-stimulatory hormones and binding proteins, did not induce cell proliferation in the system described.

Fibronectin also affects differentiation of a number of cell types in culture. Exogenous fibronectin added to cultures of myoblasts or chondrocytes inhibits differentiation *in vitro*, and alterations in the levels of synthesis and localization of fibronectin produced by myoblasts and chondrocytes correlates with observed differentiative changes in these cells in culture (Dessau *et al.*, 1978, 1981; Furcht *et al.*, 1978; Podleski *et al.*, 1979; Pennypacker *et al.*, 1979; West *et al.*, 1979; Weiss and Reddi, 1980; Yamada *et al.*, 1981). Fibronectin also inhibits the expression of differentiated function of cultured rat ovarian granulosa cells in serum-free medium, as indicated by decreased steroidogenic potential of the cultures in the presence of fibronectin, while stimulating the growth of these cells *in vitro* in the proper mixture of hormones, binding proteins, and nutrients (Orly *et al.*, 1980). In some systems, fibronectin can be demonstrated to promote rather than inhibit differentiation of cells in culture. Neural crest cells and particularly several lines of embryonal carcinoma cells in serum-free medium are reported to respond to fibronectin in this way (Sieber-Blum *et al.*, 1981; Rizzino, 1983). In most of the cases in which effects of fibronectin on differentiation of cells in culture have been observed, marked morphological effects of fibronectin are also seen, and it has been suggested that the effects of fibronectin on differentiation and proliferation may be a result of effects of this protein on cell shape (Birdwell *et al.*, 1978; Rizzino, 1983).

In addition to stimulatory and inhibitory effects of fibronectin on differentiative capacity of cultured cells, the adhesive response to fibronectin of late passage fibroblast strains in culture is altered, compared to early-passage fibroblasts (Chandrasekhar and Millis, 1979).

Differences in fibronectin-promoted adhesion of late-passage vs. early-passage cells are reported to be due to both an altered ability of late-passage cells to respond to normal fibronectin and a defective fibronectin produced by the late-passage cells. It is possible that some of the effects of fibronectin on the differentiation of some cell types may also be found to be related to production of altered forms of the fibronectin molecule or alterations in the ability of some cell types to respond to the molecule.

Synthesis and Deposition of Fibronectin by Cells in Culture

Fibronectin synthesized by cells in culture is often found deposited extracellularly in association with collagen, proteoglycans such as heparin sulfate proteoglycans and glycosaminoglycans (e.g., hyaluronic acid) (Birdwell *et al.*, 1978; Chen *et al.*, 1978b; Vaheri *et al.*, 1978; Furcht *et al.*, 1979; Murray *et al.*, 1980; Hayman *et al.*, 1982a). Fibronectin synthesized by fibroblastic cells *in vitro* is probably the mode by which these cells adhere and spread on the culture vessel in serum-free medium in the absence of added spreading-promoting factors (Grinnell and Feld, 1979). Depending on the cell type and conditions of the culture examined, fibronectin may be found intracellularly in the perinuclear region, on the apical and basal cell surfaces in diffuse form or in discrete surface clusters, or in pericellular or extracellular matrix, sometimes in filamentous structures (Mosher *et al.*, 1977; Crouch *et al.*, 1978b; Birdwell *et al.*, 1978; Chen *et al.*, 1978b; Dessau *et al.*, 1978; Jaffe and Mosher, 1978; Vaheri *et al.*, 1978; Furcht *et al.*, 1979; Hedman *et al.*, 1979; Murray *et al.*, 1980; Alitalo *et al.*, 1981; Hayman *et al.*, 1982a). The formation of extracellular structures containing fibronectin does not depend on the presence of collagen in the structures; fibronectin filaments may be observed in the matrix in the absence of collagenous protein or after collagenase treatment of the matrix (Chen *et al.*, 1978b; Hedman *et al.*, 1979). The organization or formation of extracellular matrix fibronectin does, however, require the formation of disulfide bridges between the molecules, and the matrix can be solubilized by agents such as mercaptoethanol (Chen *et al.*, 1978b). The requirement for disulfide crosslinking of extracellular fibronectin structures in the matrix may explain the inhibitory effect of reducing agents such as dithiothreitol on fibroblastic cell spreading in serum-free medium in the absence of exogenous spreading factors added to the medium (Grinnell and Feld, 1980). Fibronectin also may be detected in soluble form in the medium of many types of cultured cells (Mosher *et al.*, 1977; Crouch *et al.*, 1978b; Jaffe and Mosher, 1978; Yamada and Olden, 1978; Orly and Sato, 1979; Hayman *et al.*, 1982a). Differences in the form and amounts of fibronectin synthesized in cell cultures are related to many factors, including the cell type, age and density of the culture, exposure to some hormones, state of differentiation, type of substratum and, for individual cells, the stage of the cell growth cycle (Mosher *et al.*, 1977; Stenman *et al.*, 1977; Birdwell *et al.*, 1978; Chen *et al.*, 1978b; Furcht *et al.*, 1979; Marceau *et al.*, 1980; Dessau *et al.*, 1981).

Fibronectin is synthesized by many, if not most, untransformed cells in culture, including cells of virtually every species examined. Although fibronectin synthesis by cells in culture is common, fewer cell types have been found to deposit fibronectin on an organized fibrillar matrix, and such a phenomenon is most often seen in cultures of primary or early passage cells exhibiting fibroblastic morphology. Fibronectin is synthesized by many differentiated cell types in culture: myoblasts, chondrocytes, hepatocytes, astroglia, keratinocytes, granulosa, amniotic epithelial cells, and corneal and vascular endothelial cells (Mosher *et al.*, 1977; Birdwell *et al.*, 1978; Chen *et al.*, 1978b; Crouch *et al.*, 1978b; Dessau *et al.*, 1978, 1981; Furcht *et al.*, 1978; Jaffe and Mosher, 1978; Yamada and Olden, 1978; Orly and Sato, 1979; Marceau *et al.*, 1980; Kleinman *et al.*, 1981). Detection of fibronectin synthesis by endothelial cells (Birdwell *et al.*, 1978; Jaffe and Mosher, 1978) is of particular interest because of the relationship of these cells to the blood clotting process and the involvement of collagen, fibrin, fibronectin, and platelets in this process. Fibronectin is also synthesized *in vitro* by tumor cells of mesenchymal origin, including fibrosarcomas, leiomyosarcomas, osteosarcomas, and rhabdomyosarcomas, as well as tumor cells of epithelial origin (Alitalo *et al.*, 1981; Vartio and Vaheri, 1981). Some human cell types in culture also synthesize and secrete into the culture medium a protein of lower molecular weight that, similar to fibronectin, binds collagen, but is immunologically unrelated to fibronectin (Vartio and Vaheri, 1981).

The synthesis and deposition of fibronectin by cells in culture is observed to vary with the conditions of the culture. For instance, fibronectin production is increased in confluent cultures of corneal endothelium compared with less dense, growing cultures, and in some cases changes in localization of fibronectin from apical and basal cell surfaces to only basal surfaces are observed when cultures reach confluence (Birdwell *et al.*, 1978; Yamada and Olden, 1978). Some changes in amount or distribution of surface fibronectin are cell-cycle related (Stenman *et al.*, 1977). Although the signals and mechanisms involved in evoking these changes are poorly understood, at least some of the changes may involve responses of cells to hormonal influences in the culture environment. Dexamethasone or cortisol, but not other steroid hormones, have been found to increase deposition of fibronectin in the extracellular matrix of fibroblasts and hepatocytes (Furcht *et al.*, 1979; Marceau *et al.*, 1980), and similar effects of epidermal growth factor and thrombin have been observed in other cell culture systems (Yamada and Olden, 1978).

The Role of Fibronectin in Neoplastic Transformation

Cell surface fibronectin is often observed to be decreased on transformed cells in culture relative to levels of cell surface fibronectin on untransformed control cultures (Hynes, 1973, 1976; Yamada and Olden, 1978). In some systems it appears that a decreased amount of cell surface fibronectin or a decreased percentage of cells expressing cell surface fibronectin in the cultures are associated with increased tumorigenicity or metastatic potential of the cultures (Chen *et al.*, 1976, 1978a). As further evidence of a relationship between decreased levels of extracellular fibronectin and the transformed phenotype in culture, it has been reported that chick embryo fibroblasts transformed by a Rous sarcoma virus temperature sensitive for transformation deposit a fibronectin-containing matrix only at the temperature nonpermissive for expression of the transformed phenotype (Vaheri *et al.*, 1978). Intercellular fibronectin can be detected in these cells at both permissive and nonpermissive temperatures. When added to cultures of transformed fibroblasts, fibronectin can restore a morphology to the transformed cells that is similar to that of nontransformed cells, but cannot reverse other characteristic properties of cultures of transformed fibroblasts (Hynes, 1976; Yamada and Olden, 1978).

Decreased fibronectin on the cell surface or in the extracellular matrix associated with carcinoma cells *in vivo* or in culture also has been observed, but the relationship between neoplastic transformation and fibronectin expression in the pericellular matrix or on the surface of epithelial cells in culture is less clear than with fibroblastic cells in culture (Labat-Robert *et al.*, 1980; Yang *et al.*, 1980c; Keski-Oja *et al.*, 1982). Reduction in the amount of surface or matrix fibronectin on transformation may be the result of a number of possible alterations in the normal pattern of synthesis, processing, deposition, and degradation of the protein, and probably is the result of several such changes including increased proteolytic activity associated with transformed cells, increased secretion of fibronectin into the culture medium as a result of decreased binding of the protein to cell surfaces or altered ability to deposit the protein in the extracellular matrix, decreased synthesis of the protein, and increased degradation of the protein intracellularly or extracellularly (Hynes, 1973, 1976; Chen *et al.*, 1978a; Vaheri *et al.*, 1978; Yamada and Olden, 1978; Furcht *et al.*, 1979; Kleinman *et al.*, 1981; Hayman *et al.*, 1982a).

Laminin

Biochemical and Biological Properties of Laminin

Like fibronectin, laminin is a large glycoprotein that is a major constituent of basement membrane *in vivo* (Chung *et al.*, 1979; Timpl *et al.*, 1979, 1982; Kleinman *et al.*, 1981; Ledbetter *et al.*, 1984). Whereas fibronectin is found to a limited extent in basement membrane and is also found in the interstitial stroma of mesenchymal tissues, laminin is found ubiquitously and exclusively in basement membrane. It is likely that laminin plays a major role in cellular adhesion *in vivo* in the adult organism (Kleinman *et al.*, 1981; Grotendorst *et al.*, 1982; Timpl *et al.*, 1982). Furthermore, the protein is reported to be synthesized by early embryos and appears to be developmentally regulated, suggesting it may play a role in directing cell adhesion and migration in the maturing organism (Leivo *et al.*, 1980; Ekblom *et al.*, 1980; Howe and Solter, 1980; Hogan, 1980; Cooper *et al.*, 1981; Hogan and Tilly, 1981). Laminin is a multisubunit complex composed of one glycoprotein subunit of approximately 400,000 mol. wt. and three probably identical glycoprotein subunits, each of approximately 200,000 mol. wt. (Chung *et al.*, 1979; Timpl *et al.*, 1979, 1982; Engel *et al.*, 1981). The subunits are linked together in a cross-shaped structure and connected by disulfide bonds. The larger of the subunits can be removed from the complex without loss of the cell attachment-promoting activity of the molecule (Rohde *et al.*, 1979; Engel *et al.*, 1981; Risteli and Timpl, 1981; Ott *et al.*, 1982; Rao *et al.*, 1982). The entire complex contains about 12% to 15% carbohydrate (Chung *et al.*, 1979; Timpl *et al.*, 1979). Laminin binds type IV collagen, heparin, and some proteoglycans, but unlike fibronectin does not bind hyaluronic acid (Timpl *et al.*, 1979, 1982; Del Rosso *et al.*, 1981; Engel *et al.*, 1981).

The Use of Laminin in Cell Culture

Although laminin may be isolated from several natural sources, the highest yields are obtained from the EHS transplantable mouse tumor. This tumor produces large amounts of both type IV collagen and laminin, and has proved useful in a number of studies of the biology and biochemistry of basement membrane components (Timpl *et al.*, 1978, 1979, 1982; Ledbetter *et al.*, 1984). Laminin is purified from soluble extracts of the tumor by a series of salt precipitations and ion exchange chromatography, and can be isolated by these procedures in a form free of proteoglycans, collagen, or fibronectin (Timpl *et al.*, 1982; Ledbetter *et al.*, 1984). Laminin in a purified, concentrated form is somewhat difficult to work with because it shows a tendency to form large aggregates or gels. It has been recommended that laminin for cell culture use be stored at −70 °C in aliquots at about 1 mg/ml of protein in Tris-saline or culture medium. For use in culture dishes, the laminin solutions may be used to precoat directly the plastic dishes or to pretreat dishes containing type IV collagen gels, or may be added directly to the culture medium shortly before adding cells (Timpl *et al.*, 1982; Ledbetter *et al.*, 1984).

Procedures for purification of laminin from cell cultures or culture medium conditioned by laminin-producing cells have been developed, using heparin-agarose and lectin-agarose affinity chromatography (Sakashita *et al.*, 1980; Shibata *et al.*, 1982). Yields from these procedures generally are not large enough to make them practical for isolation of the material for routine cell culture use. Laminin is not present in serum or plasma at concentrations useful for development of isolation procedures from these fluids, and probably does not contribute significantly to the cell spreading-promoting capacity of serum in conventional culture media (Kleinman *et al.*, 1981; Grotendorst *et al.*, 1982). Laminin is less sensitive to proteolysis than fibronectin, and procedures also have been developed for isolation of fragments of laminin after limited proteolytic degradation of material containing basement membrane from placenta and kidney (Risteli and Timpl, 1981; Ott *et al.*, 1982).

Effects of Laminin on Adhesion, Proliferation, and Differentiation of Cells in Culture

Most of the studies using laminin in cell culture have involved investigations of effects on adhesion primarily of epithelial cells to type IV collagen-coated culture dishes (Timpl *et al.*, 1979; Terranova *et al.*, 1980; Kleinman *et al.*, 1981). Laminin also promotes the adhesion of some cell types to dishes treated with other collagen types, and plates treated with laminin alone also possess adhesion-promoting activity characteristic of the laminin-type IV collagen complexes. Among the cell types reported to recognize laminin as an attachment factor in culture are epidermal cell lines and epidermal cells in primary culture, intestinal epithelial cells, hepatocytes, colon carcinoma cells, and several lines of embryonal carcinoma cells (Rizzino *et al.*, 1980; Terranova *et al.*, 1980; Burrill *et al.*, 1981; Carlsson *et al.*, 1981; Johansson *et al.*, 1981; Vlodavsky and Gospodarowicz, 1981; Stenn *et al.*, 1982). Some cell types such as hepatocytes and fibroblasts, can use either fibronectin or laminin as an attachment factor, depending on the culture conditions (Johansson *et al.*, 1981; Couchman *et al.*, 1983). One interesting report suggests that hepatocytes from regenerating liver can use either laminin or fibronectin as an attachment factor, whereas hepatocytes from nongrowing liver prefer fibronectin (Carlsson *et al.*, 1981). Laminin in serum-free culture is capable of promoting growth of embryonal carcinoma cells in the presence of other medium supplements, and also is reported to affect the differentiation of embryonal carcinoma cells in serum-free culture (Rizzino *et al.*, 1980; Darmon, 1982; Rizzino, 1984).

Synthesis and Deposition of Laminin by Cells in Culture

Laminin is synthesized in culture and secreted into the medium by several embryonal carcinoma-derived parietal endodermlike cell lines and by normal parietal endoderm cells, parietal yolk sac cells, and cells from a yolk sac tumor (Chung *et al.*, 1979; Hogan, 1980; Howe and Solter, 1980; Strickland *et al.*, 1980; Sakashita and Ruoslahti, 1980; Cooper *et al.*, 1981; Hogan and Tilly, 1981; Wewer *et al.*, 1981; Leivo *et al.*, 1982). Laminin also can be detected in embryos in Reichert's membrane, and some of the endodermlike lines secrete laminin in an extracellular matrix also containing type IV collagen and fibronectin (Chung *et al.*, 1979; Howe and Solter, 1980; Hogan, 1980; Sakashita and Ruoslahti, 1980; Leivo *et al.*, 1982). The cells themselves are synthesizing the collagen in this matrix, but the fibronectin found in the matrix appears to come from the serum of the culture medium, at least in some cases (Leivo *et al.*, 1982). Some undifferentiated embryonal carcinoma cell lines can be induced to synthesize laminin by treatment with dibutryl cyclic AMP and retinoic acid (Strickland *et al.*, 1980).

Other cell types that synthesize laminin in culture include neuroblastoma, astrocytes, amniotic epithelial cells, and vascular and corneal endothelial cells (Alitalo *et al.*, 1980a,b; Gospodarowicz *et al.*, 1981; Liesli *et al.*, 1983). Both fibronectin and laminin are synthesized and secreted by endothelial cells in culture, but vascular endothelial cells secrete much more laminin than corneal endothelial cells; both cell types secrete about equal amounts of fibronectin (Gospodarowicz *et al.*, 1981). Laminin production by both cell types is decreased as cultures reach confluence, whereas fibronectin production shows no change or is increased somewhat in confluent cultures. Laminin in the extracellular matrix of cultured endothelial cells is associated with type IV collagen secreted by the cells and is found only on the basal cell surface (Gospodarowicz *et al.*, 1981). Many tumor cell types in culture also synthesize laminin and secrete it into the medium, although considerably fewer of these lines have been observed to deposit laminin in an extracellular matrix (Alitalo *et al.*, 1981; Salomon *et al.*, 1984a). Studies of extracellular matrix production by the NRK normal rat kidney cell line compared with a Kirsten sarcoma virus-transformed NRK line suggest that viral transformation results in a loss of the ability to deposit laminin and other components in an extracellular matrix, but does not result in a loss in the ability of the cells to synthesize laminin (Hayman *et al.*, 1981, 1982a).

Chondronectin

Chondronectin is glycoprotein found in serum and cartilage that mediates the attachment of chondrocytes in culture to type II collagen (Hewitt *et al.*, 1980, 1982). Chondronectin does not show a similar activity for fibroblastic or other cell types. The molecular weight of the native molecule is about 180,000 (Hewitt *et al.*, 1980; Kleinman *et al.*, 1981). On reduction, the purified material breaks down into chains of about 70,000 mol. wt. Chondronectin *in vivo* is found in cartilage in a pericellular network around chondrocytes, but not in the extracellular matrix of cartilage (Hewitt *et al.*, 1982). The binding of chondronectin to type II collagen is enhanced by proteoglycan, also present in the matrix and pericellular areas (Grotendorst *et al.*, 1982; Varner *et al.*, 1984). Under some circumstances, chondrocytes in culture and *in vivo* can also synthesize fibronectin, but differentiation to a chondrocyte phenotype is associated with chondronectin synthesis and absence of fibronectin synthesis by the cells (Dessau *et al.*, 1978, 1981; Pennypacker *et al.*, 1979; West *et al.*, 1979).

Chondronectin has been isolated from chicken serum, although other sera also contain a similar activity and probably also can be used as a source for purification of the molecule (Hewitt *et al.*, 1980, 1982). The purification procedure involves ammonium sulfate precipitation, DEAE-cellulose chromatography, Cibacron blue F3GA-agarose, and wheat germ agglutinin-sepharose chromatography (Hewitt *et al.*, 1982; Varner *et al.*, 1984). Purified material can be stored in aliquots frozen in Tris-saline for extended periods. Chondronectin is present in serum at levels of 20 μg/ml or less, so large amounts of serum must be used as the starting material (Kleinman *et al.*, 1981; Hewitt *et al.*, 1982). Although the molecule can also be isolated from cartilage or some tumor tissue, yields are much less than that obtained from serum (Hewitt *et al.*, 1980, 1982; Varner *et al.*, 1984).

Serum-Spreading Factor

Biochemical and Biological Properties of Serum-Spreading Factor

Serum-spreading factor (SF) is a glycoprotein present in human serum that promotes attachment and spreading of a large number of cell types in culture (Barnes and Sato, 1979, 1980a; Barnes *et al.*, 1980, 1981, 1982, 1983). This activity was first reported to be present in a fraction of human serum isolated by glass bead affinity chromatography (Homes, 1967; Holmes *et al.*, 1979). This fraction contains both cell spreading-promoting activity and growth-promoting activity. The spreading-promoting activity can be separated from the growth-promoting activity in these preparations, and the spreading-promoting activity in the preparation has been termed serum-spreading factor (Barnes *et al.*, 1980, 1981, 1983). The factor exists in two forms in human serum; one form migrates in sodium dodecyl sulfate polyacrylamide gel electrophoresis in a manner consistent with a nominal mol. wt. of 75,000 to 78,000 (SF 75) and the other migrates in a manner consistent with a nominal mol. wt. of 65,000 to 70,000 (SF 65) (Barnes *et al.*, 1983a). A biologically active thrombin cleavage product of mol. wt. about 57,000 (SF 57) also has been identified (Silnutzer and Barnes, 1984a). Serum-spreading factor is present in human serum at concentrations in the range of 0.1 to 0.4 mg/ml serum and is also present in amniotic fluid and urinary protein (Shaffer *et al.*, 1984). The protein is distinct from fibronectin, laminin and chondronectin based on immunological and biochemical criteria and biological activities (Barnes *et al.*, 1980, 1982, 1983; Hayman *et al.*, 1982b). Serum-spreading factor has also been found associated with platelets, and it has been suggested that this protein, as with fibronectin, may be involved in platelet adhesion and aggregation (Barnes *et al.*, 1983). Tissue-associated forms of SF also have been identified (Hayman *et al.* 1983).

The Use of Serum-Spreading Factor in Cell Culture

Serum-spreading factor may be partially purified by modifications of the glass head column procedure described by Holmes (Holmes, 1967; Barnes *et al.*, 1980, 1981) and further purified by concanavalin A-sepharose affinity chromatography, DEAE-agarose chromatography, and heparin-agarose affinity chromatography (Barnes and Silnutzer, 1983; Silnutzer and Barnes, 1984b). Preparations that are free of fibronectin can be made in the glass bead column step by using a large amount of serum relative to glass bead column volume (100 ml serum per 200 ml bead volume). If smaller amounts of serum are used (e.g., 20 ml serum per 200 ml glass bead column volume), considerably higher yields of material per milliliter of serum are obtained, but some fibronectin is present in this first step preparation (Barnes *et al.*, 1984; Silnutzer and Barnes, 1984b). The fibronectin is subsequently purified away from the serum-spreading factor in the other chromatographic steps. Plasma or serum that has been passed through a gelatin-sepharose column to remove intact fibronectin and gelatin-binding fibronectin fragments also can be employed as an early step in the purification. Serum-spreading factor does not bind to gelatin (Barnes *et al.*, 1980). Serum-spreading factor also does not bind to heparin under conditions that favor fibronectin-heparin binding (Barnes and Silnutzer, 1983). The combination of gelatin affinity column and heparin affinity column purification effectively removes any fibronectin or fragments of fibronectin from the serum-spreading factor preparations.

Separation of fibronectin from serum-spreading factor in the preparations may be important for some kinds of experiments, since both have similar effects on some cell types, but exhibit different effects on others, and both are present in serum at relatively high concentration (Barnes and Sato, 1979, 1980a; Barnes *et al.*, 1980, 1982, 1983, 1984; Shaffer *et al.*, 1984). Experimental evidence exists to suggest that both serum-spreading factor and fibronectin contribute significantly to the total spreading-promoting activity of serum in conventional culture media (Grinnell, 1978; Barnes *et al.*, 1983). Either serum-spreading factor or fibronectin can act as an attachment factor and stimulate the growth of a number of cell types in serum-free media supplemented with hormones, nutrients, and binding proteins (Barnes and Sato, 1980b,c). As with fibronectin, serum-spreading factor stimulates the growth of the RFl rat ovary line by mediating cytokinesis in serum-free culture (Barnes *et al.*, 1982). Also as with fibronectin, serum-spreading factor exhibits a strong affinity for polystyrene cell culture plastic vessels, and may be introduced into the culture system by a precoating of the vessel with solutions of the attachment factor (Barnes *et al.*, 1980; Silnutzer and Barnes, 1984b).

Effects of Serum-Spreading Factor on Adhesion, Proliferation, and Differentiation of Cells in Culture

Serum-spreading factor is capable of stimulating the attachment and spreading in culture of both fibroblastic and epithelial cell types in serum-free media, and also is capable of affecting differentiation of embryonal carcinoma cells in manner not observed for fibronectin (Barnes and Sato, 1979, 1980a,b,c; Barnes *et al.*, 1980, 1982, 1983). Serum-spreading factor also influences cell migration *in vitro* (Basara *et al.*, 1984). In some situations in which serum-spreading factor influences both cell proliferation and cell morphology in serum-free media supplemented with hormones, binding proteins, and nutrients, it can be demonstrated that the concentration dependence of the proliferative effect of serum-spreading factor is essentially the same as the concentration dependence of the morphological effect, suggesting that the ability of serum-spreading factor to promote cell proliferation in these systems may be related to the ability of the factor to affect cell shape and adhesion (Barnes *et al.*, 1980).

Epibolin

Epibolin is a glycoprotein isolated from human plasma that supports attachment, spreading, and migration of epidermal cells in culture (Stenn, 1981). Molecular weight of the purified material is approximately 65,000. Purification from human plasma involves ammonium sulfate precipitation, DEAE-agarose chromatography, isoelectric precipitation, gel filtration, and preparative electrophoresis (Stenn, 1981, 1984). Although yields of epibolin by this sequence of procedures are low, fresh or outdated human plasma generally is readily available, so access to source material is little or no problem. Maximum biological activity of epibolin on epidermal cell spreading is seen in the presence of a second factor, termed coepibolin, that has not yet been characterized biochemically (Stenn, 1981). Epibolin is immunologically and biochemically distinct from laminin, which can also promote attachment and spreading of epidermal cells in culture (Stenn *et al.*, 1982). Comparison of the biochemical and biological properties of epibolin and serum-spreading factor and preliminary immunological tests suggest that these two attachment factors may be related (Barnes *et al.*, 1983). Further experiments to clarify this point are in progress. Possible sites of synthesis or extracellular deposition of epibolin have not yet been examined.

Fetuin

Preparations of fetuin, the major protein in fetal calf serum, were shown many years ago to promote attachment, spreading, and growth of cells in culture (Fisher *et al.*, 1958). Although it was reported soon after the initial observations that the component active for cell adhesion and spreading in the fetuin preparation could be separated from fetuin (Lieberman *et al.*, 1959), the nature of the factors active for cell adhesion, spreading, and proliferation still remain a point of some confusion. It is not surprising that the biologically active fetuin preparations isolated as described originally (Fisher *et al.*, 1958) may contain a number of active contaminants because the isolation procedure involves only a series of ammonium sulfate precipitations from fetal calf serum. Chromatographic procedures have identified several active components, although the relationship of these active contaminants to each other will require additional work in the future (Lieberman *et al.*, 1959; Salomon *et al.*, 1982, 1984). Fetuin has been used as a source of attachment and cell proliferation-promoting activity in serum-free cell cultures of HeLa, rat myoblast, and mouse embryo and embryonal carcinoma cells, as well as other cell types (Fisher *et al.*, 1958; Rizzino and Sato, 1978; Florini and Roberts, 1979; Rizzino and Sherman, 1979), and it may be that more than one contaminating activity is involved for some or all of these cell types. Only in the case of myoblasts has it been reported that more highly purified fetuin preparations retain the biological activity of the material purified by ammonium sulfate precipitation (Florini and Roberts, 1979).

The amounts of the attachment-promoting activities that may exist in fetuin preparations may vary with different preparations. Fetuin preparations may contain either fetal bovine fibronectin and biologically active fibronectin fragments or fetal bovine serum-spreading factor, or both or other attachment factors. A spreading-promoting activity that is different from fibronectin and similar in some respects to serum-spreading factor has been isolated from fetal calf serum (Whateley and Knox, 1980), and may copurify with fetuin in the salt precipitation procedures. A factor apparently identical to α_2-macroglobulin that stimulates the proliferation of embryonal carcinoma cells in serum-free also been isolated from fetuin (Salomon *et al.*, 1982, 1984b). This material was isolated based on assay of mitogenic activity for attached embryonal carcinoma cells, and the attachment or spreading-promoting capacity of the isolated mitogenic factor was not examined. Although the possibility exists that one or more of the several spreading-promoting activities in fetuin may exist in extracellular matrix of the fetal organism, no data in this area have been reported.

Extracellular Matrix in Cell Culture

A more complicated and perhaps more physiologically relevant approach to providing cells with an appropriate substratum *in vitro* is to plate cells on tissue culture vessels on which extracellular matrix components have been deposited by previous growth of matrix-producing cells on the vessels (Weiss *et al.*, 1975; Gospodarowicz *et al.*, 1978, 1979, 1980, 1982; Vlodavsky *et al.*, 1980). The matrix deposited under these conditions may in some instances closely resemble the extracellular matrix of the intact organism in composition, if not in organization. Present in the matrices commonly used for these types of culture systems may be collagen, hyaluronic acid, chondroitin sulfate, heparin sulfate, dermatan sulfate, elastin, fibronectin, and laminin, although there is some evidence that the structure of the matrix may not correspond in an exact way with that seen *in vivo* with regard particularly to hyaluronate-proteoglycan interactions (Birdwell *et al.*, 1978; Culp *et al.*, 1978; Hedman *et al.*, 1979; Rollins and Culp, 1979; Gospodarowicz *et al.*, 1980, 1981; Murray *et al.*, 1980; Tseng *et al.*, 1981; Lark and Culp, 1982).

Most types of cells in culture deposit some sort of extracellular matrix on the culture vessel, and several cell types have been used for such purposes. The most widely used cell type, making the best characterized matrix, are bovine corneal endothelial cells (Birdwell *et al.*, 1978; Gospodarowicz *et al.*, 1980, 1981; Tseng *et al.*, 1981; Gospodarowicz, 1984). These cells are allowed to grow as a monolayer *in vitro*, and live cells are removed from the matrix by treatment with 0.5% triton X-100 in phosphate-buffered saline or 20 mM ammonium hydroxide in water. The latter treatment is recommended for cultures that have been maintained on the plate for extended time periods and have deposited a thick matrix on the surface of the vessel (Gospodarowicz, 1984). Because the extracellular matrix left on the plate by these procedures is complex and because the potential exists for the presence in the matrix of residual growth factors or growth-stimulatory molecules that may remain bound to one or more matrix components and maintain biological activity under these circumstances (Iverius, 1972; Smith *et al.*, 1982), it is difficult to establish which substances in the matrix are responsible for the biological activities observed to be mediated by the material remaining on the surface of the culture vessel after detergent or base treatment. It does appear, however, that the extracellular matrix is reasonably free of cellular components, as determined by examination for remaining cytoskeletal proteins in the matrix, and treatments that might be expected to inactivate some growth factors, such as exposure to reducing agents or heating to 70 °C for short periods, do not diminish the growth-promoting capacity of the matrix (Gospodarowicz, 1984; Gospodarowicz *et al.*, 1983). Furthermore, the matrix is reported to maintain equal biological activity whether it is produced by cells grown in culture medium supplemented with serum, plasma, or with defined supplements, suggesting that biological activities of the material are the result of matrix deposited by the cells during the period of culture, rather than the result of components of the culture medium that have become associated with the surface of the vessel during the period of culture (Giguere *et al.*, 1982; Gospodarowicz *et al.*, 1983).

Thus far a number of cell types have exhibited improved characteristics if cultured on extracellular matrix and compared with cells cultured on conventional cell culture substrata. For instance, bovine granulosa cells, when cultured on extracellular matrix produced by bovine corneal endothelial cells, exhibit an increased lifespan *in vitro*, and can be grown in the absence of serum in a basal nutrient medium supplemented with a small number of factors (Savion *et al.*, 1981). These cells can be grown from low cell densities on extracellular matrix, but not on untreated cell culture vessels on which no matrix has been deposited, and require FGF if plated on untreated cell culture vessels but not if plated on the extracellular matrix (Savion *et al.*, 1981). Adrenal cortex cells show similar properties; these cells can proliferate at low cell densities on extracellular matrix but not in its absence and need FGF if plated on plastic cell culture vessels that do not contain extracellular matrix (Gospodarowicz *et al.*, 1980). The elimination of some growth factor requirements is also seen for pheochromocytoma cells; these cells in short-term experiments will extend neurite outgrowths on extracellular matrix in the absence of nerve growth factor, but require NGF for this differentiative event if extracellular matrix is absent (Fujii *et al.*, 1982). Some human tumor cell types that attach poorly to conventional cell culture substrata will attach, grow at a higher rate, and exhibit a lower serum requirement if cultured on extracellular matrix compared with cultures of these cells plated on cell culture plastic with no matrix (Vlodavsky *et al.*, 1980).

Another approach to providing a matrix for cells in culture that is as similar as possible to that to which cells are exposed *in vivo* is to isolate matrix material from tissues and apply this material as a cell culture substratum to the plastic culture vessel as a substratum. Insoluble material of this kind, isolated from basement membrane and interstitial stroma, has been termed "biomatrix," and contains collagen of all types, glycosaminoglycans, fibronectin, and depending on the isolation procedure may also contain laminin, as well as other basement membrane components that may have lectinlike activity or biological activity in cell–substratum interactions (Gerfaux *et al.*, 1979; Rojkind *et al.*, 1980). Biomatrix is reported to improve the long-term survival and maintenance of hepatocyte cultures and support the growth and differentiation of mammary epithelial cultures (Rojkind *et al.*, 1980; Wicha *et al.*, 1982).

Although the effect of neoplastic transformation on extracellular matrix production as a whole has not been examined in detail, it might be expected from the available evidence that transformation leads to several different types of alterations in extracellular matrix, arising from aberrations in synthetic or degradative rates, secretion, and extracellular organization of matrix components and resulting in matrices that are improperly or incompletely organized or altered in amounts of some components relative to matrices made by the corresponding nontransformed cell type. The inability of tumor cells to make a normal matrix or the ability of tumor cells to destroy such a matrix are recognized as potentially important aspects of tumor cell behavior that may be responsible in a major way for the malignant nature of cancer cells (Liotta *et al.*, 1979, 1980; Alitalo *et al.*, 1981; Hayman *et al.*, 1981, 1982a).

Future Trends: Studies of Cell Shape and Cell–Substratum Interactions

Considerable evidence points to a relationship between the nature and manner of cell adhesion and shape of cells and the control of cell proliferation and differentiation (Folkman, 1977; Folkman and Moscona, 1978; Gospodarowicz *et al.*, 1978, 1979). For instance, anchored cells show a different quantitative serum requirement for growth, as well as changes in rates of RNA and protein synthesis, when compared with cells in suspension (Yaoi *et al.*, 1972; Ben Ze'ev *et al.*, 1980; Benecke *et al.*, 1980). Furthermore, when the shape of anchored cells is manipulated by changes in the composition of the culture substratum, it can be shown that the proliferative rate of cells, rates of macromolecular synthesis, and the nature of the proteins synthesized are related to cell shape (Folkman and Moscona, 1978). Cell shape, as controlled by the nature of the substratum, can also be shown to influence the ability of cells to respond to particular hormones or growth factors. For instance, corneal epithelial cells plated on plastic exhibit a flattened morphology and a proliferative response to FGF, whereas the same cell type plated on collagen assumes a columnar shape and shows a proliferative response to EGF (Gospodarowicz *et al.*, 1978, 1979). Similar conclusions regarding the relationship between cell shape and adhesion and regulation of proliferation and differentiation in the whole organism may be made by examination of the shape and behavior of cells of proliferating tissues *in vivo* (Folkman, 1977). Experiments both *in vivo* and *in vitro* also suggest that the normal association between cell shape and adhesion and the regulation of proliferation is not functional in transformed cells (Folkman, 1977; Folkman and Moscona, 1978; Gospodarowicz *et al.*, 1978, 1979). Future studies on the nature of the processes involved in the coupling of these phenomena in normal cells should be aided by the use of well-characterized attachment factors and defined, serum-free culture conditions and will allow a better understanding of how the relationships have changed in cancer cells.

As important as how the substratum influences cell shape, proliferation and differentiation is the question of how the cells themselves affect the composition and organization of the substratum and how these effects are regulated. Evidence reviewed in this chapter suggests that cells *in vitro*, and almost certainly *in vivo*, are influenced by the soluble components of the extracellular environment with regard to cellular synthesis, secretion, processing, deposition, and degradation of the insoluble extracellular matrix. Studies, particularly in serum-free media, on actions of peptide growth factors like EGF and FGF, steroid hormones like glucocorticoids, and nutrients or low-molecular-weight factors such as ascorbic acid or retinoic acid indicate that soluble components of the extracellular environment can produce marked indirect effects on cell–substratum interactions (Marceau *et al.*, 1980; Strickland *et al.*, 1980; Salomon *et al.*, 1981; Tseng *et al.*, 1982; Kumegawa *et al.*, 1982; Kidwell *et al.*, 1983). The intriguing relationships of the association between the cell and the extracellular martrix *in vitro* and *in vivo* are likely to remain interesting areas of investigation for some time to come, and the availability of well-characterized molecules of the cell substratum that are clearly of physiological relevance and culture techniques that allow the maintenance and growth of cells *in vitro* in defined environments should allow investigators in these areas to design experiments aimed at answering major questions regarding the role of cell shape and cell adhesion on the molecular interactions involved in the association of cell membranes with extracellular matrices and the effect of the cell substratum on elements of the cytoskeleton, membrane hormone-receptor interactions, transport and secretory processes, and ultimately cellular metabolism, proliferation, and differentiation.

References

Adamson, E. D., Gaunt, S. J., and Graham, C. F., 1979, The differentiation of teratocarcinoma stem cells is marked by the types of collagen which are synthesized, *Cell* **17**:469–476.

Alitalo, K., Kurkinen, M., and Vaheri, A., 1980a, Extracellular matrix components synthesized by human amniotic epithelial cells in culture, *Cell* **19**:1053–1062.

Alitalo, K., Kurkinen, M., Vaheri, A., Virtanen, H., Rohde, H., and Timpl, R., 1980b, Basal lamina glycoproteins are produced by neuroblastoma cells, *Nature* (*London*) **287**:465–466.

Alitalo, K., Keski-Oja, J., and Vaheri, A., 1981, Extracellular matrix proteins characterize human tumor cell lines, *Int. J. Cancer* **27**:755–761.

Alitalo, K., Kuismanen, E., Myllyla, R., Kustala, U., Askoseljavaara, S., and Vaheri, A., 1982, Extracellular matrix proteins of human epidermal keratinocytes and feeder 3T3 cells, *J. Cell Biol.* **94**:497–505.

Barnes, D. W., 1982a, Epidermal growth factor inhibits growth of A431 human epidermoid carcinoma in serum-free cell culture, *J. Cell Biol.* **93**:1–4.

Barnes, D. W., 1982b, Growth of A-431 human epidermoid carcinoma in serum-free cell culture: Inhibition by epidermal growth factor, *Cold Spring Harbor Conferences on Cell Proliferation* **9**:937–941.

Barnes, D. W., and Sato, G., 1979, Growth of a human mammary tumor cell line in a serum-free medium, *Nature* (*London*) **281**:388–389.

Barnes, D. W., and Sato, G., 1980a, Factors that stimulate proliferation of breast cancer cells *in vitro* in serum-free medium, in: *Cell Biology of Breast Cancer* (C. M. McGrath, M. J. Brennan, and M. A. Rich, eds.), Academic Press, New York, pp. 277–287.

Barnes, D. W., and Sato, G., 1980b, Methods for growth of cultured cells in serum-free medium, *Anal. Biochem.* **102**:255–270.

Barnes, D. W., and Sato, G., 1980c, Serum-free cell culture: A unifying approach, *Cell* **22**:649–655.

Barnes, D. W., and Silnutzer, J., 1983, Isolation of human serum spreading factor, *J. Biol. Chem.* **258**:12548–12552.

Barnes, D. W., Wolfe, R., Serrero, G., McClure, D., and Sato, G., 1980, Effects of a serum spreading factor on growth and morphology of cells in serum-free medium, *J. Supramol. Struct. Cell. Biochem.* **14**:47–63.

Barnes, D. W., van der Bosch, J., Masui, H., Miyazaki, K., and Sato, G., 1981, The culture of human tumor cells in serum-free medium, in: *Methods in Enzymology*, Volume 79, *Interferons*, Part B (S. Pestka, ed.), Academic Press, New York, pp. 368–391.

Barnes, D. W., Darmon, M., and Orly, J., 1982, Serum-spreading factor: Effects on RF1 rat ovary cells and 1003 mouse embryonal carcinoma cells in serum-free media, *Cold Spring Harbor Conferences on Cell Proliferation* **9:**155–167.

Barnes, D. W., Silnutzer, J., See, C., and Shaffer, M., 1983, Characterization of human serum spreading factor with monoclonal antibody, *Proc. Natl. Acad. Sci. U.S.A.* **80:**1362–1366.

Barnes, D. W., Mousetis, L., Amos, B. and Silnutzer, J., 1984b, Glass bead affinity chromatography of cell attachment and spreading promoting factors of human serum, *Analytical Biochemistry* **137:**196–204.

Basara, M. L., McCarthy, J. B., Barnes, D. W., and Furcht, L. T., 1984, Tumor cell migration to serum-spreading factor, *Fed. Proc.* (in press).

Benecke, B., Ben-Ze'ev, A., and Penman, S., 1980, The regulation of RNA metabolism in suspended and reattached anchorage-dependent 3T6 fibroblasts, *J. Cell. Physiol.* **103:**247–254.

Bensusan, H. B., Koh, T. L., Henry, K. G., Murray, B. A., and Culp, L. A., 1978, Evidence that fibronectin is the collagen receptor on platelet membranes, *Proc. Natl. Acad. Sci. U.S.A.* **75:**5864–5868.

Ben-Ze'ev, A., Farmer, S. R., and Penman, S., 1980, Protein synthesis requires cell-surface contact while nuclear events respond to cell shape in anchorage-dependent fibroblasts, *Cell* **21:**365–372.

Birdwell, C. R., Gospodarowicz, D., and Nicolson, G. L., 1978, Identification, localization, and role of fibronectin in cultures of bovine endothelial cells, *Proc. Natl. Acad. Sci. U.S.A.* **75:**3273–3277.

Burrill, P. H., Bernardini, I., Kleinman, H. K., and Kretchmer, N., 1981, Effect of serum, fibronectin, and laminin on adhesion of rabbit intestinal epithelial cells in culture, *J. Supramol. Struct. Cell. Biochem.* **16:**385–392.

Carlsson, R., Engvall, E., Freeman, A., and Ruoslahti, E., 1981, Laminin and fibronectin in cell adhesion: Enhanced adhesion of cells from regenerating liver to laminin, *Proc. Natl. Acad. Sci. U.S.A.* **78:**2403–2406.

Chandrasekhar, S., and Millis, A. J. T., 1979, Fibronectin from aged fibroblasts is defective in promoting cellular adhesion, *J. Cell. Physiol.* **103:**47–54.

Chen, L. B., Gallimore, P. H., and McDougall, J. K., 1976, Correlation between tumor induction and the large external transformation sensitive protein on the cell surface, *Proc. Natl. Acad. Sci. U.S.A.* **73:**3570–3574.

Chen, L. B., Burridge, K., Murray, A., Walsh, M. L., Copple, C. D., Bushnell, A., McDougall, J. K., and Gallimore, P. H., 1978a, Modulation of cell surface glycocalyx: Studies on large, external transformation-sensitive protein, *Ann. N.Y. Acad. Sci.* **312:**366–381.

Chen, L. B., Murray, A., Segal, R. A., Bushnell, A., and Walsh, M. L., 1978b, Studies on intercellular LETS glycoprotein matrices, *Cell* **14:**377–391.

Chung, A. E., Jaffe, R., Freeman, I. L., Vergnes, J., Braginski, J. E., and Carlin, B., 1979, Properties of a basement membrane-related glycoprotein synthesized in culture by a mouse embryonal carcinoma-derived cell line, *Cell* **16:**277–287.

Civerchia-Perez, L., Faris, B., LaPointe, G., Beldekas, J., Leibowitz, H., and Franablau, C., 1980, Use of collagen-hydroxyethylmethacrylate hydrogels for cell growth, *Proc. Natl. Acad. Sci. U.S.A.* **77:**2064–2068.

Cooper, A. R., Kurkinen, M., Taylor, A., and Hogan, B. L. M., 1981, Studies on the biosynthesis of laminin by murine parietal endoderm cells, *Eur. J. Biochem.* **119:**189–197.

Couchman, J. R., Hook, M., Rees, D. A., and Timpl, R., 1983, Adhesion, growth, and matrix production by fibroblasts on laminin substrates, *J. Cell. Biol.* **96:**177–183.

Courtoy, P. J., Kanmar, Y. S., Nynes, R. O., and Farquhar, M. G., 1980, Fibronectin localization in the rat glomerulus, *J. Cell Biol.* **87**:691–696.

Crouch, E., Bornstein, P., and Canfield, T., 1978a, Collagen synthesis by human amniotic fluid cells in culture: Characterization of a procollagen with three identical proαl(I) chains, *Biochemistry* **17**:5499–5509.

Crouch, E., Balian, G., Holbrook, K., Duksin, D., and Bornstein, P., 1978b, Amniotic fluid fibronectin: Characterization and synthesis by cells in culture, *J. Cell Biol.* **78**:701–715.

Culp, L. A., Barrett, J. R., Buniel, J., and Hitri, S., 1978, Two functionally distinct pools of glycosaminoglycan in the substrate adhesion site of murine cells, *J. Cell Biol.* **79**:788–800.

Darmon, M. Y., 1982, Laminin provides a better substrate than fibronectin for attachment, growth, and differentiation of 1003 embryonal carcinoma cells, *In Vitro* **18**:997–1003.

Del Rosso, M., Cappelletti, R., Viti, M., Vannucchi, S., and Chiarugi, V., 1981, Binding of the basement-membrane glycoprotein laminin to glycosaminoglycans, *Biochem. J.* **199**:699–704.

Dessau, W., Timpl, R., Jilek, F., and von der Mark, K., 1978, Synthesis and extracellular deposition of fibronectin in chondrocyte cultures, *J. Cell Biol.* **79**:342–355.

Dessau, W., Vertel, B. M., von der Mark, H., and von der Mark, K., 1981, Extracellular matrix formation by chondrocytes in monolayer culture, *J. Cell Biol.* **90**:78–83.

Ekblom, P., Alitalo, K., Vaheri, A., Timpl, R., and Saxen, L., 1980, Induction of a basement membrane glycoprotein in embryonic kidney: Possible role of laminin in morphogenesis, *Proc. Natl. Acad. Sci. U.S.A.* **77**:485–489.

Emerman, J. T., and Pitelka, D. R., 1977, Maintenance and induction of morphological differentiation in dissociated mammary epithelium on floating collagen membranes, *In Vitro* **13**:316–328.

Engel, J., Odermatt, E., Engel, A., Madri, J. A., Furthmayr, H., Rohde, H., and Timpl, R., 1981, Shapes, domain organizations and flexibility of laminin and fibronectin, two multifunctional proteins of the extracellular matrix, *J. Mol. Biol.* **150**:97–120.

Engvall, E., and Ruoslahti, E., 1977, Binding of soluble form of fibroblast surface protein, fibronectin, to collagen, *Int. J. Cancer* **20**:1–5.

Engvall, E., Ruoslahti, E., and Miller, E. J., 1978, Affinity of fibronectin to collagens of different genetic types and to fibrinogen, *J. Exp. Med.* **147**:1584–1595.

Fisher, H. W., Puck, T. T., and Sato, G., 1958, Molecular growth requirements of single mammalian cells: The action of fetuin in promoting cell attachment to glass, *Proc. Natl. Acad. Sci. U.S.A.* **44**:4–10.

Florini, J. R., and Roberts, S. B., 1979, A serum-free medium for the growth of muscle cells in culture, *In Vitro* **15**:983–992.

Folkman, J., 1977, Conformational control of cell and tumor growth, *Recent Adv. Cancer Res.* **1**:119–130.

Folkman, J., and Moscona, A., 1978, Role of cell shape in growth control, *Nature (London)* **273**:345–349.

Fujii, D. K., Massoglia, S. L., Savion, N., and Gospodarowicz, D., 1982, Neurite outgrowth and protein synthesis by PC12 cells as a function of substratum and nerve growth factor, *J. Neurosci.* **2**:1157–1175.

Furcht, L. T., Mosher, D. F., and Wendelschafer-Crabb, G., 1978, Immunocytochemical localization of fibronectin (LETS Protein) on the surface of L6 myoglasts: Light and electron microscopic studies, *Cell* **13**:263–271.

Furcht, L. T., Mosher, D. F., Wendelschafer-Crabb, G., and Foidart, J., 1979, Reversal by glucocorticoid hormones of the loss of a fibronectin and procollagen miatrix around transformed human cells, *Cancer Res.* **39**:2077–2083.

Gerfaux, J., Chang-Fournier, F., Bardos, P., Mun, J., and Chang, C., 1979, Lectin-like activity of components extracted from human glomerular basement membrane, *Proc. Natl. Acad. Sci. U.S.A.* **76:**5129–5133.

Giguere, L., Cheng, J., and Gospodarowicz, D., 1982, Factors involved in the control of proliferation of bovine corneal endothelial cells maintained in serum-free medium, *J. Cell. Physiol.* **110:**72–80.

Gold, L. I., and Pearlstein, E., 1980, Fibronectin-collagen binding and requirement during cellular adhesion, *Biochem. J.* **186:**551–559.

Gospodarowicz, D., 1984, Preparation of extracellular matrices produced by cultured bovine corneal endothelial cells and PF-HR-9 endodermal cells; their use in cell culture, *in: Cell Culture Methods for Molecular and Cell Biology* (D. Barnes, D. Sirbasku, and G. Sato., eds.), A. R. Liss Inc., New York (in press).

Gospodarowicz, D., Greenberg, G., Birdwell, C. R., 1978, Determination of cellular shape by the extracellular matrix and its correlation with the control of cellular growth, *Cancer Res.* **38:**4155–4171.

Gospodarowicz, D., Vlodavsky, I., Greenberg, G., and Johnson, L. K., 1979, Cellular shape is determined by the extracellular matrix and is responsible for the control of cellular growth and function, *Cold Spring Harbor Conferences on Cell Proliferation* **6:**561–592.

Gospodarowicz, D., Delgado, D., and Vlodansky, I., 1980, Permissive effect of the extracellular matrix on cell proliferation *in vitro, Proc. Natl. Acad. Sci. U.S.A.* **77:**4094–4098.

Gospodarowicz, D., Greenberg, G., Foidart, J. M., and Savion, N., 1981, The production and localization of laminin in cultured vascular and corneal endothelial cells, *J. Cell. Physiol.* **197:**171–183.

Gospodarowicz, D., Lui, G. M., and Gonzalez, R., 1982, High-density lipoproteins and the proliferation of human tumor cells maintained on extracellular matrix-coated dishes and exposed to defined medium, *Cancer Res.* **42:**3704–3713.

Gospodarowicz, D., Gonzalez, R., and Fujii, D. K., 1983, Are factors originating from serum, plasma, or cultured cells involved in the growth-promoting effect of the extracellular matrix produced by cultured bovine corneal endothelial cells?, *J. Cell. Physiol.* **114:**191–202.

Green, J., Todaro, G. J., and Goldberg, B., 1966, Collagen synthesis in fibroblasts transformed by oncogenic viruses, *Nature (London)* **209:**916–917.

Grinnell, F., 1976, Cell spreading factor, *Exp. Cell Res.* **102:**51–62.

Grinnell, F., 1978, Cellular adhesiveness and extracellular substrata, in: *International Review of Cytology*, Volume 53 (Bourne, G. H., Danielli, J. F., and Jean, K. W., eds.), Academic Press, New York, pp. 65–144.

Grinnell, F., 1980, Fibroblast receptor for cell-substratum adhesion: Studies on the interaction of baby hamster kidney cells with latex beads coated by cold insoluble globulin (plasma fibronectin), *J. Cell Biol.* **86:**104–112.

Grinnell, F., and Feld, M. K., 1979, Initial adhesion of human fibroblasts in serum-free medium: Possible role of secreted fibronectin, *Cell* **17:**117–129.

Grinnell, F., and Feld, M. K., 1980, Spreading of human fibroblasts in serum-free medium: Inhibition by dithiothreitol and the effect of cold insoluble globulin (plasma fibronectin), *J. Cell. Physiol.* **104:**321–334.

Grinnell, F., and Feld, M. K., 1982, Fibronectin adsorption on hydrophilic and hydrophobic surfaces detected by antibody binding and analyzed during cell adhesion in serum-containing medium, *J. Biol. Chem.* **257:**4888–4893.

Grinnell, F., and Hays, D. G., 1978, Cell adhesion and spreading factor, *Exp. Cell Res.* **115:**221–229.

Grinnell, F., and Minter, D., 1978, Attachment and spreading of baby hamster kidney cells to collagen substrata: Effects of cold-insoluble globulin, *Proc. Natl. Acad. Sci. U.S.A.* **75**:4408–4412.

Grotendorst, G. R., Seppa, H. E. J., Kleinman, H. K., and Martin, G. R., 1981, Attachment of smooth muscle cells to collagen and their migration toward platelet-derived growth factor, *Proc. Natl. Acad. Sci. U.S.A.* **78**:3669–3672.

Grotendorst, G. R., Kleinman, H. K., Rohrbach, D. H., Hewitt, A. T., Varner, H. H., Horigan, E. A., Hassell, J. R., Terranova, V. P., and Martin, G. R., 1982, Role of attachment factors in mediating the attachment, distribution, and differentiation of cells, in: *Growth of Cells in Hormonally Defined Media* (G. H. Sato, A. B. Pardee, and D. A. Sirbasku, eds.), Cold Spring Harbor Laboratory, Cold Spring Harbor, New York, pp. 403–413.

Harper, P. A., and Juliano, R. L., 1980, Isolation and characterization of Chinese hamster ovary cell variants defective in adhesion to fibronectin-coated collagen, *J. Cell Biol.* **87**:755–763.

Harper, P. A., and Juliano, R. L., 1981, Two distinct mechanisms of fibroblast adhesion, *Nature (London)* **290**:136–138.

Hauschka, S. D., and Konigsberg, I. R., 1966, The influence of collagen on the development of muscle clones, *Proc. Natl. Acad. Sci. U.S.A.* **55**:119–126.

Hay, E. D., 1981, Extracellular matrix, *J. Cell Biol.* **91**:205s–223s.

Hayashi, M., and Yamada, K. M., 1981, Differences in domain structures between cellular and plasma fibronectins, *J. Biol. Chem.* **256**:11292–11301.

Hayman, E. G., and Ruoslahti, E., 1979, Distribution of fetal bovine serum fibronectin and endogenous rat cell fibronectin in extracellular matrix, *J. Cell Biol.* **83**:255–259.

Hayman, E. G., Engvall, E., and Ruoslahti, E., 1981, Concomitant loss of cell surface fibronectin and laminin from transformed rat kidney cells, *J. Cell Biol.* **88**:352–357.

Hayman, E. G., Oldberg, A., Martin, G. R., and Ruoslahti, E., 1982a, Codistribution of heparin sulfate proteoglycan laminin, and fibronectin in the extracellular matrix of normal rat kidney cells and their coordinate absence in transformed cells, *J. Cell Biol.* **94**:28–35.

Hayman, E. G., Engvall, E., A'Hearn, E., Barnes, D., Pierschbacher, M., and Ruoslahti, E., 1982b, Cell attachment on replicas of SDS polyacrylamide gels reveals two adhesive plasma proteins, *J. Cell Biol.* **95**:20–23.

Hayman, E. G., Pierschbacher, M. D., Ohgren, Y., and Ruoslahti, E., 1983, Serum spreading factor (vitromectin) is present at the cell surface and in tissues, *Proc. Natl. Acad. Sci.* **80**:4003–4007.

Hedman, K., Kurkinen, M., Alitalo, K., Vaheri, A., Johansson, S., Höök, M., 1979, Isolation of the pericellular matrix of human fibroblast cultures, *Cell Biol.* **81**:83–91.

Hewitt, A. T., Kleinman, H. K., Pennypacker, J. P., and Martin, G. R., 1980, Identification of an adhesion factor for chondrocytes, *Proc. Natl. Acad. Sci. U.S.A.* **77**:385–388.

Hewitt, A. T., Varner, H. H., Silver, M. H., Dessau, W., Wilkes, C. M., and Martin, G. R., 1982, The isolation and partial characterization of chondronectin, an attachment factor for chondrocytes, *J. Biol. Chem.* **257**:2330–2334.

Hogan, B. L. M., 1980, High molecular weight extracellular proteins synthesized by endoderm cells derived from mouse teratocarcinoma cells and normal extraembryonic membranes, *Dev. Biol.* **76**:275–285.

Hogan, B. L. M., and Tilly, R., 1981, Cell interactions and endoderm differentiation in cultured mouse embryos, *J. Embryol. Exp. Morphol.* **62**:379–394.

Holmes, R., 1967, Preparations from human serum of an alpha-one protein which induces the immediate growth of unadapted cells *in vitro*, *J. Cell Biol.* **32**:297–308.

Holmes, R., Mercer, G., and Mohamed, N., 1979, Studies of α-protein in human cell cultures, *In Vitro* **15:**522–531.

Howe, C. C., and Solter, D., 1980, Identification of noncollagenous basement membrane glycopolypeptides synthesized by mouse parietal entoderm and an entodermal cell line, *Dev. Biol.* **77:**480–487.

Hynes, R. O., 1973, Alteration of cell-surface proteins by viral transformation and by proteolysis, *Proc. Natl. Acad. Sci. U.S.A.* **70:**3170–3174.

Hynes, R. O., 1976, Cell surface protein and malignant transformation, *Biochim. Biophys. Acta* **458:**73–107.

Iverius, P., 1972, The interaction between human plasma lipoproteins and connective tissue glycosaminoglycans, *J. Biol. Chem.* **247:**2607–2613.

Jaffe, E. A., and Mosher, D. F., 1978, Synthesis of fibronectin by cultured human endothelial cells, *J. Exp. Med.* **147:**1779–1791.

Johansson, S., and Höök, M., 1980, Heparin enhances the rate of binding of fibronectin to collagen, *Biochem. J.* **185:**521–524.

Johansson, S., Kjellen, L., Höök, M., and Timpl, R., 1981, Substrate adhesion of rat hepatocytes: A comparison of laminin and fibronectin as attachment proteins, *J. Cell Biol.* **90:**260–264.

John, H. A., and Lawson, H., 1980, The effect of different collagen types used as substrata on myogenesis in tissue culture, *Cell Biol. Int. Rep.* **4:**841–850.

Keski-Oja, J., Gahmberg, C. G., Alitalo, K., 1982, Pericellular matrix and cell surface glycoproteins of virus-transformed mouse epithelial cells, *Cancer Res.* **42:**1147–1153.

Ketley, J. N., Orkin, R. W., Martin, G. R., 1976, Collagen in developing chick muscle *in vivo* and *in vitro*, *Exp. Cell Res.* **99:**261–268.

Kidwell, W. R., Wicha, M. S., Salomon, D., and Liotta, L. A., 1980, Differential recognition of basement membrane collagen by normal and neoplastic mammary cells, in: *Cell Biology of Breast Cancer* (C. McGrath, M. J. Brennan, and M. A. Rich, eds.), Academic Press, New York, pp. 17–32.

Kidwell, W. R., Salomon, D. S., Liotta, L. A., Zweibel, J. A., and Bano, M., 1982, Effects of growth factors on mammary epithelial cell proliferation and basement-membrane synthesis, in: *Growth of Cells in Hormonally Defined Media* (G. H. Sato, A. B. Pardee, and D. A. Sirbasku, eds.), Cold Spring Harbor Laboratory, Cold Spring Harbor, New York, pp. 807–818.

Kidwell, W. R., Bano, M., and Salomon, D., 1984, The growth of normal mammary epithelium on collagen in serum-free medium, *in: Cell Culture Methods for Molecular and Cell Biology* (D. Barnes, D. Sirbasku, and G. Sato, eds.), A. R. Liss Inc., New York (in press).

Klebe, R. J., 1974, Isolation of a collage-dependent cell attachment factor, *Nature (London)* **250:**248–251.

Klebe, R. J., Hall, J. R., Naylor, S. L., Dickey, W. D., 1978, Bioautography of cell attachment proteins, *Exp. Cell Res.* **115:**73–78.

Klebe, R. J., Bentley, K. L., Sasser, P. J., and Schoen, R. C., 1980, Elution of fibronectin from collagen with chaotropic agents, *Exp. Cell Res.* **130:**111–117.

Kleinman, H. K., McGoodwin, E. B., Martin, G. R., Klebe, R. J., Fietzek, P. P., and Woolley, D. E., 1978, Localization of the binding site for cell attachment in the α 1 (I) chain of collagen, *J. Biol. Chem.* **253:**5642–5646.

Kleinman, H. K., McGoodwin, E. B., Rennard, S. I., and Martin, G. R., 1979a, Preparation of collagen substrates for cell attachment: Effect of collagen concentration and phosphate buffer, *Anal. Biochem.* **94:**308–312.

Kleinman, H. K., Martin, G. R., and Fishman, P. H., 1979b, Ganglioside inhibition of fibronectin-mediated cell adhesion to collagen, *Proc. Natl. Acad. Sci. U.S.A.* **76:**3367–3371.

Kleinman, H. K., Klebe, R. J., and Martin, G. R., 1981, Role of collagenous matrices in the adhesion and growth of cells, *J. Cell Biol.* **88:**473–485.

Kumegawa, M., Hiramatsu, M., Yajima, T., Hatakeyama, K., Hosoda, S., and Namba, M., 1982, Effect of epidermal growth factor on collagen formation in liver–derived epithelial clone cells, *Endocrinology* **110:**607–612.

Labat-Robert, J., Birembaut, P., Adnet, J. J., Mercantini, F., and Robert, L., 1980, Loss of fibronectin in human breast cancer, *Cell Biol. Int. Rep.* **4**(6)**:**609–616.

Lark, M. W., and Culp, L. A. 1982, Selective solubilization of hyaluronic acid from fibroblast substratum adhesion sites, *J. Biol. Chem.* **257:**14073–14080.

Ledbetter, S. R., Kleinman, H. K., Hassell, J. R., and Martin, G. R., 1984, Methods for the isolation of laminin, *in: Cell Culture Methods for Molecular and Cell Biology* (D. Barnes, D. Sirbasku and G. Sato, eds.), A. R. Liss Inc., New York, (in press).

Leivo, I., Vaheri, A., Timpl, R., Wartiovaara, J., 1980, Appearance and distribution of collagen and laminin in the early mouse embryo, *Dev. Biol.* **76:**100–114.

Leivo, I., Alitalo, K., Risteli, L., Vaheri, A., Timpl, R., Wartiovaara, J., 1982, Basal lamina glycoproteins laminin and type IV collagen are assembled into five-fibered matrix in cultures of a tetracarcinoma-derived endodermal cell line, *Exp. Cell. Res.* **137:**15–23.

Lieberman, I., Lamy, F., and Ove, P., 1959, Nonidentity of fetuin and protein growth (flattening) factor, *Science* **129:**43–44.

Liesli, P., Dahl, D., and Vaheri, A., 1983, Laminin is produced by early rat ostrocytes in primary culture, *J. Cell Biol.* **96:**920–924.

Liotta, L. A., Shigeto, A., Robey, P. G., and Martin, G. R., 1979, Preferential digestion of basement membrane collagen by an enzyme derived from a metastatic murine tumor, *Proc. Natl. Acad. Sci. U.S.A.* **76:**2268–2272.

Liotta, L. A., Tryggvason, K., Garbisa, S., Hart, I., Foltz, C. M., and Shafie, S., 1980, Metastatic potential correlates with enzymatic degradation of basement membrane collage, *Nature (London)* **284:**67–68.

Marceau, N., Gayette, R., Valet, J. P., and Deschenes, J., 1980, The effect of dexamethasone on formation of a fibronectin extracellular matrix by rat hepatocytes *in vitro*, *Exp. Cell Res.* **125:**497–502.

McDonald, J. A., and Kelley, D. G., 1980, Degradation of fibronectin by human leukocyte elastase, *J. Biol. Chem.* **255:**8848–8858.

Michalopoulos, G., and Pitot, H. C., 1975, Primary culture of parenchymal liver cells on collagen membranes, *Exp. Cell Res.* **94:**70–78.

Mosesson, M. W., and Umfleet, R. A., 1970, The cold-insoluble globulin of human plasma: I. Purification, primary characterization, and relationship to fibrinogen and other cold-insoluble fraction components, *J. Biol. Chem.* **245:**5728–5736.

Mosher, D. F., Saksela, O., Keski-Oja, J., and Vaheri, A., 1977, Distribution of a major surface-associated glycoprotein, fibronectin, in cultures of adherent cells, *J. Supramol. Struct. Cell. Biochem.* **6:**551–557.

Murakami, H., and Masui, H., 1980, Hormonal control of human colon carcinoma cell growth in serum-free medium, *Proc. Natl. Acad. Sci. U.S.A.* **77:**3464–3468.

Murray, B. A., Ansbacher, R., and Culp, L. A., 1980, Adhesion sites of murine fibroblasts on cold-insoluble globulin-adsorbed substrata, *J. Cell. Physiol* **104:**335–348.

Murray, J. C., Stingl, G., Kleinman, H. K., Martin, G. R., and Katz, S. I., 1979, Epidermal cells adhere preferentially to type IV (basement membrane) collagen, *J. Cell Biol.* **80:**197–207.

Orly, J., and Sato, G., 1979, Fibronectin mediates cytokinesis and growth of rat follicular cells in serum-free medium, *Cell* **17:**295–305.

Orly, J., Sato, G., and Erickson, G. F., 1980, Serum suppresses the expression of hormonally induced functions in cultured granulosa cells, *Cell* **20:**817–827.

Ott, U., Odermatt, E., Engel, J., Furthmay, H., and Timpl, R., 1982, Protease resistance and conformation of laminin, *Eur. J. Biochem.* **123:**63–72.

Pennypacker, J. P., Hassell, J. R., Yamada, K. M., and Pratt, R. M., 1979, The influence of an adhesive cell surface protein on chondrogenic expression *in vitro*, *Exp. Cell Res.* **121:**411–415.

Pierschbacher, M. D., Hayman, E. G., and Ruoslahti, E., 1981, Location of the cell attachment site in fibronectin using monoclonal antibodies and proteolytic fragments of the molecule, *Cell* **26:**259–267.

Pierschbacher, M. D., Hayman, E. G., and Ruoslahti, E., 1983, A synthetic peptide with the cell attachment activity of fibronectin, *Proc. Natl. Acad. Sci. U.S.A.* **80:**1224–1227.

Podleski, T. R., Greenberg, I., Schlessinger, J., and Yamada, K. M., 1979, Fibronectin delays the fusion of L_6 myoblasts, *Exp. Cell Res.* **122:**317–326.

Postlethwaite, A. E., Seyer, J. M., and Kang, A. H., 1978, Chemotactic attraction of human fibroblasts to type I, II, and III collagens and collagen-derived peptides, *Proc. Natl. Acad. Sci. U.S.A.* **75:**871–875.

Rao, C. N., Margulies, I. M. K., Tralka, T. S., Terranova, V. P., Madri, J. A., and Liotta, L. A., 1982, Isolation of a subunit of laminin and its role in molecular structure and tumor cell attachment, *J. Biol. Chem.* **257:**9740–9744.

Rennard, S. I., Berg, R., Martin, G. R., Foidart, J. M., Robey, P. G., 1980, Enzyme-linked immunoassay (ELISA) for connective tissue components, *Anal. Biochem.* **104:**205–214.

Risteli, L., and Timpl, R., 1981, Isolation and characterization of pepsin fragments of laminin from human placental and renal basement membranes, *Biochem. J.* **193:**749–755.

Rizzino, A., 1983, Two multipotent embryonal carcinoma cell lines irreversibly differentiate in defined media, *Dev. Biol.* **95:** 126–136.

Rizzino, A., 1984, The growth and differentiation of embryonal carcinoma cells in defined and serum-free media, *in: Cell Culture Methods for Molecular and Cell Biology* (D. Barnes, D. Sirbasku, and G. Sato, eds.), A. R. Liss Inc., New York (in press).

Rizzino, A., and Crowley, C., 1980, Growth and differentiation of embryonal carcinoma cell line F_9 in defined media, *Proc. Natl. Acad. Sci. U.S.A.* **77:**457–461.

Rizzino, A., and Sato, G., 1978, Growth of embryonal carcinoma cells in serum-free medium, *Proc. Natl. Acad. Sci. U.S.A.* **75:**1844–1848.

Rizzino, A., and Sherman, M. I., 1979, Blastocysts in serum-free medium, *Exp. Cell Res.* **121:**221–233.

Rizzino, A., Terranova, V., Rohrbach, D., Crowley, C., and Rizzino, H., 1980, The effects of laminin on the growth and differentiation of embryonal carcinoma cells in defined media, *J. Supramol. Struct. Cell. Biochem.* **13:**243–253.

Rohde, H., Wick, G., and Timpl, R., 1979, Immunochemical characterization of the basement membrane glycoprotein laminin, *Eur. J. Biochem.* **102:**195–201.

Rojkind, M., Gatmaitan, Z., MacKensen, S., Giambrone, M., Ponce, P., and Reid, L. M., 1980, Connective tissue biomatrix: Its isolation and utilization for long-term cultures of normal rat hepatocytes, *J. Cell Biol.* **87:**225–263.

Rollins, B. J., and Culp, L. A., 1979, Preliminary characterization of the proteoglycans in the substrata adhesion sites of normal and virus-transformed murine cells, *Biochemistry* **18:**5621–5629.

Rubin, K., Höök, M., Öbrink, B., Timpl, R., 1981, Substrate adhesion of rat hepatocytes: Mechanism of attachment to collagen substrates, *Cell* **24:**463–470.

Ruoslahti, E., and Engvall, E., 1980, Complexing of fibronectin glycosaminoglycans and collagen, *Biochim. Biophys. Acta* **631:**350–358.

Ruoslahti, E., and Vaheri, A., 1974, Novel human serum protein from fibroblast plasma membranes, *Nature (London)* **248:**789–791.

Ruoslahti, E., Hayman, E. G., Engvall, E., Cothran, W. C., and Butler, W. T., 1981, Alignment of biologically active domains in the fibronectin molecule, *J. Biol. Chem.* **256:**7277–7281.

Ruoslahti, E., Hayman, E. G., Pierschbacher, M., and Engvall, E., 1982, Fibronectin: Purification, immunochemical properties, and biological activities, *Methods Enzymol* **82:**803–831.

Russo, J., Soule, H. D., McGrath, C., and Rich, M. A., 1976, Re-expression of the original tumor pattern by a human breast carcinoma cell line (MCF-7) in sponge culture, *J. Natl. Cancer Inst.* **56:**279–282.

Sakashita, S., and Ruoslahti, E., 1980, Laminin-like glycoproteins in extracellular matrix of endodermal cells, *Arch. Biochem. Biophys.* **205:**283–290.

Sakashita, S., Engvall, E., and Ruoslahti, E., 1980, Basement membrane glycoprotein laminin binds to heparin, *FEBS Lett.* **116:**243–246.

Salomon, D. S., Liotta, L. A., and Kidwell, W. R., 1981, Differential response to growth factor by rat mammary epithelium plated on different collagen substrata in serum-free medium, *Proc. Natl. Acad. Sci. U.S.A.* **78:**382–386.

Salomon, D. S., Bano, M., Smith, K. B., and Kidwell, W. R., 1982, Isolation and characterization of a growth factor (embryonin) from bovine fetuin which resembles α_2-macroglobulin, *J. Biol. Chem.* **257:**14093–14101.

Salomon, D. S., Liotta, L. A., Panneerselvam, M., Terranova, V. P., Sahai, A., And Feknel, P., 1984a, Analysis of basement membrane synthesis and turnover in mouse embryonal and human A431 epidermoid carcinoma cells in serum-free medium, *in: Cell Culture Methods for Molecular and Cell Biology* (D. Barnes, D. Sirbasku, and G. Sato, eds.), A. R. Liss Inc., New York (in press).

Salomon, D. S., Smith, K. B., Losonczy, I., Bano, M., Kidwell, W. R., Alessandri, G., and Gullino, P. M., 1984b, α_2-Macroglobulin, a contaminant of commercially prepared Pedersen fetuin: Isolation, characterization and biological activity, *in: Cell Culture Methods for Molecular and Cell Biology* (D. Barnes, D. Sirbasku, and G. Sato, eds.), A. R. Liss Inc., New York (in press).

Sandmeyer, S., Smith, R., Kiehn, D., Bornstein, P., 1981, Correlation of collagen synthesis in procollagen messenger RNA levels with transformation in rat embryo fibroblasts, *Cancer Res.* **41:**830–838.

Savion, N., Lui, G., Laherty, R., and Gospodarowicz, D., 1981, Factors controlling proliferation and progesterone production by bovine granulosa cells in serum-free medium, *Endocrinology* **109:**409–420.

Shaffer, M. C., Foley, T. P., and Barnes, D. W., 1984, Quantitation of spreading factor in human biological fluids, *J. Lab. Clin. Med.* (in press).

Shibata, S., Peters, B., Roberts, D., Goldstein, I., and Liotta, L., 1982, Isolation of laminin by affinity chromatography on immoblized Griffonia Simplicifolia I Lectin, *FEBS Lett.* **142:**194–198.

Sieber-Blum, M., Sieber, F., and Yamada, K. M., 1981, Cellular fibronectin promotes adrenergic differentiation of quail neural crest cells *in vitro*, *Exp. Cell Res.* **133:**285–395.

Silnutzer, J., and Barnes, D. W., 1984a, A biologically active thrombin cleavage product of human serum spreading factor, *Biochem. Biophys. Res. Comm.* **118:**339–343.

Silnutzer, J., and Barnes, D. W., 1984b, Human serum spreading factor: preparation, use and assay in cell culture, *in: Cell Culture Methods for Molecular and Cell Biology* (C. D. Barnes, D. Sirbasku and G. Sato, eds.),A. R. Liss, Inc., New York (in press).

Sirica, A. E., Richards, W., Tsukada, Y., Sattler, C. A., and Pitot, H. C., 1979, Fetal phenotypic expression by adult rat hepatocytes on collagen gel/nylon meshes, *Proc. Natl. Acad. Sci. U.S.A.* **76:**283–287.

Smith, J. C., Singh, J. P., Lillquist, J. S., Goon, D. S., and Stiles, C. D., 1982, Growth factors adherent to cell substrate are mitogenically active in situ, *Nature* (*London*) **296:**154–156.

Stathakis, N. E., and Mosesson, M. W., 1977, Interactions among heparin, cold-insoluble globulin, and fibrinogen in formation of the heparin-precipitable fraction of plasma, *J. Clin. Invest.* **60:**855–865.

Stenman, S., Wartiovaara, J., and Vaheri, A., 1977, Changes in the distribution of a major fibroblast protein, fibronectin, during mitosis and interphase, *J. Cell Biol.* **74:**453–467.

Stenman, T. S., Vaheri, A., Timpl, R., 1979, Changes in the matrix proteins, fibronectin and collagen, during differentiation of mouse tooth germ, *Dev. Biol.* **70:**116–126.

Stenn, K. S., 1981, Epibolin: A protein of human plasma, that supports epithelial cell movement, *Proc. Natl. Acad. Sci. U.S.A.* **78:**6907–6911.

Stenn, K. S., 1984, Purification of epibolin from human plasma, *in*: *Cell Culture Methods for Molecular and Cell Biology* (D. Barnes, D. Sirbasku, and G. Sato, eds.), A. R. Liss Inc., New York (in press).

Stenn, K. S., Madri, J. A., Roll, F. J., 1979, Migrating epidermis produces AB_2 collagen and requires continual collagen synthesis for movement, *Nature* (*London*) **277:**229–232.

Stenn, K. S., Madri, J. A., Tinghitella, T., and Terranova, V. P., 1982, Multiple mechanisms of dissociated epidermal cell spreading, *J. Cell Biol.* **96:**63–67.

Strickland, S., Smith, K. K., Marotti, K. R., 1980, Hormonal induction of differentiation in teratocarcinoma stem cells: Generation of parietal endoderm by retinoic acid and dibutyryl cAMP, *Cell* **21:**347–365.

Strom, S. C., and Michalopoulos, G., 1982, Collagen as a substrate for cell growth and differentiation, *Methods Enzymol.* **82:**544–555.

Sun-Yang, N., Kube, D., Park, C., Furmanski, P., 1981, Growth of human mammary epithelial cells on collagen gel surfaces, *Cancer Res* **41:**4093–4100.

Terranova, V. P., Rohibach, D. H., and Martin, G. R., 1980, Role of laminin in the attachment of PAM 212 (epithelial) cells to basement membrane collagen, *Cell* **22:**719–726.

Timpl, R., Martin, G. R., Bruckner, P., Wick, G., and Wiedemann, H., 1978, Nature of the collagenous protein in a tumor basement membrane, *Eur. J. Biochem.* **84:**43–52.

Timpl, R., Heilwig, R., Robey, P. G., Rennard, S. I., Foidart, J., and Martin, G. R., 1979, Laminin: A glycoprotein from basement membranes, *J. Biol. Chem.* **254:**9933–9937.

Timpl, R., Rohde, H., Risteli, L., Ott, U., Roby, P. G., and Martin, G. R., 1982, Laminin, *Methods Enzymol.* **82:**831–838.

Tseng, S. C. G., Savion, N., Gospodarowicz, D., and Stern, R., 1981, Characterization of collagens synthesized by cultured bovine corneal endothelial cells, *J. Biol. Chem.* **256:**3361–3365.

Tseng, S. C. G., Savion, N., Stern, R., and Gospodarowicz, D., 1982, Fibroblast growth factor modulates synthesis of collagen in cultured vascular endothelial cells, *Eur. J. Biochem.* **122:**355–360.

Vaheri, A., Kurkinen, M., Lehto, V.-P., Linder, E., Timpl, R., 1978, Codistribution of pericellular matrix proteins in cultured fibroblasts and loss of transformation: Fibronectin and collagen, *Proc. Natl. Acad. Sci. U.S.A.* **75:**4944–4948.

Van der Bosch, J., 1984, Primary tissue cultures of human colon carcinomas in serum-free medium: An *in vitro* system for tumor analysis and therapy experiments, *in*: *Cell Culture Methods for Molecular and Cell Biology* (D. Barnes, D. Sirbasku, and G. Sato, eds.), A. R. Liss, Inc., New York, (in press).

Varner, H. H., Hewitt, A. T., and Martin, G. R., 1984, Methods for the isolation of chondronectin, *in*: *Cell Culture Methods for Molecular and Cell Biology* (D. Barnes, D. Sirbasku, and G. Sato, eds.), A. K. Liss, Inc., New York, (in press).

Vartio, T., and Vaheri, A., 1981, A gelatin-binding 70,000-dalton glycoprotein synthesized distinctly from fibronectin by normal and malignant adherent cells, *J. Biol. Chem.* **256:**13085–13090.

Vlodavsky, I., and Gospodarowicz, D., 1981, Respective roles of laminin and fibronectin in adhesion of human carcinoma and sarcoma cells, *Nature* (*London*) **289:**304–306.

Vlodavsky, I., Lui, G. M., and Gospodarowicz, D., 1980, Morphological appearance, growth behavior and migratory activity of human tumor cells maintained on extracellular matrix versus plastic, *Cell* **19:**607–616.

Vuento, M., Salonen, E., Salminen, K., Pasanen, M., and Stenman, U., 1980, Immunochemical characterization of human plasma fibronectin, *Biochem. J.* **191:**719–727.

Weinstein, R., Hoover, G. A., Stemerman, M. B., van der Spek, J., and Maciag, T., 1982, Fibronectin dependence for attachment *in vitro*: Smooth-muscle cell versus fibroblast, in: *Growth of Cells in Hormonally Defined Media* (G. H. Sato, A. B. Pardee, and D. A. Sirbasku, eds.), Cold Spring Harbor Laboratory, Cold Spring Harbor, New York, pp. 145–154.

Weiss, L., Poste, G., MacKearnin, A., and Willett, K., 1975, Growth of mammalian cells on substrates coated with cellular microexudates, *J. Cell Biol.* **64:**135–145.

Weiss, R. E., and Reddi, A. H., 1980, Synthesis and localization of fibronectin during collagenous matrix-mesenchymal cell interaction and differentiation of cartilage and bone *in vivo*, *Proc. Natl. Acad. Sci. U.S.A.* **77:**2074–2078.

West, C. M., Lanza, R., Rosenbloom, J., Lowe, M., Holtzer, H., and Avdalovic, N., 1979, Fibronectin alters the phenotypic properties of cultured chick embryo chondroblasts, *Cell* **17:**491–501.

Wewer, U., Albrechtsen, R., and Ruoslahti, E., 1981, Laminin, a noncollagenous component of epithelial basement membrane synthesized by a rat yolk sac tumor, *Cancer Res.* **41:**1518–1524.

Whateley, J. G., and Knox, P., 1980, Isolation of a serum component that stimulates the spreading of cells in culture, *Biochem. J.* **185:**349–354.

Wicha, M. S., Liotta, L. A., Garbisa, S., Kidwell, W. R., 1979, Basement membrane collagen requirements for attachment and growth of mammary epithelium, *Exp. Cell Res.* **124:**181–190.

Wicha, M. S., Liotta, L. A., Vonderhaar, B. K., and Kidwell, W. R., 1980, Effects of inhibition of basement membrane collagen deposition on rat mammary gland development, *Dev. Biol.* **80:**253–266.

Wicha, M. S., Lowrie, G., Kohn, E., Baganandoss, P., and Mahn, T., 1982, Extracellular matrix promotes mammary epithelial growth and differentiation *in vitro*, *Proc. Natl. Acad. Sci. U.S.A.* **79:**3213–3217.

Yamada, K. M., and Olden, K., 1978, Fibronectins-adhesive glycoproteins of cell surface and blood, *Nature* (*London*) **275:**179–184.

Yamada, K. M., Kennedy, D. W., Kimata, K., and Pratt, R. M., 1980, Characterization of fibronectin interactions with glycosaminoglycans and identification of active proteolytic fragments, *J. Biol. Chem.* **255**:6055–6063.

Yamada, K. M., Kennedy, D. W., Grotendorst, G. R., and Momoi, T., 1981, Glycolipids: Receptors for Fibronectin? *J. Cell. Physiol.* **109**:343–351.

Yamada, K. M., Kennedy, D. W., and Hayashi, M., 1982, Fibronectin in cell adhesion, differentiation and growth, in: *Growth of Cells in Hormonally Defined Media* (G. H. Sato, A. B. Pardee, and D. A. Sirbasku, eds.), Cold Spring Harbor Laboratory, Cold Spring Harbor, New York, pp. 131–143.

Yang, J., Guzman, R., Richards, J., Jintoft, V., DeVault, M. R., Wellings, S. R., Nandi, S., 1980a, Primary culture of human mammary epithelial cells embedded in collagen gels, *J. Natl. Cancer Inst.* **65**:337–341.

Yang, J., Richards, J., Buzman, R., Imagawa, W., Nandi, S., 1980b, Sustained growth in primary culture of normal mammary epithelial cells imbedded in collagen gels, *Proc. Natl. Acad. Sci. U.S.A.* **77**:2088–2092.

Yang, N. S., Kirkland, W., Jorgensen, T., and Furmanski, P., 1980c, Absence of fibronectin and presence of plasminogen activator in both normal and malignant human mammary epithelial cells in culture, *J. Cell Biol.* **84**:120–130.

Yaoi, Y., Onoda, T., Takahashi, H., 1972, Inhibition of mitosis in suspension culture of chick embryo cells, *Nature* (*London*) **237**:285–286.

CHAPTER 10

Cell Culture Studies Using Extracts of Extracellular Matrix to Study Growth and Differentiation in Mammalian Cells

LOLA M. REID and DOUGLAS M. JEFFERSON

Introduction

With the evolution of metazoans came the evolution of multiple, specialized cell types that coordinate their activities with each other and in response to the organism's environment. The regulation of growth and differentiation in such specialized cell types, and its loss in certain diseases such as malignancy, have been the focus of many major fields of scientific interest. This review will discuss the state of the art of specific culture conditions that optimize either growth or maintenance of the differentiated state in "committed" mammalian cells, that is, cells that have differentiated during embryological development to a specific cell type such as a fibroblast, hepatocyte, or neuron. Thus, we are distinquishing two aspects of differentiation: (1) the "committment" process, occurring during embryogenesis and resulting in multiple specialized cell types, and (2) regulation of gene expression within a committed cell.

LOLA M. REID and DOUGLAS M. JEFFERSON • Department of Molecular Pharmacology, Albert Einstein College of Medicine, Bronx, New York 10461.

Once mammalian cells are committed, they can be influenced only to vary the level of gene expression for the specialized functions of that cell type. These two aspects of differentiation are shown in schematic form in Fig. 1. In the studies to be presented here, all dealing with committed cells, the goals are to maintain the gene expression of a particular cell type at levels approximating those *in vivo.*

Regulation of growth and differentiation in committed cells has been shown to result from complex cell–cell interactions (Green, 1979; Bissell, 1981; Bissell *et al.*, 1982; Vaughan and Bernstein, 1980; Zanjani and Rinehart, 1980). Although in any metazoan tissue there may be a number of cell types involved in these interactions, those between the epithelial cell and a mesenchymal cell derivative are thought to be the most important. Grobstein in 1955 described the importance of this cell–cell interaction. Its relevance was further clarified by the studies of Zwilling (1955), Saunders and Gasseling (1957), Wessells (1964,1970), Cunha (1972), Kratochwil (1964), and others. In Fig. 2 is shown a schematic diagram of the epithelial–mesenchymal relationship and several versions of it found in mammalian tissues. As indicated in this figure and in Fig. 1, there must be at least two major forms of the epithelial–mesenchymal relationship: (1) one found in tissues with stem cells (e.g., skin or colon) in which the epithelial cells bound to the basement membrane are in a proliferative state but can terminally differentiate, losing their viability by detachment from the basement membrane and loss of association with the mesenchymal cells; and (2) one found in systems in which the epithelial cells, bound to the basement membrane, are maintained in a differentiative state but can convert to a proliferative state while still remaining bound to the basement membrane. In the latter types of tissues, the cells are plastic and can go back and forth from a proliferative, less differentiated state to an amitotic, fully differentiated condition.

The complex cellular interactions effecting differentiation have been reduced, in part, to identifiable variables and factors influencing the gene expression of cells: nutrients; hormones and growth factors; soluble signals from neighboring cell–cell interactions (e.g., signals that pass between cells via gap junctions, conditioned media signals); and substrata of tissue-specific extracellular matrix. Since elsewhere in this book there are excellent reviews discussing many of these variables, this review will focus on studies of extracellular matrix and its known relevance to regulation of growth and differentiation when acting alone or in combination with hormones as in defined media or with serum supplemented media. Furthermore, since most studies of matrix biology have used individual components of matrix and since elsewhere in this

I. "COMMITMENT" (also called determination)

A unidirectional process resulting in the generation of cell types and involving the restriction of the genetic expression to that for one cell type. Thereafter, such cells cannot change their commitment even in diseased states such as malignancy.

Categories of cells

Embryonic or pluripotent cells

Hepatocytes Neurons Fibroblasts

II. REGULATION OF EXPRESSION OF COMMITTED CELLS

A. Conversion of Stem Cells to Terminally Differentiated Cells

Stem cells are bound to a form of basement membrane and in association with other cell types; the interactions resulting in proliferation. Detachment from the basement membrane and loss of association with the other cell types results in terminal differentiation and loss of viability.

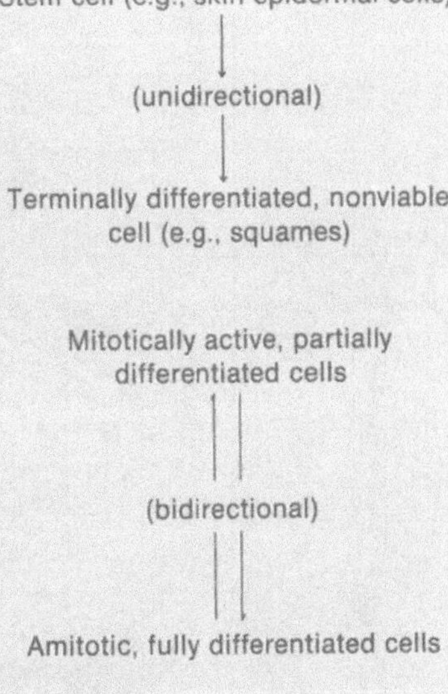

B. Maintenance of Differentiated Cells

In tissues such as liver or lung, the cells are maintained stably in an amitotic, differentiated state. Under specific conditions (e.g., partial hepatectomy), the tissue is stimulated to regenerate. All of the cells, in association with each other, grow in a coordinated fashion. During growth, the differentiated functions are reduced somewhat (the extent varies with the tissue). Once a specific tissue mass is achieved, the cells stop growing and then return to the fully differentiated state. In malignant tissue, the well-known phenomenon of dedifferentiation (Uriel, 1979) may reflect that tumor cells are stuck in the mitotically active, partially differentiated state.

Figure 1. Types of mechanisms resulting in differentiation of cells.

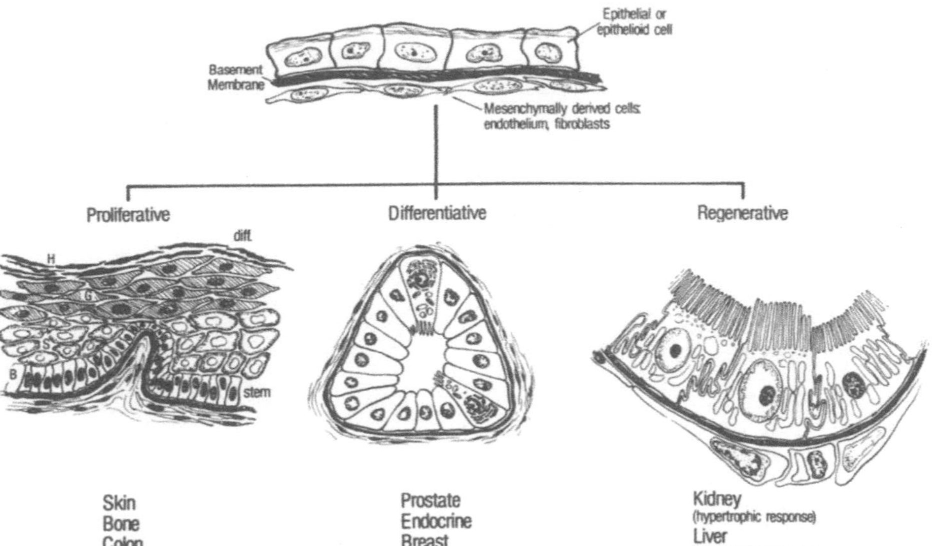

Figure 2. Schematic representation of tissue architectural paradigm. Differentiation of metazoan tissues is inextricably coupled to cellular interactions. Metazoan tissues can be represented by a tissue architectural paradigm as presented schematically in this figure showing the epithelial-mesenchymal cell relationship, among the first heterotypic cell–cell relationships to evolve in metazoans, and ubiquitous in all metazoan tissues. The relationship consists of an epithelial or epithelioid cell attached to basement membrane material, and in turn, associated with a mesenchymally derived cell, most usually fibroblasts or endothelial cells. Almost all cells *in vivo* capable of proliferation or of long-term survival are found in association with basement membranes. The basement membrane is not a membrane (a misnomer from studies of early anatomists, who saw an acellular layer at the "basement" of the epithelium and called it a membrane), but a complex,

collageneous secretion produced by both the epithelium and the mesenchymal cell and located between them. The interactions between the epithelium and the mesenchymal cells consist of (1) soluble signals to be found in conditioned medium from primary cultures, and (2) insoluble signals to be found in basement membranes.

There are *at least* two classes of tissue forming subdivisions of the tissue architectural paradigm in adult tissues. The two classes are (1) stem cell tissues, and (2) tissue maintained in the differentiated state. Each of these classes is distinct in that the cellular interactions effect proliferation or maintain differentiation or the ability to switch between the two. The distinctions should result in different expectations for culture models as clarified below.

Stem Cells are represented by bone marrow, skin, or colon. Stem cells are associated with the basement membrane and with fibroblasts. The stem cells proliferate as long as they are attached to the basement membrane. Daughter cells of the stem cells become detached from the basement membrane (skin, bone marrow) or slide to new regions of presumably chemically distinct basement membrane (colon) and in so doing lose their mitotic potential. Thus, the communications between the epithelium and the mesenchymal cells are concerned primarily with maintenance of the proliferative state. Success in mimicking the proliferative tissues should produce stem cells anchored to substrata of basement membrane. The stem cells should produce floating cells that undergo differentiation. Soluble factors in the medium should dictate the kinetics of differentiation.

Differentiated Tissues are represented by prostate, breast, liver, or endocrine tissues. Differentiated epithelium or epitheloid cells are bound to basement membrane and associated with either fibroblasts or (in tissues with extensive transport or secretion) endothelium. Although some proliferation of these cells is possible, it is limited and involves the concomitant growth of all cell types in association with the basement membrane. Thus, the primary concern of the cellular interactions is the maintenance of specialized functions of the epithelial cells. Simulation of the epithelial-mesenchymal relationship should produce nongrowing differentiated epithelial cells. These tissues can undergo growth during regenerative responses. A few tissues have an unusual capacity to regenerate either by hyperplasia (liver) or by hypertrophy (kidney). Under certain conditions, such as a partial hepatectomy or partial nephrectomy, the epithelial-mesenchymal relationship can switch back to a proliferative state. Most tissues have some restricted capacity to regenerate. Successful simulation of the cellular interactions should produce an extended proliferative or differentiative response depending on the soluble signals (the hormones) and the chemical composition of the basement membrane.

volume there are reviews of these studies, this review will concentrate on more recent studies using crude extracts of matrix that give a more general perspective of the roles of matrix.

Background on Extracellular Matrix

Early Studies Indicating Relevance of Extracellular Matrix

Studies in Whole Organisms and Tissues

Interest in the extracellular matrix has increased dramatically in recent years due to increasing awareness of its relevance in the regulation of growth and differentiation. Several recent reviews and books, especially those of Yamada (1983), Hay (1981a,b,c), Bissell (1981), Bissell *et al.* (1982), and Kleinman *et al.* (1981), provide excellent coverage of the past and present studies on extracellular matrix. What should become apparent from these reviews is that ironically the observation that extracellular material is pertinent to differentiation has existed for decades in the fields of embryology and developmental biology. The extensive pioneering efforts of developmental biologists (Banerjee *et al.*, 1977; Bernfield and Banerjee, 1982; Bernfield *et al.*, 1972; Hay, 1981a,b,c) laid the groundwork for the realization of matrix's significance and for many of the fundamental principles guiding the rash of current investigations in this field.

However, even though matrix has long been known to contain signals directing differentiation, progress in studying its influences has been hampered until recently due to its chemical complexity and insolubility. The current renaissance of interest in matrix biology is in part a reflection of the recent availability of purified matrix components and in part a reflection of the technical and theoretical developments in molecular biology enabling one to analyze molecular mechanisms regulating gene expression in cells. The final necessary technological breakthroughs are the development of methodologies for maintenance of cells, whether normal or diseased, in culture in a physiological state approximating that *in vivo*. Until now, there have been extreme difficulties in culturing cells with retention of their differentiated state when the cells are isolated from tissues. Disassociation of cells from tissues frequently has led to loss of function or death of the cells. There has been the dilemma of studying differentiation in tissues or whole organisms where the numerous variables make interpretation of experiments difficult or impossible, or to use isolated cell types such as clonal cell lines that empirically were found to lose or have reduced levels of the cell's differentiated functions. Recent breakthroughs in cell culture technology are providing methods (many of them described in this book) by which to establish individual cell types, including epithelial cells, in culture with retention of their differentiated phenotype. These breakthroughs involve the elimination of serum (now shown to be toxic and/or inhibitory to many epithelial cells), the development of serum-free, hormonally defined media designed for each cell type, and the use of substrata of tissue-specific extracellular matrix.

A number of approaches have guided the investigations of extracellular matrix. Matrix was characterized in tissues by morphological and histochemical methods or individual components were isolated, purified, and chemically characterized (Miller, 1976; Grotendo *et al.*, 1980; Bernfield and Banerjee, 1982; Hauschka and Konigsberg, 1966; Hay, 1981a,b,c). These studies have been expanded by new technologies, especially the use of monoclonal antibodies, permitting isolation and identification of ever more matrix components. However, studies on matrix biology have suffered from two inherent limitations in past scientific approaches:

1. Much of the matrix is inherently insoluble due to extensive crosslinking of its components. Therefore, many of its components are difficult to isolate intact (for example, type IV collagen) and often require extraction methods that destroy parts of the molecules. Most of the matrix components studied to date (e.g., type I collagen, fibronectin) are those that are amenable to solubilization by standard methods of chemical isolation and purification. There are presumed to be matrix components that have not been isolated and purified because of the inherent insolubility of matrix.
2. The influence of the matrix on cells is the net result of a complex mixture of molecules, not just one or two. The tendency of scientists to analyze one component by itself, albeit an essential approach, cannot exclude the need for studies that acknowledge that the matrix is a mixture of components operating in synchrony.

A wholly different approach has been adopted in recent years to assess the biological role of matrix. Crude extracts of cell layers or of whole tissues are made with protocols designed to select for the insoluble components in extracellular matrix. The crude extracts are then used as cell culture substrata for clonal cell lines and/or primary cultures of tissues. Biological activity in terms of attachment, survival, growth, or levels of tissue-specific functions has been assessed under these new conditions. The early studies using these extracts are revealing fundamental new appreciations for the relevance of extracellular matrix to cellular physiology. Moreover, those investigations using purified hormones, especially the serum-free hormonally defined media, in combination with these extracts suggest complicated synergies between matrix and hormones in regulating cell function (see section on synergies between hormones and matrix, p. 267).

Cell Culture Technology: A Critical Assessment

Studies of normal and diseased epithelial cells are often complicated by variables that are difficult to control in whole animals. The hope has long been to develop culture systems of normal or diseased cells with retention of cellular phenotypes observed *in vivo*. Therefore, the need for cell culture model systems is apparent as a means to minimize the number of variables affecting the property being analyzed. The use of genetically homogeneous populations of normal or mutant cells permits one to more precisely analyze mechanisms affecting cellular physiology. However, under standard culture conditions, normal cells die or lose tissue-specific functions (see reviews by Bissell and Guzelian, 1981; Bissell, 1981; Jakoby and Pastan, 1979). The usual methods of preparing cells for culture start with the disruption of the tissue's architecture. If the culture method preserves some of the cell–cell relationships as in organ cultures and Wolf cultures, the differentiative potential of the cells is also preserved (Wolf and Wolf, 1975; Kratochwil, 1964). However, for cultures in which the cells are cloned to derive homogeneous populations, the specialized functions can be totally lost. A compromise solution has been to resort to minimally deviant neoplastic cells carried as transplantable tumor lines in syngeneic or immunosuppressed hosts. Such tumor lines have been adapted more readily to cell culture than normal cells. Using this approach, many partially differentiated cell cultures have been established (Buonassisi *et al.*, 1962).

It has been proposed that the traditional culture conditions have been inadequate primarily because they destroy critical cell–cell relationships in the tissues (Green, 1979; Guguen-Guillouzo *et al.*, 1983; Reid, 1982; Reid and Rojkind, 1979). Malignant tumors, known to contain one (or at most two or three) cell types proliferating in an unregulated fashion, would therefore be cells that are qualitatively or quantitatively autonomous to the cellular interactions. Their autonomy would give them a selective advantage *in vitro* and make them more easily adaptable for cell culture conditions. The corollary is that development of culture conditions for normal cells should require simulation of critical cell–cell relationships, especially the epithelial–mesenchymal interaction (Grobstein, 1953). In classic cell culture technologies, the epithelial-mesenchymal relationship was simulated, in part, by the use of feeder layers of cells, a technique found valuable for primary cultures of many types of differentiated epithelial cells (Jakoby and Pasten, 1979). For recent reviews on the epithelial–mesenchymal relationship see those by Bissell *et al.* (1982) and Yamada (1983).

Epithelial Cell Cultures

Of all the cell types, epithelial cells have proved to be one of the most difficult to maintain in culture (Bissell and Guzelian, 1981; Bissell, 1981; Buonassisi *et al.*, 1962; Green, 1979; Jefferson *et al.*, 1980; Michalopoulos and Pitot, 1975; Montgomery and Guzelian, 1980). When plated onto tissue culture plastic, the epithelial cells rapidly undergo morphological and biochemical deterioration resulting in a loss within two weeks of markers of differentiation and within three to four weeks of viable cells. Extensive studies (review by Bissell, 1981) of the variables contributing to the inability to culture normal epithelial cells have revealed the usual problems of requirements for carefully defined nutrients (Ham, 1965; Ham and McKeehan, 1979), gas exchange, efficiency of nutrient and waste exchange, and requirements for specific hormones and growth factors (Barnes and Sato, 1980; Ambesiim *et al.*, 1980; Bottenstein *et al.*, 1979). The importance of collagenous substrata has long been noted (Hauschka and Konigsberg, 1966). Innovations in the use of collagenous substrata (floating collagen gels) were introduced with remarkable enhancement in the survival and differentiative state of the cells associated with more three-dimensional shapes. The floating collagen gels have proven a boon for various normal epithelial cell types, including mammary cells, skin epithelium, colon epithelium, and others (Benya and Shaffer, 1982; Chambard *et al.*, 1980; Chazov *et al.*, 1981; Guido and Bernfield, 1981; Guzman *et al.*, 1982; Emmerman and Pitelka, 1977; Michalopoulos and Pitot, 1975; Michalopoulos *et al.*, 1976, 1978). Morphologically, the cells were distinctly different when plated onto collagen gels in contrast to tissue culture plastic. On plastic the cells spread to form a squamouslike shape, appeared increasingly translucent and agranular, and rapidly lost their granular endoplasmic reticulum. On collagen gels, the cells assumed polygonal shapes that became cuboidal or columnar over time in culture. The histology of the cells and ultrastructural studies indicated formation of cellular junctions, the development of polarity of the cells, and well-developed endoplasmic reticulum and Golgi bodies.

Until the late 1970s and early 1980s, virtually all cell culture studies using collagenous substrata used type I collagen, primarily due to its ease of isolation and preparation. However, increasingly it has become apparent that specific cell types attach and function preferentially with specific collagen types (Huw and Lawson, 1980; Murray *et al.*, 1980) (see Table I). Thus, the potential for collagens to influence cells in culture will require re-evaluation once investigations have been done with appropriate collagen types.

Chemical Composition of the Extracellular Matrix

General Considerations

Extracellular matrix is produced and secreted by all metazoan cells. Its composition in various tissues consist of a complex mixture of collagens, elastin, proteoglycans, glycoproteins, microfibrillar proteins, and anchorage proteins such as fibronectin or laminin (Kleinman *et al.*, 1981; Gerfaux *et al.*, 1981; Hascall and Sajdera, 1969; Hascall and Hascall, 1981; Hassell *et al.*, 1980; Hay, 1981a,b,c; Hewitt *et al.*, 1982; Iozzo and Wight, 1982; Kuettner *et al.*, 1982; Risteli and Risteli, 1981; Risteli and Timpl, 1981; Risteli *et al.*, 1981; Timpl *et al.*, 1978,1979,1981). Although detailed studies of the chemistry, biosynthesis, deposition, localization, and functions of some forms of collagen (Hay, 1981a,b,c; Kleinman *et al.*, 1981; Miller 1976, 1982; Timpl *et al.*, 1978,1979,1981), elastin (Hay, 1981a,b,c; Lansing *et al.*, 1952), fibronectin (Huw and Lawson, 1980; Hynes and Destree, 1978; Klebe, 1974; Yamada, 1981,1983; Yamada *et al.*, 1981; Yamada and Olden, 1978), and laminin (Hirata *et al.*, 1983; Risteli and Risteli, 1981; Risteli and Timpl, 1981; Risteli *et al.*, 1981) are available, other components of the extracellular matrix are less well characterized. It is also known that the chemical composition and functions of matrix can change depending on a variety of exogenous stimuli such as hormones, nutrients, and neural signals. It can also change in certain diseased states such as malignancy. Bauer *et al.* (1982), Weiss *et al.* (1980), Tseng *et al.* (1980), Liotta *et al.* (1976,1977,1983), Jones and DeClerck (1980), Unkeless *et al.* (1973), and Kramer *et al.* (1982) have shown that tumor cells secrete enzymes that degrade the extracellular matrix permitting the tumor cells to be invasive. Yamada *et al.* (1981,1983) and Hynes (1976,1979,1981) have shown that fibronectin is often absent or reduced in neoplastically transformed cells.

Table I. Families of Extracellular Matrix Present in Adult Mammalian Tissues

	Primary components		
Cell type	Collagen type	Anchorage protein	Proteoglycans (major ones)
Fibroblast	I	Fibronectin	Dermatan sulfate Chondroitin sulfate
Cartilage cell	II	Chondronectin	Chondroitin sulfate
Reticular cell fibroblast	III	Fibronectin	Heparan sulfate
Epithelial cell	IV, V	Laminin	Heparan sulfate
Muscle	V	—	Heparan sulfate

Summarized from various reviews and articles (Hassell *et al.*, 1980; Kleinman *et al.*, 1981; Timpl *et al.*, 1978, 1979, 1981; Yamada and Olden, 1978; Hay, 1981). The extracellular matrix is constructed in the paradigm indicated below.

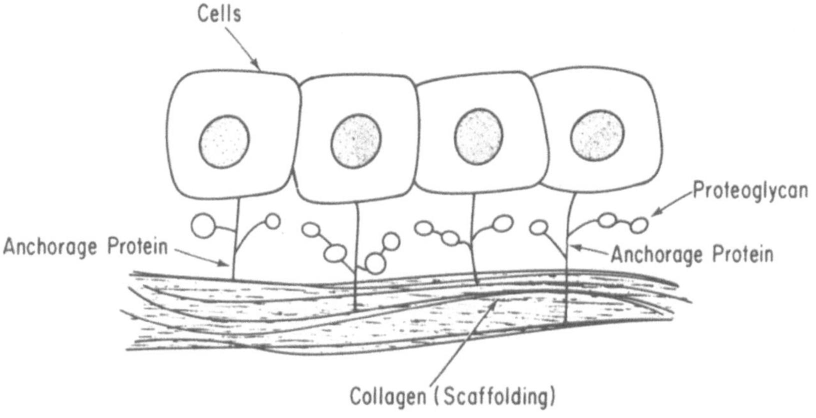

The extracellular matrix has classically been assigned the role of structural or skeletal support for tissues. The elasticity and tensile strength of collagens and elastin confer appropriate structural support to each tissue (Lansing *et al.*, 1952; Miller, 1976,1982). The specific mixture of collagen types and elastin in a given tissue is thought to be critically important in the specific functions of the tissue (Miller, 1976,1982). For example, in the skin, type I collagen predominates. Since type I collagen forms dense sheets of hydrogen-bonded collagen fibrils, tissues containing a significant amount of this collagen type in the extracellular matrix have greater tensile strength and skeletal properties. In contrast, tissues requiring elasticity such as ligaments contain significant amounts of elastin (Lansing *et al.*, 1952; Foster,

1982). Not only are the collagen types, the presence or absence of elastin, and their ratios tissue-specific, but so also are the associated proteins, glycoproteins, and proteoglycans. Laminin and heparan sulfate are usually found complexed with type IV collagen in basement membranes, whereas fibronectin and chondroitin sulfate are found complexed with type I collagen (Kleinman *et al.*, 1981; Sakashita *et al.*, 1980; Timpl *et al.*, 1978). This gives rise to matrix types or "families," each containing a specific collagen type with its complex of proteoglycans, proteins, and glycoproteins. In the schematic diagram associated with Table I is shown a paradigm governing the known relationships of matrix components to each other and with a cell. The cells are bound by anchorage proteins to a scaffolding, which in animals consists of collagen molecules. Associated with the anchorage protein are one or more other proteins, glycoproteins, or proteoglycans. Some proteoglycans have been implicated as transmembrane molecules connecting the cytoskeletal proteins to the anchorage protein (Koda and Bernfield, 1982). The matrix families can be categorized by the collagen types, and each is produced by and associated with a specific cell type. In Table I is given the known composition of the most common types of matrix in adult mammalian tissues and the cell type which produces it.

The tissue specificity of a matrix derives in part from whether it is, for example, a type I vs. type II matrix and in part from components that are unique to matrix of a specific tissue. Proteoglycans may be one of the tissue-specificity factors, since the protein cores of a class of proteoglycans can vary from tissue to tissue. Therefore, heparan sulfate in the skin may be distinct, both chemically and functionally, from heparan sulfate in the lungs (Hassell *et al.*, 1981,1980; Hascall and Sajdera, 1969; Hascall and Hascall, 1981).

Although Table I is an overgeneralization, it does convey the principles that certain components are universally present in one of the families of matrices (e.g., type I collagen is always present in type I matrices), that certain components are unique in a specific tissue matrix (e.g., the heparan sulfate found in the skin is different from that found in other tissues), and that specific cell types produce and are in association with specific types of matrix. Therefore, the development of cell culture substrata of matrix must take these principles into account and use them to guide experimental designs. One's expectations for specific cell responses are predicated on the type of matrix, whether embryonic versus adult, whether derived from growing or nongrowing tissue, whether type I versus type III, and whether derived from normal versus diseased tissue. For example, one should see a preferential attachment by epithelial cells over fibroblasts, but should not expect a strict tissue-specific response in epithelial cells if they are plated onto pure type IV collagen, since that matrix component is present in all forms of type IV matrices.

Known Functions of Matrix Proteins

Relevance of Cell Shape to Differentiation

Cells seeded onto matrix or its components become more cuboidal or columnar than they do on tissue culture plastic on which they tend to exhibit a more squamouslike appearance. Characterization of ultrastructural markers in cells when plated onto mixtures of components of matrix or even when suspended in solutions containing matrix components has shown that the cytoskeleton is significantly altered depending on the substratum chemistry (Sugrue and Hay, 1981). On a plastic substratum the cells underwent blebbing, whereas on collagen gels or on mixtures of collagen gels with laminin or fibronectin, Sugrue and Hay (1981) have shown that the cells assumed smooth contours more similar to their morphology *in vivo*. The loss of blebbing in the presence of collagens and anchorage proteins, whether bound to a substratum or in solution, correlated with the development of a basal cell cortex containing a highly organized mat of actin filaments that could be decorated with myosin S1 fragments. Sugrue and Hay postulated the existence of cell surface receptors for matrix molecules which in turn interacted with the cytoskeleton.

The influence of the extracellular matrix on the phenotypic expression of cells may be due, in part, to its influence on the cell shape and on the specific cytoskeletal patterns (Ben-Ziev *et al.*, 1979,1980; Condeelis, 1980,1981; Condeelis *et al.*, 1981; Woloeswick and Porter, 1979; Woloeswick *et al.*, 1980). The cytoskeleton, containing structural components such as microtubules, intermediate filaments, and actin, and a force-producing component containing actin and myosin, has been implicated in the spatial integration of the cytoplasm in nonmuscle cells (Lazarides, 1980; Risteli and Risteli, 1981; Taylor and Condeelis, 1979; Woloeswick and Porter, 1979). The cytoskeleton as spatial integrator is responsible for the placement of organelles within the cytoplasm, regulation of cell shape, and surface architecture (Condeelis *et al.*, 1981; Woloeswick *et al.*, 1980), that is, the morphological phenotype of differentiated cells. Recent evidence suggests that this spatial organization may actually influence the metabolic activity of cells (Ben-Ziev *et al.*, 1979,1980).

Influence of Matrix Chemistry on Cell Response: General Comments

Although the roles of matrix include its influence on cell shape, it may also act via interactions between factors at the cell surface and the chemical components of the matrix. Thus, the presence of certain matrix components might selectively stabilize one or another hormone receptor in the cell membrane, alter the physical properties of a membrane channel for nutrients, or might present certain molecular-sieving properties determining the diffusion rate of nutrients, waste products, hormones, and so forth into or out of the cell or tissue. These aspects of matrix biology have been implicated in studies on the composition of glomerular basement membranes in normal versus nephrotic animals indicating the relevance of matrix composition in regulating filtration of proteins from the plasma into the urinary space as an ultrafiltrate (Brenner *et al.*, 1978; Caulfield and Farquhar, 1978). Some of the findings using crude extracts of matrix in cell culture have also confirmed these principles: the chemistry of the matrix can alter hormone responsiveness (Gatmaitan *et al.*, 1983; Gospodarowicz *et al.*, 1980), migration patterns of cells (Hume and Potten, 1980), or growth responses (Gospodarowicz *et al.*, 1980; Vlodavsky *et al.*, 1980; Enat *et al.*, 1984). Furthermore, it is likely that the matrix may bind and stabilize hormones or growth factors, in a selective fashion, resulting in a set of signals that are stably presented to cells. Thus, the well-known phenomenon in endocrinology that hormones have a rapid turnover may not apply or applies differently if those hormones are bound to a matrix. This area of matrix biology is largely in its infancy. It indicates, however, the extraordinary complexity possible in considering both the numerous components of matrix, its synergies with hormones, and the turnover of hormones and/or matrix components under given physiological conditions.

Anchorage Proteins

A review of the known anchorage proteins, their chemical characterization and known functions, is presented elsewhere in this volume and will not be reviewed again here.

Collagens

Collagens were the first matrix components isolated and characterized (see review by Linsenmayer, 1981; Miller and Gay, 1982). They are macromolecular proteins composed largely of proline, hydroxyproline, and glycine. Since collagens are among the only proteins containing hydroxyproline, this amino acid is often used to assess the presence or to track synthesis and turnover of collagens (Miller, 1976; Miller and Gay, 1982). For an excellent review of the fields of collagen chemistry and synthesis, see the reviews by Linsenmayer (1981), Hay (1981a,b,c), Olsen (1981), Gross (1981), and Cunningham and Frederiksen (1982).

The influence of collagens in cell culture has long been known to facilitate establishment of primary cultures especially of epithelial cells (Hauschka and Konigsberg, 1966; Jakoby and Pastan, 1979). As noted earlier, all studies, until recently, used type I collagen, a fact that has determined in part the success or lack of success with various cell types. The influence of collagens on cells is indirect in that they serve as chemical scaffoldings for other matrix components that contact the cell directly. Therefore, the chemical composition of a specific collagen type dictates in turn what can bind to it and thereby determines many of the functions of that collagen. The best-known roles of collagens are structural, providing the tensile strength, elasticity, and support for all types of tissue throughout metazoan organisms. In its role as structural support, collagen's unusually flexible nature permits cell shape changes associated with tissue-specific functions and motility. This flexibility stems from the unusual chemical composition of collagen in that every third amino acid residue is glycine.

A detailed review of the many types of collagens will not be presented here, but can be found in the reviews by those of Linsenmayer (1981), Hay (1981a,b,c), Miller (1976), and Miller and Gay (1982).

Proteoglycans

Proteoglycans are macromolecules composed of a protein core and at least one (usually there are many) long unbranched polysaccharide chain (for reviews see Dorfman, 1981; Hascall and Hascall, 1981; Hascall and Kimura, 1982). The polysaccharide chains, commonly known as glycosaminoglycans are covalently attached to the core protein at a serine residue and account for at least 90% of the proteoglycan weight. The proteoglycans are categorized according to the complexity of the number, length, and composition of the attached glycosaminoglycan chains. The number of core proteins that have been isolated and characterized is small, but for a given type of proteoglycan (e.g., heparan sulfate), the protein core may vary from tissue to tissue. Therefore, the heparan sulfate in one tissue (e.g., kidney basement membranes) may be distinct from the one in another tissue. The functions of proteoglycans are poorly understood at present. Using organ cultures of salivary glands, Bernfield and his associates (Banerjee *et al.*, 1977; Bernfield and Banerjee, 1982; Bernfield *et al.*, 1972; Cohn *et al.*, 1977) have found that the proteoglycans associated with the cell surface influence the branching morphogenesis of salivary gland epithelial cells. They and others (Furukawa and Bhavanandan, 1982) have also shown an influence of the proteoglycans on cell division. Most recently, Koda and Bernfield (1982) reported evidence suggesting that some proteoglycans are transmembrane molecules connecting cytoskeletal components to anchorage proteins.

Cell Culture Methods Using Crude Extracts of Extracellular Matrix

Over the last decade, studies on the biology and chemistry of extracellular matrix have entered a new era in which extensive analyses of its chemistry and function on cultured cells are expanding those insights developed previously by chemists and by embryologists studying organ cultures or whole organisms. The excitement is that many fields (matrix biology, endocrinology, oncology, molecular biology), long separate, are coming together to provide a fundamental understanding of growth regulation and differentiation.

The predominant approach for studies on matrix was, until the late 1970s, to characterize changes in the matrix by histochemistry during various physiological changes or to isolate and purify a single matrix component and then to test it in culture for its influences on cells. However, this approach is now being complemented by use of crude extracts of extracellular matrix prepared from cultured cells or from whole tissues. Why would anyone want to use a crude extract if they can use a purified component (a question posed by many chemists to the biologists)? The reasons are fundamental to an understanding of matrix biology. These include the following considerations:

Matrix is a complex mixture of proteins, glycoproteins, and proteoglycans (and other factors?), not all of them yet identified or purified, working in synergy with each other to exert the biological effects of matrix. Thus, as with the findings with hormones, one's understanding of the influence of matrix is biased, sometimes in erroneous directions, if one analyzes only one component at a time. This does not negate the need for experiments using purified components one at a time but argues the necessity of also including experiments that use mixtures of matrix components.

One of the most notable aspects of the matrix is that it is largely insoluble and therefore difficult to handle until one adapts one's approaches to take that aspect of its nature into account. Matrix's insolubility compounds the problems of its multiplicity of components working in synergy in that it makes it difficult to isolate and purify the factors without destroying their activity and makes it difficult to find ideal ways of "presenting" these components to cells.

All of these problems—complex mixture of factors, some identified and some not, and the problems resulting from their insolubility—are overcome by use of crude extracts prepared by negative selection protocols (i.e., protocols that select for insoluble components as opposed to a positive selection in which the active factor is solubilized by an extraction solution).

The protocols developed in the three approaches described below are all similar in that they use negative selection protocols that isolate all or most of a cell layer's or tissue's matrix components (usually more than 90% of the collagens and their associated proteins, glycoproteins, and proteoglycans) and do not prejudge which components might be active. Certainly, the use of crude extracts to study the biology of matrix is expected to be a temporary phase scientifically. The future will include undoubtedly purification and characterization of more and more biologically active factors and then reconstitution of these factors into a "defined matrix." Recently, Oldberg and Ruoslahti (1982) described a "defined matrix" prepared from type I collagen, chondroitin sulfate and fibronectin. A distinct "defined matrix" was prepared by Kleinman, *et al.* (1983) from type IV collagen, laminin, and heparan sulfate.

The main purpose of this chapter is to review the use of different types of extracellular substrata that influence cultured cells to express specific differentiated functions in a fashion more characteristic of cells *in vivo*. However, we would be remiss if we did not acknowledge the *in vivo* use of extracellular matrix by Reddi and his co-workers (for review, see Reddi, 1983). They have used an extracellular matrix to study the regulation of localized differentiation of cartilage and bone. Bone matrix is prepared from the shafts of long bones of the rat using a negative selection process that dissolves away undesirable components. Powders of bone are demineralized with HCl and extracted with water, ethanol, and ether leaving the insoluble extracellular matrix behind (Reddi and Huggins, 1972). Following subcutaneous implantation of the extracellular bone matrix, which is primarily collagenous in nature, into rats, sequential development of endochondral bone is initiated, remodeled, and finally populated with bone marrow cells (Reddi, 1981).

Extracellular matrix prepared from different sources, including tendon and skin, did not induce bone formation. The biological properties of bone extracellular matrix have been shown to be sensitive to alkali, proteases, and reducing agents. Recently, two interesting proteins have been isolated from the bone extracellular matrix. One of these proteins demonstrates potent chemotactic activity (Somerman *et al.*, 1982) and the other a mitogenic component (Sampath *et al.*,1982), which may be important in the sequence of events that lead to endochondral bone formation. The use of extracellular matrix *in vivo* as a model system may offer important insights into the regulation of differentiated function for a number of different tissues.

Extracellular Matrix

Denis Gospodarowicz introduced one of the first protocols for the isolation of a crude extract of extracellular matrix (Gospodarowicz *et al.*, 1976,1977,1978,1979,1980,1981; Gospodarowicz and Greenberg, 1979,1981; Savion and Gospodarowicz, 1980; Vlodavsky and Gospodarowicz, 1981). He and his associates described the isolation of extracellular matrix (ECM), a cellular exudate or residue left behind on cell culture dishes after extraction of a monolayer cell culture with dilute alkali solutions or dilute detergents followed by rinsing with saline or serum-free medium solutions. Most of the early studies used monolayers of bovine endothelial cells. The choice of endothelial cells was fortuitous, since such simple extractions of endothelial cells do leave behind the matrix as a residue on the dish without causing the cells plus their matrix components to detach. More recently, the protocol has been adapted for other cell types, both epithelial cells or other mesenchymal cell types (Hawrot, 1980; Crickard and Golbus, 1982; Bissell, 1981; Hellerquist, 1982). The success in the isolation of an ECM from cultures of these cells has been variable, depending on the cell type and especially depending on whether or not the cells are malignant. Isolation of ECM from normal cells has proved more successful than from malignant cells, perhaps reflecting the fact that malignant cells produce enzymes that degrade the matrix components (Liotta *et al.*, 1976,1977,1980,1981,1982; Kramer *et al.*, 1982; Jones and DeClerck, 1980; DeClerck and Jones, 1980).

Chemical characterization of ECM from endothelial cells indicates that it is composed of collagens types I, III, IV, and V at ratios of 3:16:1 (Tseng *et al.*, 1981), anchorage proteins including fibronectin and laminin (Gospodarowicz *et al.*, 1981), and the proteoglycans heparan sulfate and chondroitin sulfate (Gospodarowicz and Greenburg, 1981).

Functionally, endothelial ECM substrata have proved far superior to regular tissue culture dishes for improving cell attachment, clonal growth, survival, and differentiation. Even cells, such as normal epithelial cells, which have long been the most notoriously difficult cell types to establish in culture, responded favorably to ECM. Attachment efficiencies, which for normal epithelial cells plated onto tissue culture plastic, might be as low as 0.1% to 10% (highly dependent on seeding density and cell type) increased to 90% to 100% for most cell types, both normal and neoplastic, and for both primary cultures and established cell lines (Gospodarowicz and Greenburg, 1981; Gospodarowicz *et al.*, 1980). Similarly, clonal growth efficiency (a measure of the ability of single cells to form colonies of cells when in isolation from any neighboring cells), which for normal epithelial cells on tissue culture dishes is usually less than 0.1%, was increased to above 50% if cells were plated onto ECM. The length of survival of the primary cultures, usually a week or two for normal epithelial cells plated onto tissue culture plastic, increased dramatically to many weeks or months on ECM. Almost all of the studies with endothelial ECM have suggested that it is not species or tissue specific, a distinction with the findings using crude extracts of whole tissues which will be discussed below. The lack of tissue or species specificity may be due to the fact either that the endothelium used to prepare the ECM is an established cell line or that it is derived from embryonic endothelium (certain types of embryonic tissues, e.g., amniotic membranes, appear to have matrix with no or little tissue specificity).

The initial studies with ECM spawned many more both by Gospodarowicz *et al.* (1981) and others (Fujii *et al.*, 1982; Crickard and Golbus, 1982; Collins, 1980; Hawrot, 1980). Besides ECM's dramatic influence on attachment and survival of cells, its most notable influence has been on the growth of cells. Gospodarowicz and co-workers have shown that cells plated onto ECM grow better and with a reduction in the serum requirement or an alteration in the hormone requirements for growth (Gospodarowicz *et al.*, 1980,1981; Gospodarowicz and Lui, 1981). For example, vascular smooth muscle cells plated onto tissue culture plastic require serum for growth. If plated onto ECM, the cells will grow in either plasma or serum. The known distinctions between plasma and serum are primarily that serum contains an abundance of platelet-derived growth factor (PDGF), whereas plasma does not (Gospodarowicz and Ill, 1980). One interpretation for these experiments is that vascular smooth muscle cells, known to be dependent for their growth on PDGF, can become autonomous to this factor if grown on ECM either because ECM regulates or stabilizes the same mechanisms as PDGF or because PDGF causes the cells to produce ECM. However, a different interpretation, perhaps as likely, is that ECM contains PDGF bound to it, since the endothelial cells used to prepare the ECM were grown in a serum-supplemented medium and since recent studies (Smith *et al.*, 1982) have shown that PDGF tenaciously sticks to dishes and to various forms of matrix. If the latter interpretation is true, these studies implicate a possibly major role for matrix as a solid-state material that can either stabilize or destabilize (depending on the chemistry of the matrix) hormones and growth factors. Thus, matrix may be a reservoir of signals (both matrix and hormones) that are stably presented to specific cell types.

Extracellular matrix, derived from endothelial cell lines, is remarkably potent in certain but not all aspects of tissue-specific function. Its influence is primarily with respect to attachment, survival, and growth behavior including such phenomena as outgrowth of neurites from normal or neoplastic neural tissue (Fujii *et al.*, 1982), but it has limited capacity to augment other tissue-specific functions such as albumin synthesis by hepatocytes or insulin secretion by pancreatic islet cells (Muschel *et al.*, unpublished data; Reid, 1982; Reid *et al.*, 1983).

The advantages of ECM are clearly the simplicity in preparing it, the fact that the matrix exudate left on the dish is in a form that is chemically, stoichiometrically and structurally (i.e., the chemical anatomy) intact and native, and that it is derived from only one cell type, simplifying the number of variables to be identified. Its limitations are the other side of the coin of these same attributes. Since *in vivo* the chemical composition of the matrix produced by a single cell type is influenced by cell–cell interactions (mediated by both matrix and soluble factors) within the tissues and by systemic cellular interactions via hormones and/or neural transmitters, the matrix produced by the cells under culture conditions is likely to be different. Secondly, since the amount of material secreted by cells on the dish is small, the ability to do chemistry is limited. The paucity of the material on the dishes does not hamper identification of the matrix components, but it does preclude purification, in bulk, of some matrix components, especially those that are minor components in the matrix or are in an unusually insoluble (cross-linked) form.

ECM has been and is likely to continue to be used for many cell types due to its ease of preparation, its lack of tissue or species specificity, and its potent mitogenic influence on many cell types. Given its limited influence on differentiative functions, ECM will probably be used less for studies of gene regulation of adult, tissue-specific functions. For these types of studies, other types of matrix substrata are proving more efficacious.

Biomatrix

The Development of the Method for Isolation of Biomatrix

Whereas ECM is a crude extract of extracellular matrix derived by alkali or detergent extractions of clonal cell cultures, "biomatrix" is a crude matrix extract derived by negative selection protocols of whole tissue (plant or animal; normal or diseased). In Fig. 3 is given a flow diagram of current protocols for isolation of biomatrix from either tumor or normal tissue and for isolation of a biomatrix conducive to either growth or differentiation of cells. The crude extract is referred to as biomatrix to indicate that it is a mixture of all forms of the tissue matrix, not any one form (e.g. basement membrane).

The first reports of isolation of a matrix extract from tissues came from the work of Meezan *et al.* (1977). They developed a protocol involving a sequence of solutions to dissolve and thereby eliminate undesired components, leaving a crude insoluble extract of matrix from kidney glomerular tissue. The sequence of solutions to dissolve the "undesirable" components were water, low-ionic strength salt buffers, and detergents. The soluble and insoluble fractions were separated by centrifugation. In an improvement on Meezan *et al.*'s method, Rojkind *et al.* (1980) and Rojkind and Ponce (1982) showed that using the same protocol, one could enrich for matrix components if the soluble and insoluble fractions were separated by means of polyester filters of specific pore diameter to select for matrix fibers of given length in a specific tissue; that is, the filters permitted the isolation of insoluble fibers greater than the length of the pore diameter of filter used. The fibers consisted primarily of collagens, elastins, and various proteins complexed with them. Analogous procedures have now been developed for various tissues and organs (Day and Lenhoff, 1980; Duhamel *et al.*, 1981; Murray *et al.*, 1980; Schleich *et al.*, 1981). Rojkind's method was further adapted and modified by Reid and associates (1980,1981,1982) to isolate extracellular matrix for use as a cell culture substratum for differentiated epithelial cells. Such extracts, referred to as "biomatrix," were found to significantly enhance the survival and the retention of the differentiated state of various normal and neoplastic epithelial cells.

Chemical characterization of biomatrix indicated tissue and species specificity in composition. Biomatrix derived from rat liver, pancreas,

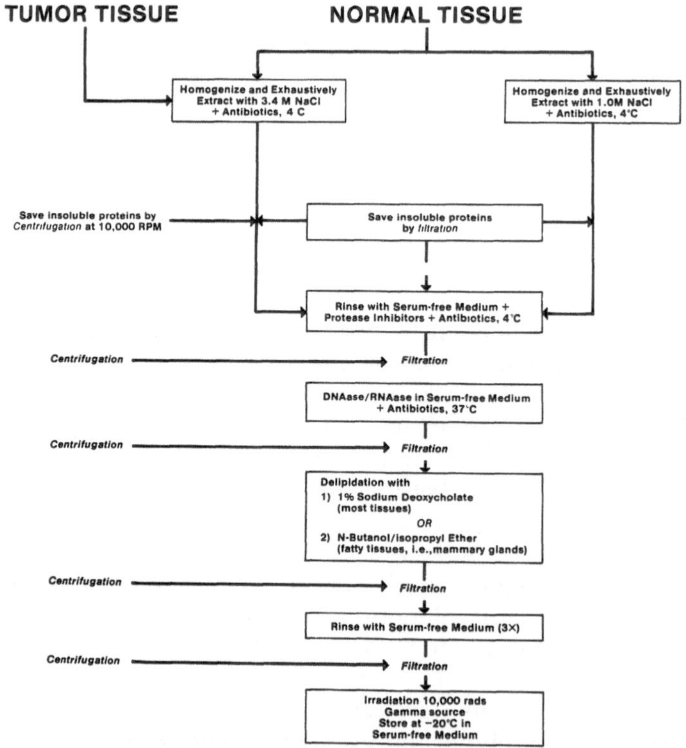

Figure 3. Protocols for isolation of biomatrix. Biomatrix is a crude extract, enriched in extracellular matrix, and derived from any type of tissue. All the protocols for isolation of biomatrix operate on the principle of negative selection. For more details on these methods see Rojkind *et al.* (1980) and Reid (1982). The tissue is extracted sequentially with solutions that dissolve presumed undesired components, and a constant selection is made for insoluble proteins. Selection can be either by centrifugation (at 10,000 RPM for 10–15′) or by filtration through a cloth filter of given pore diameter. The results from using matrix prepared by centrifugation or filtration can be distinct, but both forms of extract show biological activity on cell growth and differentiation. One must use centrifugation for preparing biomatrix from tumors or any type of tissue in which the matrix is not fully cross-linked or in which the matrix does not contain interstitial collagens.

Two distinct protocols are shown: tissues can be initially extracted with 3.4 M NaCl or 1.0 M NaCl. The use of the former results in an extract more conducive to differentiation and the latter is permissive for growth of normal epithelial cells.

Although the studies are incomplete, it is suspected that proteoglycans may be more abundant in the 3.4 M NaCl extract and depleted in the 1.0 M NaCl extract and are among the factors relevant to the extracts' functional distinctions.

Biomatrix extracts from a tumor line, the EHS tumor, a mouse tumor line shown to constitutively secrete basement membrane proteins (Orkin *et al.*, 1977), are prepared by a shortened version of the protocols in the figure. The tumors are dissected free from mice, pressed through a collector (Bellco Labs, Vineland, N.J.) with a No. 2 coarse grid and prostate all contained collagens types I, III, and IV (approximate ratios 1:20:3) and various anchorage proteins, including fibronectin and laminin (Reid *et al.*, 1981; Rojkind and Ponce, 1981). However, biomatrix from each tissue type contained unique components recognizable on SDS gels when the biomatrix was extracted with either acetic acid solutions (Ponce *et al.*, 1981) or with guanidine HCl (Reid *et al.*, 1981; Reid, 1982). This tissue and species specificity in the chemical compositions has proved relevant biologically to the matrice's influence. This will be discussed in more detail below.

Influence of Biomatrix Substrata on Differentiation

Tissue-specific substrata of biomatrix in combination with serum-supplemented media were found to dramatically enhance attachment efficiencies, plating efficiencies, and survival and to better maintain differentiated functions that were evaluated in cultures of rat liver, rat pancreatic islet cells, rat prostatic epithelial cells (Reid *et al.*, 1980,1981; Reid, 1982; Rojkind *et al.*, 1980), and in rat mammary epithelial cells (Wicha *et al.*, 1982). Rat liver cultures were maintained for more than six months with retention of albumin synthesis (levels were lower than those observed *in vivo*). In subsequent studies a number of tissue-specific functions were systematically evaluated over time in culture: for rat liver (albumin, glutathione-S-transferase, tyrosine aminotransferase), for rat pancreatic islet cells (insulin), and for rat prostastic epithelial cells (prostatic binding protein). Similar studies were done by Wicha *et al.* (1982) with mammary epithelial cells and evaluating various tissue specific functions such as lactalbumin synthesis. It was realized that for all the types of tissues studied (Reid, 1982; Gatmaitan *et al.*, 1983; Wicha *et al.*, 1982), the cultures start off in their first week at levels of differentiation comparable to those seen with cultures on collagen gels. Gradually the cells' differentiative functions increased, peaking after three to four weeks in culture and maintained at steady levels thereafter.

(to eliminate the stromal capsule around the tumor), and the tumor cells collected in serum-free medium. The cells are pelleted and brought up and homogenized in 10X volume: volume ratio with 3.4 M NaCl + antibiotics (penicillin/streptomycin) all at 4°C. The homogenate is centrifuged at 10,000 RPM for 15 minutes. The pellet is again brought up in the 3.4 M NaCl buffer and stirred for an hour. it is centrifuged and rinsed twice with serum-free medium containing antibiotics and protease inhibitors. If cells to be plated onto EHS biomatrix are to be used for mRNA studies, the biomatrix is extracted with DNase only. If not, then both RNase and DNase are used. After nuclease extraction, the insoluble proteins are rinsed twice with serum-free medium and sterilized by gamma irradiation.

Wicha (1982) reported that mammary epithelial cells differentiate to a greater extent on biomatrix that is free floating in the dish as compared to matrix attached to the dish a finding thought to be due to the ability of the cells to assume a more cuboidal shape on the floating matrix. By trying many new protocols in the isolation of biomatrix, extracts were isolated that cause a faster response with respect to differentiation. The essential changes in the methods of isolating the biomatrix were: (1) the use of buffer solutions with high ionic strength to maintain in an insoluble form various components including proteoglycans that appear to be critical for differentiation; (2) to minimize the use of detergents that bind irreversibly to the biomatrix and make it toxic; and (3) to use a liquid nitrogen pulverizer to convert the tough, fibrous biomatrix into a powder that could be spread over a cell culture dish to make a homogeneous substratum. When comparing the biological activity of the two extracts, one prepared with initial extractions with 3.4 M NaCl (biomatrix 3.4) vs. those with 1.0 M NaCl (biomatrix-1.0), it was found that they differed in their influence on growth and differentiation of normal rat hepatocytes. Hepatocytes (Enat *et al.*, 1983) or mammary epithelium (Wicha *et al.*, 1982) grew on biomatrix-1.0 (in combination with serum-free defined media) but were amitotic and more differentiated on biomatrix-3.4. Investigations are ongoing to ascertain the chemical distinctions of these two extracts. Among the known differences are larger amounts of laminin, of proteoglycans and of other, guanidine-HCl-extractable proteins in the biomatrix-3.4. Although these studies are incomplete, it is probable that proteoglycans and/or glycosaminoglycans are among the relevant components dictating the distinctions in biological activity. Banerjee *et al.* (1977), Hascall and Hascall (1981), Kramer *et al.* (1982), and others have shown that abundance of specific proteoglycans (e.g., heparan sulfate) are correlated with inhibition of cell growth and maintenance of the differentiated state, and their loss results in conditions permissive for cell division.

The most influential form of biomatrix substratum, with respect to differentiation of epithelial cells, has been that derived from EHS tumors, a transplantable mouse tumor line shown to constitutively secrete basement membrane proteins (Orkin *et al.*, 1977). When EHS biomatrix was used in combination with serum-free, hormonally defined media, tissue-specific functions of normal rat liver were maintained at levels approximately 70% or more of those *in vivo*. Furthermore, the response by the cells occurred within a few days after plating the cells, a fact quite distinct from the findings with biomatrix from normal tissue. However, the stability of the cultures was less than that observed when normal tissue biomatrix was used. Depending on the cell type, the cultures lasted only a few weeks. Thus, the differentiative response of normal epithelial cells to EHS biomatrix was faster but more labile than that seen with normal tissue biomatrix. In contrast, use of normal tissue biomatrix resulted in cells with maximal response after a week or two (with biomatrix 3.4) or after up to four weeks (with biomatrix 1.0), but once achieved the responses were stable for months. These observations can be interpreted as implicating type IV matrix (basement membrane) components as the biologically active factors with respect to differentiation. Thus, EHS biomatrix, which contains only type IV matrix components, gives a greater response than do biomatrices from normal tissues in which type IV matrix components constitute 10% to 15% of the proteins. It is likely that the instability of cellular response to EHS biomatrix is due to the reduced crosslinking in this tumor matrix resulting from release by the tumor cells of degradative enzymes. Since the biomatrix from normal tissue is largely interstitial collagens and associated proteins, the normal cells plated onto the normal tissue biomatrix response less (approximately 25% to 40% of normal, *in vivo* levels of tissue-specific functions). However, the gradual restoration of differentiative function probably correlates with secretion and deposition of extracellular matrix by the cells and stabilization by the biomatrix of the newly synthesized type IV matrix components. Once the cultures of normal cells have restored their differentiated function, they then remain stable for months.

Influence of Biomatrix on Growth of Normal Epithelial Cells

The only systematic study done to date has been with normal rat hepatocytes. Enat *et al.* (1983) showed that stable proliferation of normal hepatocytes required specific forms of liver biomatrix: either biomatrix-3.4 M derived from regenerating liver or biomatrix-1.0 M. In addition, proliferation of the cells depended on the use of serum-free media, a low-seeding densities for the cells and a defined medium containing insulin, glucagon, EGF, growth hormone, prolactin, and trace elements. If hepatocytes were plated onto EHS biomatrix or onto rat liver biomatrix-3.4 M, the cells would not divide even if the rest of the culture conditions were appropriate. Thus, some forms of biomatrix are permissive to growth and some are not, and hormones and matrix operate in synergistic fashion to influence, either growth or differentiation.

Matrix and Neoplastically Transformed Cells

Malignant transformation, regardless of the transforming event, induces a cascade of changes in the cell that disrupts its homeostatic relationship with the host, including the cellular environment and neighboring cells (Hart and Fidler, 1981). These alterations include those of cell morphology with concomitant changes in growth behavior, surface structural components, surface antigenicity, membrane transport mechanisms, and enzymatic composition, and incomplete synthesis of surface glycoproteins (Hynes, 1976; Yamada, 1981,1983). These observations are noted along with the phenomenon of a loss or reduction of the normal cellular response to control mechanisms for proliferation (Nicolson and Poste, 1976). This finding may be antecedent to or concomitant with observed cellular alterations, permitting the tumor cell to escape from the constraints of growth and behavior to which normal cells are subject. Tumor cell growth is less dependent on cell–cell contact interactions and is increasingly autonomous to regulation by serum factors, hormones, and matrix components (Cherington *et al.*, 1979; Reid *et al.*, 1979).

Biomatrix has been demonstrated to enable tumor cells to express morphological and growth properties that resemble those of their normal counterparts *in vivo* to a more faithful extent than is allowed by culture on plastic substrates (Reid, 1982; Muschel *et al.*, unpublished data), thus suggesting that some regulatory constraints are extant for neoplastic cells and that they can still respond to modulating factors. On biomatrix, tumor cells attach and grow more efficiently with a reduced requirement for serum supplementation for growth, suggesting a finely tuned synergism between hormones and substrates in the modulation of cell proliferation (Gatmaitan *et al.*, 1983). Similarly, in the development of defined media for hepatoma cells, it was found that collagenous and biomatrix substrata reduced the hormone requirements for growth both qualitatively and quantitatively (see Table II).

Whereas normal epithelial cells respond stably to a matrix, tumor cells have responses that change over time. In studies with rat insulinoma cells (Muschel *et al.*, unpublished data) and hepatoma cells (Jefferson *et al.*, 1983b), the cells were found to respond differently to biomatrix if cultured in serum supplemented medium (SSM) as opposed to a serum-free, hormonally defined medium (DM).

Growth rates of the hepatomas (similarly the insulinomas) were equivalent if the cells were plated onto tissue culture plastic in SSM or DM or plated onto biomatrix in SSM. However, if they were plated onto biomatrix in DM, the cells underwent growth arrest after one or

Table II. Constituents of "Defined Media" for Optimal Clonal Growth of Normal and Neoplastic Rat Hepatocytes

		Rat hepatoma	
Factor	Normal adult rat hepatocyte	Tissue culture plastic and DM	Collagen type I and DM
Epidermal growth factor	50 ng/ml	Not required	Not required
Insulin	10 μg/ml	100 μg/ml	1–5 μg/ml
Glucagon	10 μg/ml	10 μg/ml	5 μg/ml
Hydrocortisone	Not required	1×10^{-8} M	Not required
Transferrin	Not required	10 μg/ml	Not required
Linoleic acid[a]	5 μg/ml	10 μg/ml	5 μg/ml
Triiodothyronine	Not required	1×10^{-9} M	Not required
Prolactin	20 mU/ml	2 mU/ml	2 mU/ml
Growth hormone	10 mU/ml	10 uU/ml	10 uU/ml
Trace elements			
Zinc	1×10^{-10} M	1×10^{-10} M	1×10^{-10} M
Copper	1×10^{-7} M	1×10^{-7} M	1×10^{-7} M
Selenium	3×10^{-10} M	3×10^{-10} M	3×10^{-10} M

[a] Linoleic acid must be present in combination with fatty acid free bovine serum albumin (Pentax, Inc.) at a 1:1 molar ratio. If added alone, it is quite toxic to the cells.

Pituitary hormones were purchased from commercial sources when possible. An alternate source was through the NIH hormone distribution service.

The trace elements were also required on tissue culture plastic. However, the concentration required for optimal clonal growth is unknown. The concentrations used were the same as for cell cultures plated on collagenous substrata. The listed concentrations of the required trace elements are the minimum concentrations necessary to give maximal growth. However, growth curves at various concentrations of the trace elements, show that there is a range of concentrations which are optimal for each trace element. For selenium, it is a broad range from 3×10^{-10} M to 10^{-7} M, for copper, it is from 10^{-7} to 10^{-6} M and for zinc, it is from 10^{-10} to 10^{-8} M.

two divisions. They remained in growth arrest for 10 to 12 days in the case of the rat and human hepatomas and for 18 to 20 days in the case of the rat insulinoma. During the growth arrest, tissue-specific functions were augmented 5- to 50-fold above that observed under any other culture condition. The phase of growth arrest accompanied by augmented tissue-specific functions was transient for all tumor cell lines studied. The growth arrest and augmentation of tissue-specific functions lasted 3 to 4 days for HeLa cells, 10 to 12 days for human hepatoma cells, and 18 to 20 days for the rat insulinoma cells. At the end of growth arrest, regions of the dishes contained cells growing in piles on top of one another. The cells in the piles eventually detached and floated into the medium. If the floating cells or cells in the piles were transferred to dishes with fresh medium and biomatrix, they again went into growth arrest.

The response of tumor cells to the biomatrix and DM conditions indicates both that tumor cells are capable of responding to regulatory signals and that the cells are able (most probably through the secretion of enzymes degradative to matrix and hormones) to escape regulation by these signals.

Metastatic Site Potential and Matrix

The ability of tumor cells to escape regulation by hormones and matrix signals appears also to apply to tumor metastasis. Tumor metastasis is a complex phenomenon that encompasses multifactorial events and conditions in which an interaction between host and tumor factors results in local invasiveness and successful tumor spread (Hart and Fidler, 1981). The host does not exist merely as a "culture medium" permitting neoplastic proliferation passively, but instead mounts an array of defense maneuvers that include mechanical processes as well as highly specific humoral and cell-mediated immune responses. Should the tumor cell successfully elude these forces, it is then able to circulate and home in to sites of metastasis. Traditionally, the metastatic process has been thought to include a series of steps, the veracity and exact order of which are speculative, whereby tumor cells grow, become vascularized, lose their anchorage to surrounding cells and tissues, embolize, and extravasate, all of these occurring in the presence of continuing host defense maneuvers. Most of the studies on metastasis have focused on features of the cell and on host immunologic responses with very few studies addressing the role, if any, of the host tissues, whether merely permissive or actively involved, in the final development of the metastatic focus.

The complexity of the interaction between tumor cells and extracellular matrix, the model for the basement membranes that tumor cells are thought to invade *in vivo*, is amplified by the finding that tumor cells, possibly as a contingent property of their transformed state, secrete enzymes that degrade multiple components of basement membranes (Unkeless *et al.*, 1973; Liotta *et al.*, 1977; Liotta *et al*, 1982). Plasminogen-specific protease has been noted in virally transformed fibroblasts (Unkeless *et al.*, 1973), and has been thought by some investigators (Sisskin *et al.*, 1980) to correlate with tumorigenicity. Other enzymes, virtually a battery of proteases that can degrade most of the known components of basement membrane including elastin, collagen, glycoproteins, and proteoglycans, have been reported to be present in different types of tumor cell lines (Sisskin *et al.*, 1980; Jones and DeClerck, 1980; Kramer *et al.*, 1982). This finding suggests that the transformed cell has a built-in mechanism to ensure its successful invasion through tissue barriers in order to propagate.

Some intriguing data have emerged from studies investigating the production of enzymes by tumor cells. Attachment assays done in parallel with *in vivo* tumorigenicity studies suggest that attachment of tumor cells to endothelial cell matrix appears to correlate with the production of enzymes that degrade glycoproteins and proteoglycans, permitting attachment of cells to subendothelial matrix (Kramer *et al.*, 1982) that in turn seems to correlate with the efficiency of metastasis formation *in vivo* (Nicolson and Poste, 1976). Moreover, whereas normal cells are both tissue specific and to some extent species specific in their growth properties on biomatrix extracted from different tissues, malignant cells appear to have lost species specificity and to have an altered tissue-specificity (Chiuten *et al.*, 1983; Doerr *et al.*, 1983). At high-seeding densities, tumor cells showed no tissue specificity, attaching and growing at equal rates and growing to equal saturation densities on biomatrix from a number of tissues. At low-seeding densities appropriate for clonal growth, some tissue specificity remained. Clonal growth occurred on some but not all types of biomatrix. The more anaplastic the tumor cell, the more types of biomatrix were permissive for clonal growth. The tissue specificity that remained appeared to correlate with the metastatic site potential of the tumor *in vivo*. Although untested as yet, it may be that the ability of tumors to grow at clonal seeding densities correlates with their ability to secrete the proper repertoire of enzymes to degrade relevant growth inhibitors (proteoglycans?) in the matrix and that in turn dictates their ability to colonize a specific tissue. This hypothesis is currently under investigation.

Synergies Between Hormones and Matrix

The significance of serum-free, hormonally defined media (DM) has been amply documented by reviews presented throughout this book. In most of the investigations reported in these reviews, the DM was used with cells plated on tissue culture plastic. Use of DM in conjunction with matrix substrata, either individual matrix components or crude extracts, gives cellular responses that strongly suggest synergies in the effects of the two sets of regulators. The composition of the DM for growth of hepatoma cells was found to be dependent on the type of substratum on which the cells were sitting. The hepatoma cells required more hormones for clonal growth on tissue culture plastic than they did on collagenous or biomatrix substrata (see Table II) (Gatmaitan *et al.*, 1983; Enat *et al.*, 1983). Although multiple mechanisms

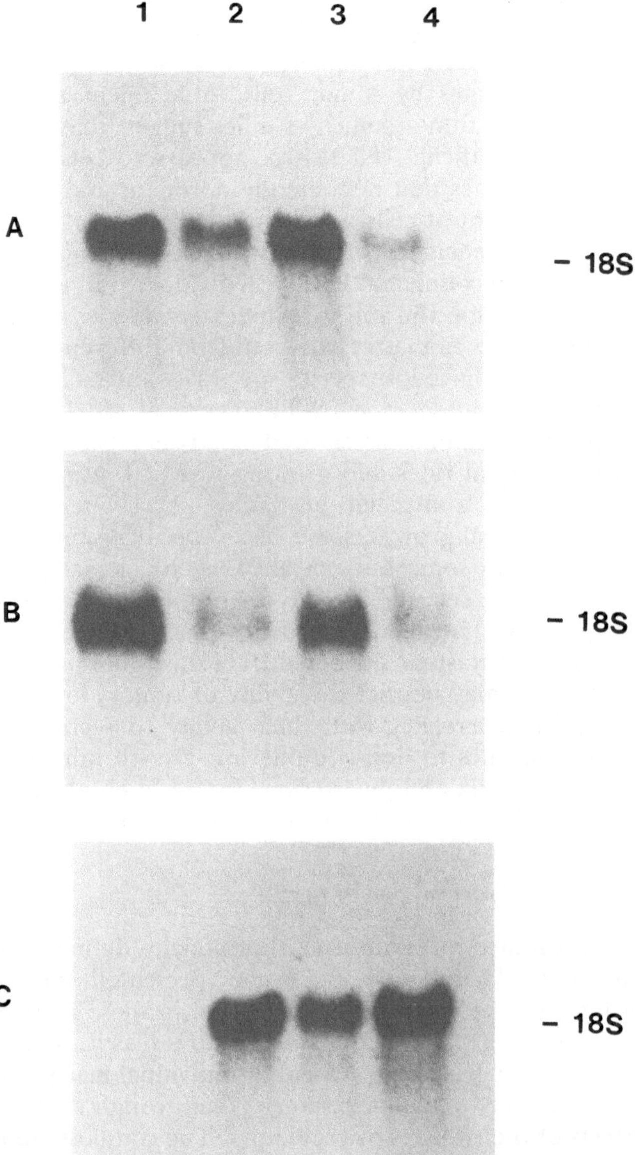
1
2
3
4
A
- 18S
B
- 18S
C
- 18S

may be involved to produce these results, it indicates that hormones do not operate alone in regulating cells; rather they operate in an integrated fashion with matrix components.

This principle of matrix and hormones interacting to effect a cellular response was also shown in the work of Enat *et al.* (1983) in which normal rat hepatocytes could be induced to grow by using a DM in combination with a liver biomatrix, but only if the biomatrix was depleted of certain growth inhibitors. Thus, there are forms of matrix that make the cells permissive to respond to certain hormones. Clearly, the evaluation of hormone effects will require knowledge of the matrix influence (potentiation or inhibition).

In analyzing differentiation, it is important to assess how exogenous factors might be affecting the expression of tissue-specific functions. Appreciation of this comes from the studies of Jefferson *et al.* (1983a) where the relative influence of DM on liver-specific functions as measured by analysis of the steady-state level of mRNAs by the Northern blot technique. Primary cultures of normal adult rat hepatocytes were prepared and plated onto tissue culture plastic. Cells were grown in either serum-supplemented medium (SSM), DM, or a mixture of DM + SSM for five days (Enat *et al.*, 1983). Cells grown in SSM exhibit a dramatic decrease in the steady-state levels for both albumin (panel A, lane 4), and glutathione-S-transferase (panel B, lane 4) compared with *in vivo* levels (see Fig. 4). However, these two liver-specific functions were maintained at near *in vivo* levels when cells were maintained in DM (panel A, B, lane 1, 3). The transcription rates for the liver-specific functions were not equivalent in cells cultured in DM or SSM but in both conditions these rates were significantly lower than the *in vivo* rates. Therefore, the influence of DM on the steady-state levels of these tissue-specific (there are other liver-specific mRNAs that are not maintained as well as albumin and gluthathione-S-transferase) mRNAs must be by the stabilization of existing mRNA. Additionally, β-actin, which

←

Figure 4. Northern blot analysis of housekeeping and liver-specific functions in primary rat hepatocyte cultures. The autoradiographic patterns in this figure represent the relative abundance of particular mRNAs. Confluent primary cultures of adult rat hepatocytes were grown for 5 days on tissue culture plastic dishes in one of the following conditions: lane 1, *in vivo* rat liver control; lane 2, mixture of serum-supplemented medium and Enat's hormonal defined medium (Enat *et al.*, 1983); lane 3, hormonally defined medium; lane 4, serum-supplemented medium. RNA was isolated from the cells, electrophoresized, blotted, and hybridized with specific 32P-cDNA probes. The resulting autoradiographic patterns represent the relative abundance of the particular mRNA. The panels were hybridized using the following 32P-cDNA probes for: A, albumin; B, glutathione-S-transferase and C- β-actin.

exists in low abundance *in vivo*, is increased (100 X) under culture conditions (panel C) with only a small increase (4 X) in the transcription rate. Studies are ongoing at present to ascertain if biomatrix affects transcription rates or if, similar to some hormones, it affects such posttranscriptional processes as mRNA stability.

In summary, biomatrix substrata have proved applicable for establishing normal, adult cells in a state of growth or one of differentiation. Achievement of an optimal response from the cells requires both an appropriate biomatrix substratum and an appropriate defined medium. Primary cultures of normal cells, which in the past were difficult to impossible to achieve, are now achievable with many of the cells' tissue-specific functions intact.

Tumor cells also respond to biomatrix, especially if it is used in combination with defined medium. However, their response, even if dramatic, is transient due to tumor cell mechanisms (most probably degradative enzyme secretion) that permit the tumor cells to escape regulation by the hormonal and matrix components.

The advantages of biomatrix are that the protocol for its isolation is applicable to any tissue in any stage of development, any physiological state, or in any diseased state. Since it is prepared from whole tissues, it can be made in bulk facilitating chemical isolation and purification of biological active components. From studies already completed using biomatrix, it is apparent that cellular responses *in vitro* can be maintained to be similar or equivalent to those *in vivo*.

1. Normal cells require a period of "recovery" for their tissue-specific functions of one to several weeks since biologically active factors have been eliminated from the biomatrix by the current protocols. The recovery period is assumed to be synthesis and deposition of those factors into the biomatrix, which then stabilizes them.
2. The biomatrix is dispersed by pulverization and painting of the material onto the dish, resulting in a disruption of the natural relationship of matrix components.
3. The protocols for biomatrix select for all matrix types resulting in a mixture of components from all the types of matrix in that tissue. The influence of matrix components from one type could negate or alter the influence of other matrix components.

In summary, the advantages and disadvantages of using biomatrix extracts are primarily that one is using not just one matrix type at a time but a mixture. The responses of cells to biomatrix may be complicated and at times difficult to interpret due to this limitation in the system. Clearly biomatrix must give way to a defined matrix produced by purifying components from biomatrix and reconstituting them in an appropriate way to yield the maximal response from cells. Even so, the current studies using these crude extracts have given many insights into regulation of cells by matrix components and the interaction of these components with the regulatory influences of hormones.

Crude Extracts from Amniotic Membranes

Liotta *et al.* (1980) first described the isolation of a crude extract of basement membrane derived from amnions. The protocols they used were similar to those for ECM involving dilute alkali or detergent extraction of the amnion followed by rinses with saline or medium solutions.

The extracted amnions consisted of a type IV matrix (basement membrane) on one side and a type III (stromal) on the other. Since the amnion is used intact after extraction (as a sheet of material), its components retain the correct stoichiometry and chemistry. The cell culture studies using these extracted amniotic membranes were for assays of degradative enzyme production by tumor cells (Liotta *et al.*, 1981). Tumor cells were assessed for invasive potential by seeding the cells onto extracted amnions and watching for colony formation on a dish under the amniotic membrane as well as assaying the culture medium for enzymatic activity. Controls included a variety of normal cells, both fibroblasts and epithelial cells. The amniotic membrane extract did not show species or tissue specificity in any aspect of cellular behavior.

Studies using amniotic membrane as a cell culture substratum are limited. Among the best of these are the studies by Madri *et al.* (1980,1982) and Madri and Williams (1983), who used the treated amnions for cultures of endothelial cells. They showed that the cells produced a distinctly different behavior if plated onto the basement membrane surface side vs. the stromal side. On the basement membrane side, the cells flattened, formed tight junctions, gap junctions, stopped any migratory behavior, and formed tube-like structures. On the stromal side, the cells proliferated, migrated into the stromal element, and only formed tube-like structures after prolonged time in culture.

These studies suggest great potential for using amniotic membrane for many types of cells, especially perhaps proliferative stem cells such as skin, intestinal tract, or bone marrow. Since the amnion is an embryonic tissue, its basement membrane shows little or no tissue-specificity.

The Future

Use of matrix substrata is still in its infancy. However, the studies with crude extracts of cell layers or tissues indicate revolutionary changes in our abilities to culture normal cells or in our abilities to achieve specific physiological or pathological states in cells in culture. However, at present one must realize that many matrix components have yet to be identified, purified, and characterized. The creation of a "defined matrix" is essential but will require many years. Not only is there limited knowledge of what constitutes all of the matrix components, but there is also limited knowledge of how to assemble them to elicit optimal cellular responses in culture. At least, though, the directions are clear: Differentiation of normal cells requires a complicated interplay between soluble regulators such as hormones and conditioned medium factors and regulators in the matrix substrata. It is hardly any wonder that the culturing of normal cells, especially epithelial cells, took so long to be achieved. There were so many factors (both soluble and insoluble ones) to be identified. Fortunately, the field of matrix biology is expanding rapidly with many laboratories focusing their diverse areas of expertise on the central issue of understanding the complex nature of extracellular matrix and how the matrix and cells functionally interact.

ACKNOWLEDGMENTS. We would like to thank Rosina Passela, Mary Joan Vaccaro-Olko, Clifford Liverpool, and Dorothy Occhino without whose help and patience this chapter could not have come to fruition. Additionally, a few paragraphs of the metastatic site potential and matrix section were written by Diana Chiuten for which we are grateful. Some of the studies presented in this review were supported in part by NIH grants (CA33164, CA30117) and an ACS grant (BC-439C).

References

Ambesi-Impiobato, F. S., Parks, L. A. M., and Coon, H. G., 1980, Culture of hormone-dependent functional epithelial cells from rat thyroids, *Proc. Natl. Acad. Sci. U.S.A.* **77**:3455–3459.

Banerjee, S. D., Cohn, R. H., and Bernfield, M. R., 1977, Basal lamina of embryonic salivary epithelia, *J. Cell Biol.* **73**:445–463.

Barnes, D., and Sato, G., 1980, Serum-free medium. A Review, *Cell* **22**:649–655.

Bauer, E. A., Seltzer, J. L., and Eisen, A. Z., 1982, Inhibition of collagen degradative enzymes by retinoic acid *in vitro*, *J. Am. Acad. Dermatol.* **6**:603–607.

Benya, P. D., and Shaffer, J. D., 1982, Dedifferentiated chondrocytes reexpress the differentiated collagen phenotype when cultured in agarose gels, *Cell* **30**:215–224.

Ben-Ziev, A., Dunn, A., Solomon, F., and Penman, S., 1979, The outer boundary of the cytoskeleton: A lamina derived from plasma membrane proteins, *Cell* **17**:859–865.

Ben-Ziev, A., Farmer, S. R., and Penman, S., 1980, Protein-synthesis requires cell-surface contact while nuclear events respond to cell-shape in anchorage-dependent fibroblasts, *Cell* **21**:365–372.

Bernfield, M., and Banerjee, S. D., 1982, The turnover of basal lamina glycosaminoglycan correlates with epithelial morphogenesis, *Dev. Biol.* **90**:291–305.

Bernfield, M. R., Banerjee, S. D., and Cohn, R. H., 1972, Dependence of salivary epithelial morphology and branching morphogenesis upon acid mucopolysaccharide-protein (proteoglycan) at the epithelial surface, *J. Cell Biol.* **52**:674–689.

Bissell, D. M., and Guzelian, P. S., 1981, Phenotypic stability of adult rat hepatocytes in primary monolayer culture, *Ann. N.Y. Acad. Sci.* **349**:85–98.

Bissell, M. J., 1981, The differentiated state of normal and malignant cells or how to define a "normal" cell in culture, *Int. Rev. Cytol.* **70**:27–100.

Bissell, M. J., Hall, G. H., and Parry, G., 1982, How does the extracellular matrix direct gene expression?, *J. Theor. Biol.* **99**:31–68.

Bottenstein, J., Hayashi, I., Hutchings, S., Masui, H., Mather, J., McClure, D., Ohasa, S., Rizzino, A., Sato, G., Serrero, G., Wolfe, R., and Wu, R., 1979, The growth of cells in serum-free hormone-supplemented media, *Methods Enzymol.* **58**:94–109.

Brenner, B. M., Hosteller, T. H., and Humes, H. D., 1978, Molecular basis of proteinaria of glomerular origin, *N. Engl. J. Med.* **298**:826–833.

Buonassisi, V., Sato, G., and Cohen, A., 1962, Hormone producing cultures of adrenal and pituitary tumor origin, *Proc. Natl. Acad. Sci. U.S.A.* **48**:1184–1190.

Caulfield, J. P., and Farquhar, M. G., 1978, Loss of anionic sites from the glomerular basement membrane in amino nucleoside nephrosis, *Lab. Invest.* **39**:505–512.

Chambard, M., Gabrion, J., and Mauchamp, J., 1980, Reorganization in culture of thyroid follicular cells-influence of collagen on the orientation of cell polarity, *C.R. Seances Acad. Sci. Ser. D.* **291**:79–81.

Chazov, E. I., Alexeev, A. V., Antonov, A. S., Koteliansky, V. E., Leytin, V. L., Lyubimova, E. V., Repin, Sviridov, D. D., Torchilin, V. P., and Smirnov, V. N., 1981, Endothelial cell culture on fibrillar *collagen*: Model to study platelet adhesion and liposome targeting to intercellular collagen matrix, *Proc. Natl. Acad. Sci. U.S.A.* **76**:5603–5607.

Cherington, P. V., Smith, B. L., and Pardee, A. B., 1979, Loss of epidermal growth factor requirement and malignant transformation, *Proc. Natl. Acad. Sci. U.S.A.* **76**:3937–3941.

Chiuten, D. D., Lan, S., Kaplan, B. H., and Reid, L. M., 1983, Neoplastic transformation alters tissue specificity and regulation of epithelial cells by extracellular matrix, *Cancer Res.* **24**:29.

Cohn, R. H., Banerjee, S. D., and Bernfield, M. R., 1977, Basal lamina of embryonic salivary epithelial: Nature of glycosaminoglycan and organization, *J. Cell Biol* **73**:464–478.

Collins, F., 1980, Neurite outgrowth induced by the substrate associated material from nonneuronal cells, *Dev. Biol.* **79**:247–252.

Condeelis, J., 1980, Reciprocal interactions between the actin lattice and cell membrane, *Neurosci. Res. Bull.* **19**:83–99.

Condeelis, J., 1981, Microfilament-membrane interactions in cell shape and surface architecture, in: *International Cell Biology* (H. G. Schweiger, ed.), Springer-Verlag, New York, pp. 306–320.

Condeelis, J., Salisbury, J., and Fujiwara, K., 1981, A new protein that gels F actin in the cell cortex of Dictyostelium discoideum, *Nature* (*London*) **292:**163.

Crickard, K., and Golbus, M. S., 1982, Influence of extracellular matrix on the proliferation of human amniotic fluid cells *in vitro, Prenatal. Diagn.* **2:**89–95.

Cunha, G. R., 1972, Tissue interactions between epithelium and mesenchyme of urogenital and integumental origin, *Anat. Rec.* **172:**529–542.

Cunningham, L. and Frederiksen, D., 1982, Structural and Contractile Proteins. Part A. Extracellular Matrix, *Methods Enzymol.* **82**.

Day, R. M., and Lenhoff, H. M., 1980, Hydra extracellular matrix: a substrate for cell attachment and spreading, *J. Cell Biol.* **87:**119a.

DeClerck, Y. A., and Jones, P. A., 1980, Effect of ascorbic acid on the resistance of the extracellular matrix to hydrolysis by tumor cells, *Cancer Res.* **40:**3228–3231.

Doerr, R., Chiuten, D., Jefferson, D. M., Gatmaitan, Z., Enat, R., and Reid, L., 1983, Assay for predicting tumorigenic and metastatic potential, *Hepatology* **3:**86.

Dorfman, A., 1981, Proteoglycan biosynthesis, in: *Cell Biology of Extracellular Matrix* (E. Hay, ed.), Plenum Press, New York, pp. 115–138.

Duhamel, R. C., Meezan, E., and Brendel, K., 1981, Basement-membranes from bovine retinal microvessels—Extraction of non-collagenous and collagenous polypeptides into separate fractions by non-enzymatic solubilization, *Fed. Proc.* **40:**343.

Emmerman, J. T., and Pitelka, D. R., 1977, Maintenance and induction of morphological differentiation in dissociated mammary epithelium on floating collagen membranes, *In Vitro* **13:**346–348.

Enat, R., Jefferson, D. M., Ruiz-Opazo, N., Gatmaitan, Z., Leinwand, and Reid, L. M., 1984, Hepatocyte proliferation *in vitro*: Its dependence on the use of serum-free, hormonally defined medium and substrata of extracellular matrix, *Proc. Natl. Acad. Sci. U.S.A.* (in press).

Foster, J. A., 1982, Elastin Structure and Biosynthesis: An Overview, *Methods Enzymol.* **82:**559–570.

Fujii, D. K., Massoglia, S. L., Savion, N., and Gospodarowicz, D., 1982, Neurite outgrowth of substratum and nerve growth factor, *J. Neurosci.* **2:**1157–1175.

Furukawa, K., and Bhavanandan, V. P., 1982, Influence of glycosaminoglycans on endogenous DNA synthesis in isolated normal and cancer cell nuclei, *Biochim. Biophys. Acta* **697:**344–352.

Gatmaitan, Z., Jefferson, D. M., Ruiz-Opazo, N., Biempica, L., Arias, I., Dudas, G., Leinwand, L., and Reid, L. M., 1983, Regulation of growth and differentiation of a rat hepatoma cell line by the synergistic interactions of hormones and collagenous substrata, *J. Cell Biol.* **97:**1179–1190.

Gerfaux, J., Lanson, M., and Chany-Fournier, F., 1981, Chemical characterization and functional role of human glomerular basement membrane components, *Renal Physiol.* **4:**67–73.

Gospodarowicz, D., and Greenberg, G., 1979, The coating of bovine and rabbit corneas denuded of their endothelium with bovine corneal endothelial cells, *Exp. Eye Res.* **28:**249–265.

Gospodarowicz, D., and Greenberg, G., 1981, The role of growth factors and extracellular matrices in the control of mammalian cell proliferation, in: *The Biology of Normal Human Growth* (M. Ritzen, A. Aperia, K. Hall, A. Larsson, A. Zetterberg, and R. Zetterstrom, eds), Raven Press, New York, pp. 1–21.

Gospodarowicz, D., and Ill, C. R., 1980, Plasma versus serum: Is there a difference in their ability to promote cell growth *in vitro* or *in vivo*?, *Proc. Natl. Acad. Sci. U.S.A.* **77**:2726–2730.

Gospodarowicz, D., and Lui, G. M., 1981, Effect of substrate and fibroblast growth factor on the proliferation *in vitro* of bovine aortic endothelial cells, *J. Cell. Physiol.* **109**:69–81.

Gospodarowicz, D., Moran, J., Braun, D., and Birdwell, C. R., 1976, Clonal growth of bovine endothelial cells in tissue culture: Fibroblast growth factor as a survival agent, *Proc. Natl. Acad. Sci. U.S.A.* **73**:4120–4124.

Gospodarowicz, D., Mescher, A. L., and Birdwell, C. R., 1977, Stimulation of corneal endothelial cells *in vitro* by fibroblast and epidermal growth factors, *Exp. Eye Res.* **25**:75–89.

Gospodarowicz, D., Greenberg, G., Bialecki, H., and Zetter, B., 1978, Factors involved in the modulation of cell proliferation *in vivo* and *in vitro*: The role of fibroblast and epidermal growth factors in the proliferative response of mammalian cells, *In Vitro* **14**:85–118.

Gospodarowicz, D., Vlodavsky, I., and Savion, N., 1979, The extracellular matrix and the control of proliferation of vascular endothelial and vascular smooth muscle cells, *J. Supramol. Struct. Cell. Biochem.* **10**(3):181.

Gospodarowicz, D., Delgado, D., and Vlodavsky, I., 1980, Permissive effect of the extracellular matrix on cell proliferation *in vitro, Proc. Natl. Acad. Sci. U.S.A.* **77**:4094–4098.

Gospodarowicz, D., Greenberg, G., Foidart, J.-M., and Savion, N., 1981, The production and localization of laminin in cultured vascular and corneal endothelial cells, *J. Cell. Physiol.* **107**:173–183.

Green, H., 1979, The keratinocyte as differentiated cell type: A Review or Bibliography, *Harvey Lect.* **74**:101–139.

Grobstein, C., 1954, Tissue interaction in the morphogenesis of mouse embryonic rudiments *in vitro*, in: *Aspects of Synthesis and Order in Growth* (D. Rudrik, ed.), Princeton University Press, New Jersey, pp. 233–256.

Grobstein, C., 1953, Inductive epithelio-mesenchymal interactions in cultured organ rudiments of the mouse, *Science* **118**:128–137.

Gross, J., 1981, An essay on biological degradation of collagen, in: *Cell Biology of Extracellular Matrix* (L. E. Hay, ed.), Plenum Press, New York, pp. 217–258.

Grotendo, G., Hewitt, A. T., Kleinman, H. K., Martin, G. R., Rohrbach, D. H., Terranov, V. P., Varner, H. H., and Wilkes, C. M., 1980, Collagen-specific attachment proteins for fibroblasts, chondrocytes, epithelial cells and smooth-muscle cells, *Eur. J. Cell Biol.* **22**:419.

Guguen-Guillouzo, C., Clement, B., Baffet, G., Beaumont, C., Morel-Chany, E., Glaise, D., and Guillouzo, A., 1983, Maintenance and reversibility of active albumin secretion by adult rat hepatocytes co-cultured with another liver epithelial cell type, *Exp. Cell Res.* **143**:47–54.

Guido, D., and Bernfield, M., 1981, Type I collagen reduces the degradation of basal lamina proteoglycan by mammary epithelial cells, *J. Cell Biol.* **91**:281–286.

Guzman, R. C., Osborn, R. C., Yang, J., DeOme, K. B., and Nandi, S., 1982, Transplantation of mouse mammary epithelial cells grown in primary collagen gel cultures, *Cancer Res.* **42**:2376–2383.

Ham, R. G., 1965, Clonal growth of mammalian cells in a chemically-defined synthetic medium, *Proc. Natl. Acad. Sci. U.S.A.* **53**:288–293.

Ham, R., and McKeehan, W., 1979, Media and growth requirements, *Methods Enzymol.* **58**:44–93.

Hart, I. R., and Fidler, I. J., 1981, The implications of tumor heterogeneity for studies on the biology and therapy of cancer metastasis, *Biochem. Biophys. Acta* **651**:37–50.

Hascall, V. C. and Kimura, J. H., 1982, Proteoglycans, *Methods Enzymol.* **82**:769–800.

Hascall, V. C., and Hascall, G. K., 1981, Proteoglycans, in: *Cell Biology of Extracellular Matrix* (E. Hay, ed.), Plenum Press, New York, pp. 39–64.

Hascall, V. C., and Sajdera, S. W., 1969, Proteinpolysaccharide complex from bovine nasal cartilage: The function of glycoprotein in the formation of aggregates, *J. Biol. Chem.* **244**:2384–2396.

Hassell, J. R., Robey, P. G., Barrach, H. J., Wilczek, J., Rennard, S. I., and Martin, G. R., 1980, Isolation of a heparan sulfate-containing proteoglycan from basement-membrane, *Proc. Natl. Acad. Sci. U.S.A.* **77**:4494–4498.

Hassell, J. R., Newsome, D. A., and Martin, G. R., 1981, Isolation and characterization of the proteoglycans and collagens synthesized by cells in culture, *Vision Res.* **21**:49–53.

Hauschka, S. D., and Konigsberg, R., 1966, The influence of collagen on the development of muscle clones, *Proc. Natl. Acad. Sci. U.S.A.* **55**:119–126.

Hawrot, E., 1980, Cultured sympathetic neurons: Effects of cell-derived and synthetic substrata on survival and development, *Dev. Biol.* **74**:136–151.

Hay, E., 1981a, Collagen in embryonic development, in: *Cell Biology of Extracellular Matrix* (E. Hay, ed.), Plenum Press, New York, pp. 379–409.

Hay, E., 1981b, Extracellular Matrix, *J. Cell Biol.* **91**:205s–223s.

Hay, E. (ed.), 1981c, *Cell Biology of Extracellular Matrix,* Plenum Press, New York.

Hellerqvist, C. G., 1982, Biosynthetic matrices from cells in culture, *Methods Enzymol.* **82**:530–544.

Hewitt, A. T., Varner, H. H., Silver, M. H., Dessau, W., Wilkes, C. M., and Martin, G. R., 1982, The isolation and partial characterization of chondronectin, and attachment factor for chondrocytes, *J. Biol. Chem.* **257**:2330–2334.

Hirata, K., Yoshida, Y., Siramatsu, K., Freeman, A. E., and Hayasaka, H., 1983, Effects of laminin, fibronectin and type IV collagen on liver cell cultures, *Exp. Cell Biol.* **51**:121–129.

Hume, W. J., and Potten, C. S., 1980, Changes in proliferative activity as cells move along undulating basement-membranes in stratified squamous epithelium, *Br. J. Dermatol.* **103**:499–504.

Huw, A. J., and Lawson, H., 1980, The effect of different collagen types used as substrata on myogenesis in tissue culture, *Cell Biol. Int. Rep.* **4**:841–850.

Hynes, R. O., 1976, Cell-surface proteins and malignant transformation, *Biochim. Biophys. Acta* **458**:73–107.

Hynes, R. O. (ed.), 1979, *Surfaces of Normal and Malignant Cells,* John Wiley and Sons, New York.

Hynes, R. O., 1981, Fibronectin and its relation to cellular structure and behavior, in: *Cell Biology of Extracellular Matrix* (E. Hay, ed.), Plenum Press, New York, pp. 295–334.

Hynes, R. O., and Destree, A. T., 1978, Relationships between fibronectin (LETS protein) and actin, *Cell* **15**:875–886.

Iozzo, R. V., and Wight, T. N., 1982, Isolation and characterization of proteoglycans synthesized by human colon and colon carcinoma, *J. Biol. Chem.* **257**:11135–11144.

Jakoby, W., and Pastan, I. (eds.), 1979, *Methods in Enzymology,* Volume 58, Academic Press, New York.

Jefferson, D. M., Cobb, M. H., Gannaro, J. F., and Scott, W. N., 1980, Transporting renal epithelium: Culture in hormonally defined serum-free medium, *Science* **210**:912–914.

Jefferson, D. M., Clayton, D. F., Darnell, J. E., Jr., and Reid, L. M., 1983a, Maintenance of differentiated function in cultured hepatocytes by mRNA stabilization, *J. Cell Biol.* **97:**141a.

Jones, P. A., and DeClerck, Y. A., 1980, Destruction of extracellular matrices containing glycoproteins, elastin and collagen by metastatic human tumor cells, *Cancer Res.* **40:**3222–3227.

Klebe, R. J., 1974, Isolation of a collagen-dependent cell attachment factor, *Nature (London)* **250:**248–251.

Kleinman, H. K., Klebe, R. J., and Martin, G. R., 1981, Role of Collagenous matrices in the adhesion and growth of cells, *J. Cell Biol.* **88:**473–485.

Kleinman, H. K., McGarvey, M. L., Hassell, J. R., and Martin, G. R., 1983, Formation of a supramolecular complex is involved in the reconstitution of basement membrane components, *Biochemistry,* **22:**4969–4974.

Koda, J. E., and Bernfield, M., 1982, Heparin sulfate proteoglycan binding to collagen involves the native fibril, *J. Cell Biol.* **95:**126a.

Kramer, R. H., Vogel, K. G., and Nicolson, G. L., 1982, Solubilization and degradation of subendothelial matrix glycoproteins and proteoglycans by metastatic tumor cells, *J. Biol. Chem.* **257:**2678–2686.

Kratochwil, K., 1964, Organ specificity in mesenchymal induction demonstrated in the embryonic development of the mammary gland of the mouse, *Dev. Biol.* **20:**46–71.

Kuettner, K. E., Pauli, B. U., Gall, G., Memoli, V. A., and Schenk, R. K., 1982, Synthesis of cartilage matrix by mammalian chondrocytes in vitro. I: Isolation, culture characteristics, and morphology, *J. Cell Biol.* **93:**743–750.

Lansing, A. I., Rosenthal, T. B., Alex, M., and Dempsey, E. W., 1952, The structure and chemical characterization of elastin fibers as revealed by elastase and by electron microscopy, *Anat. Rec.* **114:**555–576.

Lazarides, E., 1980, Intermediate filaments as mechanical integrators of cellular space, *Nature (London)* **283:**249–256.

Linsenmayer, T. F., 1981, Collagen, in: *Cell Biology of Extracellular Matrix* (E. Hay, ed.), Plenum Press, New York, pp. 5–38.

Liotta, L. A., 1982, Tumor extracellular matrix. *Lab. Inv.* **47:**112–113.

Liotta, L. A., Abe, S., Robey, P. G., and Martin, G. R., 1976, Preferential digestion of basement membrane collagen by an enzyme derived from a metastatic murine tumor, *Proc. Natl. Acad. Sci. U.S.A.* **76:**2268–2272.

Liotta, L. A., Kleinerman, J., Catanzaro, P., and Rynbrandt, D., 1977, Degradation of basement membrane by murine tumor cells, *J. Natl. Cancer Inst.* **58:**1427–1431.

Liotta, L. A., Lee, C. W., and Morakis, D. J., 1980, New method for preparing large surfaces of intact human basement membrane for tumor invasion studies, *Cancer Lett.* **11:**141–152.

Liotta, L. A., Goldfarb, R. H., Brundage, R., Siegal, G. P., Terranova, V., and Garbisa, S., 1981, Effect of plasminogen activator (urokinase), plasmin, and thrombin on glycoprotein and collagenous components of basement membrane, *Cancer Res.* **41:**4629–4636.

Madri, J. A., and Williams, S. K., 1983, Capillary endothelial cells cultures: Phenotypic modulation by matrix components, *J. Cell Biol.* **97:**153–165.

Madri, J. A., Roll, F. J., Furthmayr, H., and Foidart, J.-M., 1980, Ultrastructural localization of fibronectin and laminin in the basement membranes of the murine kidney, *J. Cell Biol.* **86:**682–687.

Madri, J. A., Foellmer, H. G., and Furthmayr, H., 1982, Type V collagens of the human placenta: Trimer α-chain composition, ultrastructural morphology and peptide analysis, *Collagen Relat. Res.* **2:**19–29.

Meezan, E., Hjelle, T., Brendel, K., and Carson, E., 1977, A simple, nondisruptive method for the isolation of morphologically and chemically pure basement membranes from several tissues, *Life Sci.* **17:**1721–1732.

Michalopoulos, G., and Pitot, H. C., 1975, Primary culture of parenchymal liver cells on collagen membranes, *Exp. Cell Res.* **94:**70–78.

Michalopoulos, G., Sattler, G. L., and Pitot, H. C., 1976, Maintenance of microsomal cytochromes B-5 and P-450 in primary cultures of parenchymal liver cells on collagen membranes, *Life Sci.* **18:**1139–1144.

Michalopoulos, G., Sattler, G. L., O'Connor, L., and Pitot, H. C., 1978, Unscheduled DNA synthesis induced by procarcinogens in suspensions and primary cultures of hepatocytes on collagen membranes, *Cancer Res.* **38:**1866–1871.

Miller, E. J., 1976, Biochemical and biological significance of the genetically distinct collagens, *Mol. Cell. Biochem.* **13:**165–192.

Miller, E. J. and Gay, S., 1982, Collagen: an overview, *Methods Enzymol.* **82:**3–32.

Montgomery, B. D., and Guzelian, P. S., 1980, Phenotypic stability of adult rat hepatocytes in primary monolayer culture, *Ann. N.Y. Acad. Sci.* **349:**85–98.

Murray, J. C., Stingl, G., Kleinman, H. K., Martin, G. R., and Katz, S. I., 1980, Epidermal cells adhere preferentially to type IV (basement membrane) collagen, *J. Cell Biol.* **80:**197–202.

Nicolson, G. L., and Poste, G., 1976, The cancer cell: Dynamic aspects and modifications in cell-surface organization, *N. Engl. J. Med.* **295:**197–203.

Oldberg, A. and Ruoslahti, E., 1982, Interactions between chondroitin sulfate proteoglycan, fibronectin and collagen, *J. Biol. Chem.* **257:**4859–4863.

Olsen, B. R., 1981, Collagen biosynthesis, in: *Cell Biology of Extracellular Matrix* (E. Hay, ed.), Plenum Press, New York, pp. 139–178.

Orkin, R. W., Gehron, P., McGoodwin, E. B., Martin, G. R., Valentine, T., and Swarm, R., 1977, A murine tumor producing a matrix of basement membrane, *J. Exp. Med.* **145:**204–220.

Ponce, P., Cordero, J., and Rojkind, M., 1981, A noncollagenous matrix for attachment of rat hepatocytes in culture, *Hepatology* **1:**204–210.

Reddi, A. H., 1981, Cell biology and biochemistry of endochondral bone development, *Collagen Relat. Res.* **1:**209–226.

Reddi, A. H., 1983, Regulation of local differentiation of cartilage and bone by extracellular matrix: A cascade type mechanism, *Prog. Clin. Biol. Res.* **100B:**261–268.

Reddi, A. H., and Huggins, C. B., 1972, Biochemical sequences in the transformation of normal fibroblasts in adolescent rats, *Proc. Natl. Acad. Sci. U.S.A.* **69:**1601–1605.

Reid, L. M., 1982, Regulation of growth and differentiation of mammalian cells by hormones and extracellular matrix, in: *From Gene to Protein: Translation into Biotechnology* (F. Ahmad and W. Whelan, eds.), Academic Press, New York, pp. 53–73.

Reid, L. M., and Rojkind, M., 1979, New techniques for culturing differentiated cells: Reconstituted basement membrane rafts, *Methods Enzymol.* **58:**263–278.

Reid, L. M., Morrow, B., Jubinsky, P., Schwartz, E., and Gatmaitan, Z., 1981, Regulation of growth and differentiation of epithelial cells by hormones, growth factors, and substrates of extracellular matrix, *Ann. N.Y. Acad. Sci.* **372:**354–370.

Reid, L. M., Stiles, C. D., Saier, M. H., Jr., and Rindler, M. J., 1979, Growth of nontumorigenic cells in millipore diffusion chambers implanted in mice and implications for *in vivo* growth regulation, *Cancer Res.* **39:**1467–1473.

Reid, L. M., Gaitmaitan, Z., Arias, I., Ponce, P., and Rojkind, M., 1980, Long-term cultures of normal rat hepatocytes on liver biomatrix, *Ann. N.Y. Acad. Sci.* **349:**70–76.

Reid, L. M., Jefferson, D. M., Ruiz-Opazo, N., Gatmaitan, Z., Enat, R., and Leinwand, L., 1983, Regulation of differentiation by hormones and extracellular matrix, *In Vitro* **19:**259.

Risteli, J., Wick, G., and Timpl,R., 1981, Immunological characterization of the 7-S domain of type IV collagens, *Collagen Relat. Res.* **1:**419–432.

Risteli, L., and Risteli, J., 1981, Basement-membranes are certainly not thixotropic gels, *Lancet* **2:**583.

Risteli, L., and Timpl, R., 1981, Isolation and characterization of pepsin fragments of laminin from human placental and renal basement membranes, *Biochem. J.* **193:**749–755.

Rojkind, M., and Ponce, P., 1982, The extracellular matrix of the liver, *Collagen Relat. Res.* **2:**151–175.

Rojkind, M., Gatmaitan, Z., Mackensen, S., Giambrone, M., Ponce, P., and Reid, L. M., 1980, Connective tissue biomatrix: Its isolation and utilization for long-term cultures of normal rat hepatocytes, *J. Cell Biol.* **87:**255–263.

Sakashita, S., Engvall, E., and Ruoslahti, E., 1980, Basement-membrane glycoprotein laminin binds to heparin, *FEBS Lett.* **116:**243–246.

Sampth, T. K., DeSimone, D. P., and Reddi, A. H., 1982, Extracellular bone matrix-derived growth factor, *Exp. Cell Res.* **142:**460–464.

Saunders, J. W., and Gasseling, M. T., 1957, The origin of pattern and feather germ tract specificity, *J. Exp. Zool.* **135:**503–527.

Savion, N., and Gospodarowicz, D., 1980, Patterns of cellular peptide synthesis by cultured bovine granulosa cells, *Endocrinology* **107:**1798–1807.

Schleich, A., Tchao, R., Frick, M., and Mayer, A., 1981, Interaction of human carcinoma cells with an epithelial layer and the underlying basement-membrane—A new model, *Arch. Geschwulstforsch.* **51:**40–44.

Sisskin, E. E., Weinstein, I. B., Evans, C. H., and DiPaolo, J. A., 1980, Plasminogen activator synthesis accompanying chemical carcinogen-induced *in vitro* transformation of Syrian hamster and guinea pig fetal cells, *Int. J. Cancer* **26:**331–335.

Smith, J. C., Singh, J. P., Lillquist, J. S., Goon, D. S., and Stiles, C. D., 1982, Growth factors adherent to cell substrates are mitogenically active *in situ* , *Nature* (*London*) **296:**154–156.

Somerman, M., Hewitt, A. T., Reddi, A. H., Seppa, A., Varner, H., Termine, J. D., and Schiffman, E., 1982, The role of chemotaxis in bone induction, in: *Proceedings of the 5th International Conference on Calcified Tissues*, International Conference Series No. 589, (M. Silberman and H. C. Slavkin, eds.), Excerpta Medica, Princeton, New Jersey, pp. 56–59.

Sugrue, S. P., and Hay, E. D., 1981, Response of basal epithelial cell surface and cytoskeleton to solubilized extracellular matrix molecules, *J. Cell Biol.* **91:**45–54.

Taylor, D., and Condeelis, J., 1979, Cytoplasmic structure and contractility in amoeboid cells, *Int. Rev. Cytol.* **56:**57–143.

Timpl, R., Martin, G. R., Bruckner, P., Wick, G., and Wiedemann, H., 1978, Nature of the collagenous protein in a tumor basement membrane, *Eur. J. Biochem.* **84:**43–52.

Timpl, R., Rohde, H., Robey, P. G., Rennard, S. I., Foidart, J.-M., and Martin, G. R., 1979, Laminin a glycoprotein from basement membranes, *J. Biol. Chem.* **254:**9933–9937.

Timpl, R., Wiedemann, H., Vandelde, V., Furthmay, H., and Kuhn, K., 1981, A network model for the organization of type IV collagen molecules in basement membranes, *Eur. J. Biochem.* **120:**203–211.

Tseng, S. C. G., Savion, N., Gospodararowicz, D., and Stern, R., 1980, Fibroblast growth-factor maintains the phenotypic-expression of collagen-synthesis in cultured bovine vascular endothelial-cells, *Eur. J. Cell* **22**:418.

Tseng, S. C. G., Savion, N., Gospodarowicz, D., and Stern, R., 1981, Characterization of collagens synthesized by bovine corneal endothelial cell cultures, *J. Biol. Chem.* **256**:3361–3365.

Unkeless, J. C., Tobia, A., Ossowski, L., Quigley, J. P., Rifkin, D. B., and Reich, E., 1973, An enzymatic function associated with transformation of fibroblasts by oncogenic viruses. I. Chick embryo fibroblasts transformed by avian RNA tumor viruses, *J. Exp. Med.* **137**:85–111.

Uriel, J., 1979, Retrodifferentiation and the fetal patterns of gene expression in cancer, *Adv. Cancer Res.* **29**:127–174.

Vaughan, F., and Bernstein, I., 1980, Molecular aspects of control in epidermal differentiation, *Mol. Cell. Biochem.* **12**:171–179.

Vlodavsky, I., and Gospodarowicz, D., 1981, Respective involvement of laminin and fibronectin in the adhesion of human carcinoma and sarcoma cells, *Nature* (*London*) **289**:304–306.

Vlodavsky, I., Lui, G. M., and Gospodarowicz, D., 1980, Morphological appearance, growth behavior and migrating activity of human tumor cells maintained on extracellular matrix versus plastic, *Cell* **19**:607–616.

Weiss, R. E., Reddi, A. H., and Lenk, E. V., 1980, Insulin regulates the fibronectin mediated collagenous matrix-cell interaction, *J. Cell Biol.* **87**:A124.

Wessells, N. K., 1964, Substrate and nutrient effects upon epidermal basal cell orientation and proliferation, *Proc. Natl. Acad. Sci. U.S.A.* **52**:252–259.

Wessells, N. K., 1970, Mammalian lung development: Interactions in formation and morphogenesis of tracheal buds, *J. Exp. Zool.* **175**:455–466.

Wicha, M. S., 1982, Growth and differentiation of rat mammary epithelium on mammary gland extracellular matrix, in: *Extracellular Matrix* (S. Hawkes and J. Wang, eds.), Academic Press, New York. pp. 309–314.

Wicha, M. S., Lawrie, G., Kohn, E., Bagavandross, P., and Mahn, T., 1982, Extracellular matrix promotes mammary epithelial growth and differentiation *in vitro, Proc. Natl. Acad. Sci. U.S.A.* **79**:3213–3217.

Wolf, Et., and Wolf, Em., 1975, Current research with organ cultures of human tumors, in: *Human Tumor Cells In Vitro* (J. Fogh, ed.), Plenum Press, New York, pp. 207–240.

Woloeswick, J., and Porter, K. R., 1979, Microtrabecular lattice of the cytoplasmic ground substance: Artifact or reality, *J. Cell Biol.* **82**:114–139.

Woloeswick, J., Becker, R., and Condeelis, J., 1980, The appearance of microtubules: The cytoplasmic ground substance in resinless sections of cells, in: *International Symposium on Microtubules and Microtubule Inhibitors* (M. DeBrabander, ed.), pp. 3–16.

Yamada, K. M., 1981, Fibronectin and other structural proteins, in: *Cell Biology of Extracellular Matrix* (E. Hay, ed.), Plenum Press, New York. pp. 95–114.

Yamada, K. M. (ed.), 1983, *Cell Interactions and Development*, Molecular Mechanisms, Wiley Interscience, New York.

Yamada, K. M., and Olden, K., 1978, Fibronectins, adhesive glycoproteins of cell surface and blood, *Nature* (*London*) **275**:179–184.

Yamada, K. M., Kennedy, D. W., Grotendorst, G. R., and Momoi, T., 1981, Glycolipids: Receptors for fibronectin?, *J. Cell. Physiol.* **109**:343–351.

Zanjani, E. D., and Rinehart, J. J., 1980, Role of cell–cell interaction in normal and abnormal erythropoiesis, *Am. J. Pediatr. Hematol.* **2**:233–244.

Zwilling, E., 1955, Ectoderm–mesoderm relationship in the development of the chick embryo limb bud, *J. Exp. Zool.* **55**:423–441.

Index